T0335977

Ziling Wang
Lishu Zhang

Essential Computing Skills for Biologists

Ziling Wang
Beijing Jiaotong University
Beijing, China
zlw@bjtu.edu.cn

Lishu Zhang
Beijing Jiaotong University
Beijing, China

图书在版编目（C I P）数据

生物信息计算基础 = Essential Computing Skills for Biologists : 英文 / 王字玲，张丽姝主编 . -- 北京 : 高等教育出版社，2018. 1
ISBN 978-7-04-049018-3

Ⅰ. ①生… Ⅱ. ①王… ②张… Ⅲ. ①生物信息论 - 英文 Ⅳ. ① Q811.4

中国版本图书馆 CIP 数据核字（2018）第 008093 号

策划编辑 冯 英　　责任编辑 冯 英　　封面设计 张 楠　　版式设计 童 丹
插图绘制 杜晓丹　　责任校对 高 歌　　责任印制

出版发行 高等教育出版社
社 址 北京市西城区德外大街4号
邮政编码 100120
印 刷
开 本 787mm×1092mm 1/16
印 张 19
字 数 480 千字
购书热线 010-58581118
咨询电话 400-810-0598

网 址 http://www.hep.edu.cn
http://www.hep.com.cn
网上订购 http://www.hepmall.com.cn
http://www.hepmall.com
http://www.hepmall.cn
版 次 年 月第 1 版
印 次 年 月第 次印刷
定 价 89.00 元

物 料 号 49018-00

Preface

Over the past few decades, developments in genomic, molecular research, as well as information technologies have produced a tremendous amount of information related to molecular biology. This made computing skills become essential in molecular biology research. Computational methods have evolved to analyze and interpret various types of data, such as nucleotide and amino acid sequences, protein domains, and protein structures, etc. Bioinformatics is the name given to these mathematical and computing approaches used for understanding biological processes and becomes an important part of many areas of biology. Bioinformatics not only provides tools we can use to understand the basic aspect of biology, including development, metabolism, adaptation to the environment, genetic variations of individual, and evolution, but also facilitates our understanding of the diseases processes through the analysis of molecular sequence data.

This book is a handbook of methods and protocols for biologists. It aims at undergraduate, graduate students and researchers originally trained in biological or medical sciences who need to know how to access the data archives of genomes, proteins, metabolites, gene expression profiles and the questions these data and tools can answer, for example, how to draw inferences from data archives and how to make connections among them to deduce useful predictions. For each chapter, the conceptual and experimental background is provided, together with specific guidelines for handling raw data, including preprocessing and analysis.

The book is structured in three parts. Part I introduces the basic knowledge about popular bioinformatics tools, databases and web resources, including online sequence databases, sequence alignment, predicting DNA and protein function from sequence, protein structure prediction and analysis, molecular phylogeny and evolution. Part II presents examples of Omics bioinformatics applications, including genetic variation and human disease, gene expression profile and data management, qualitative and quantitative proteomics, bioinformatics for metabolomics, and integrating Omics data for pathways and interaction networks. Part III provides basic statistical analysis skills and programming skills needed to handle and analyze Omic datasets.

Bioinformatics is an interdisciplinary field involving molecular biology and genetics, computer science, mathematics, and statistics. It is too far broad to be understood by one person. Thus, this book is written by multiple authors, each of whom brings a deeper knowledge of the subject. I wish to express my gratitude to all authors for their dedication in providing excellent chapters, some of the data and examples presented in the book are the results of their own research. As for any omissions or errors, the responsibility is mine.

In preparing the book, we read many textbooks and published papers and viewed many websites, we sincerely apologize to those authors and researchers whose work we did not cite.

Enjoy reading.

Wang Ziling
College of Life Sciences and Bioengineering,
Beijing Jiaotong University, Beijing
October, 2017

Contents

PART I DATABASES AND BIOINFORMATICS TOOLS

PART I

DATABASES AND BIOINFORMATICS TOOLS

Chapter 1 Online Sequence Database

Yong Liu[1], Lishu Zhang[2]

In recent years, with the rapid development of computer and network technology, a large number of biological information resources can be retrieved through the Internet. Such a large number of biological databases, software resources and Internet connection are making life science research more convenient and efficient.

The major objectives of biological databases are not only to store, organize and share data in a structured and searchable manner with the aim to facilitate data retrieval and visualization for humans, but also to provide web application programming interfaces (APIs) for computers to exchange and integrate data from various database resources in an automated manner.

According to the report of 2016 database issue of *Nucleic Acids Research*, there are 1685 databases that are publicly accessible online. Various databases cover all areas of life sciences, and in this chapter, we will focus on some of the commonly used biology databases or online resources, including, ① nucleic acid sequence database, such as GenBank, EMBL, DDBJ, etc.; ② protein database uniprot; ③ protein three-dimensional structure of the database PDB; ④ online human genome resources: UCSC genome browser, Ensembl genome browser and NCBI Map Viewer.

1.1 Nucleic Acid Sequence Database

There are three well-known large-scale nucleic acid sequence databases. They are GenBank (maintained by National Center for Biotechnology Information, NCBI), EMBL (maintained by The European Bioinformatics Institute, EBI) and DDBJ (maintained by National Institute of Genetics, NIG). In 2005, GenBank, EMBL and DDBJ announced the International Nucleotide Se-

1. Liu Yong, College of Life Sciences and Bioengineering, School of Science, Beijing Jiao Tong University, Beijing, China, 100044.

2. Zhang Lishu, College of Life Sciences and Bioengineering, School of Science, Beijing Jiao Tong University, Beijing, China, 100044.

quence Database Collaboration (INSDC). According to the agreement, these three databases each collects the nucleic acid sequence data published around the world, and shared their sequence data daily, to ensure that a uniform and comprehensive collection of sequence information is available worldwide. That means the sequence information underlying DDBJ, EMBL-Bank, and GenBank is equivalent. Let's begin with GenBank.

1.1.1 GenBank

GenBank is a comprehensive public database of nucleotide sequences and supporting bibliographic and biological annotation. GenBank is built and distributed by the National Center for Biotechnology Information (NCBI), a division of the National Library of Medicine (NLM), located on the campus of the US National Institutes of Health (NIH) in Bethesda, MD, USA. NCBI builds GenBank primarily from the submission of sequence data from authors and from the bulk submission of expressed sequence tag (EST), genome survey sequence (GSS), whole-genome shotgun (WGS) and other high-throughput data from sequencing centers. The US Patent and Trademark Office also contributes sequences from issued patents. GenBank data is available at no cost over the Internet, through FTP and a wide range of Web-based retrieval and analysis services. The website of GenBank is https://www.ncbi.nlm.nih.gov/genbank/, and its homepage is shown in Figure 1.1.

1. The source in GenBank

Virtually all records enter GenBank as direct electronic submissions (http://www.ncbi.nlm.nih.gov/genbank/), with the majority of authors using the BankIt or Sequin programs. Many journals require authors with sequence data to submit the data to a public sequence database as a condition of publication. GenBank staff can usually assign an accession number to a sequence submission within two working days of receipt, and do so at a rate of ~3500 per day. The accession number serves as confirmation that the sequence has been submitted and provides a means for readers of articles in which the sequence is cited to retrieve the data. Direct submissions receive a quality assurance review that includes checks for vector contamination, proper translation of coding regions, correct taxonomy and correct bibliographic citations. A draft of the GenBank record is passed back to the author for review before it enters the database.

Authors may ask that their sequences be kept confidential until the time of publication. Since GenBank policy requires that the deposited sequence data be made public when the sequence or accession number is published, authors are instructed to inform GenBank staff of the publication date of the

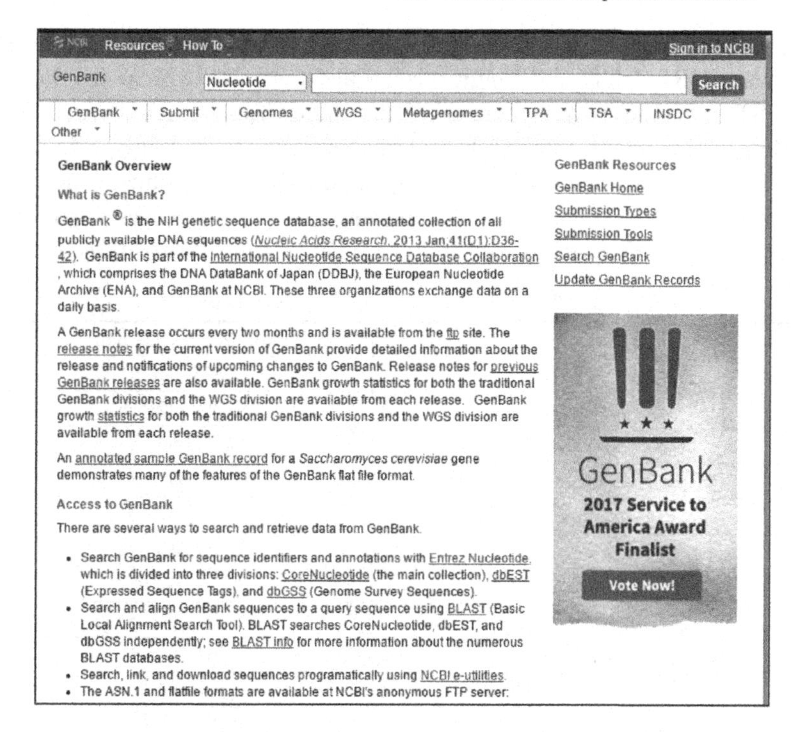

Fig. 1.1 The homepage of GenBank

article in which the sequence is cited in order to ensure a timely release of the data. Although only the submitter is permitted to modify sequence data or annotations, all users are encouraged to report lags in releasing data or possible errors or omissions to GenBank at update@ncbi.nlm.nih.gov.

NCBI works closely with sequencing centers to ensure timely incorporation of bulk data into GenBank for public release. GenBank offers special batch procedures for large scale sequencing groups to facilitate data submission, including the program tbl2asn, described at http://www.ncbi.nlm.nih.gov/genbank/tbl2asn2.html.

2. Organization of the database

(1) Sequence-based taxonomy

Database sequences are classified and can be queried using a comprehensive sequence-based taxonomy (http://www.ncbi.nlm.nih.gov/taxonomy/). Over 300 000 formally described species are represented in GenBank, and the top species in the non-WGS GenBank divisions are listed in Table 1.1.

(2) GenBank divisions

GenBank assigns sequence records to various divisions based either on the source taxonomy or the sequencing strategy used to obtain the data. There are 12 taxonomic divisions (BCT, ENV, INV, MAM, PHG, PLN, PRI, ROD, SYN, UNA, VRL, VRT) and 5 high-throughput divisions: expressed sequence tags (EST), genome survey sequences (GSS), high-throughput cDNA (HTC), high-throughput genomic (HTG), and the sequence-tagged site (STS). Finally, the PAT division contains records supplied by patent offices, the transcriptome shotgun assembly (TSA) division contains sequences from TSA projects and the WGS division contains sequences from WGS projects.

Table 1.1 Top organisms in GenBank

Organism	Non-WGS base pairs
Homo sapiens	17 575 474 103
Mus musculus	9 993 232 725
Rattus norvegicus	6 525 559 108
Bos taurus	5 391 699 711
Zea mays	5 079 812 801
Sus scrofa	4 894 315 374
Danio rerio	3 128 000 237
Triticum aestivum	1 925 428 081
Solanum lycopersicum	1 764 995 265
Hordeum vulgare	1 617 554 059
Strongylocentrotus purpuratus	1 435 261 003
Macaca mulatta	1 297 237 624
Oryza sativa Japonica Group	1 265 215 013
Xenopus tropicalis	1 249 788 384
Nicotiana tabacum	1 200 025 462
Arabidopsis thaliana	1 165 816 533
Drosophila melanogaster	1 155 228 906
Vitis vinifera	1 071 458 039
Glycine max	1 020 646 789
Pan troglodytes	1 010 316 029

Expressed sequence tags.

ESTs is a major source of sequence records and gene sequences. The top organisms represented in the EST division are *H. sapiens*, *M. musculus*, *S. scrofa*, *Arabidopsis thaliana*, *B. Taurus*, *Z. mays* and *D. rerio*. As part of its daily processing of GenBank EST data, NCBI identifies through BLAST searches all homologies for new EST sequences and incorporates that information into the companion database, dbEST (www.ncbi.nlm.nih.gov/dbEST/ index.html). The data in dbEST are processed further to produce the UniGene database (www.ncbi.nlm.nih.gov/ sites/entrez?db=unigene) of millions of gene-oriented sequence clusters.

Sequence-tagged sites, GSSs and ENV.

The sequence-tagged site (STS) division of GenBank (www.ncbi.nlm.nih.gov/dbSTS/index.html) contains anonymous STSs based on genomic sequences as well as gene-based STSs derived from the 3'-ends of genes and ESTs. These STS records usually include mapping information.

The GSS division of GenBank (www.ncbi.nlm.nih.gov/dbGSS/index.html) grew rapidly over the past years. GSS sequences are the products of as many as 80 different experimental techniques, including meta-genomic surveys of sequences arising from biological communities. However, more than one-quarter of all GSS records are single reads from bacterial artificial chromosomes ('BAC-ends') used in a variety of genome sequencing projects. The most highly represented species in the GSS division, including meta-genomic surveys, are marine meta-genome, *M. musculus*, *Z. mays* and *H. sapiens*. The human data have been used (www.ncbi.nlm.nih.gov/projects/genome/clone/) along with the STS records in tiling the BACs for the Human Genome Project.

The ENV division of GenBank accommodates non-WGS sequences obtained via environmental sampling methods in which the source organism is unknown. Many ENV sequences arise from meta-genome samples derived from various animal tissues, such as the gut or skin, or from particular environments, such as freshwater sediment, hot springs or areas of mine drainage. Records in the ENV division contain 'ENV' in the keyword field and use an '/environmental_sample' qualifier in the source feature.

HTG and HTC sequences.

The HTG division of GenBank (www.ncbi.nlm.nih.gov/HTGS/) contains unfinished large-scale genomic records, which are in transition to a finished state. These records are designated as Phase 0–3 depending on the quality of the data, with Phase 3 being the finished state. Upon reaching Phase 3, HTG records are moved into the appropriate organism division of GenBank. The HTC division of GenBank accommodates HTC sequences, which are of draft quality but may contain 5'-UTRs and 3'-UTRs, partial-coding regions and introns. HTC sequences which are finished and of high quality are moved to the appropriate organism division of GenBank.

WGS sequences.

WGS sequences appear in GenBank as a set of WGS contigs, many of them bearing annotations originating from a single sequencing project. These sequences are issued accession numbers consisting of a four-letter project ID, followed by a two-digit version number and a six-digit contig ID. Hence, the WGS accession number 'AAAA01072744' is assigned to contig number '072744' of the first version of the project 'AAAA'. WGS project contigs for *H. sapiens*, *Pan trodlodytes*, *Macacca mulatta*, *Equus caballus*, *Canis familiaris*, *Drosophila*, *Saccharomyces* and 800 other organisms and environmental samples are available. For a complete list of WGS projects with links to the data, see www.ncbi.nlm.nih.gov/projects/WGS/WGSprojectlist.cgi. Although WGS project sequences may be annotated, many low-coverage genome projects do not contain annotation. Because these sequence projects are ongoing and incomplete, these annotations may not be tracked from one assembly version to the next and should be considered preliminary.

TSA sequences.

In recent years, a growing number of sequencing traces have been deposited in the NCBI Trace Archive (TA). Given the advent of next-generation sequencing technologies, including those from Roche-454 Life Sciences, Illumina Solexa and Applied Biosystems SOLiD, NCBI deployed a Short Read Archive (SRA) in 2007. Neither of these archives is a part of GenBank, but beginning with release 166, GenBank added a new TSA division for TSA sequences, which are shotgun assemblies of sequences deposited in TA, SRA and the EST division of GenBank. TSA records (e.g. EZ000001) have 'TSA' as their keyword and a primary block that provides the base ranges and identifiers of the sequences used in the TSA assembly.

(3) Special record types

Third-party annotation (TPA).

TPA records are sequence annotations published by someone other than the original submitter of the primary sequence record in DDBJ/ENA/GenBank (http://www.ncbi.nlm.nih.gov/genbank/TPA). Each TPA record falls into one of three categories: experimental, in which case there is direct experimental evidence for the existence of the annotated molecule; inferential, in which case the experimental evidence is indirect; and assembly, where the focus is on providing a better assembly of the raw reads. TPA sequences may be created by assembling a number of primary sequences. The format of a TPA record (e.g. BK000016) is similar to that of a conventional GenBank record but includes the label 'TPA exp:', 'TPA inf:' or 'TPA asm:' at the beginning of each definition line as well as corresponding keywords. TPA experimental and inferential records also contain a primary block that provides the base ranges and identifier for the sequences used to build the TPA. TPA sequences are not released to the public until their accession numbers or sequence data and annotation appear in a peer reviewed biological journal. TPA submissions to GenBank may be made using either BankIt or Sequin.

Contig (CON) records for assemblies of smaller records.

Within GenBank, CON records are used to represent very long sequences, such as a eukaryotic chromosome, where the sequence is not complete but consists of several contig records with uncharacterized gaps between them. Rather than listing the sequence itself, CON records contain assembly instructions involving the several component sequences. An example of such a CON record is CM000663 for human chromosome 1.

3. Retrieving GenBank data

There are several ways to search and retrieve data from GenBank. Search GenBank for sequence identifiers and annotations with Entrez Nucleotide, which is divided into three divisions: CoreNucleotide (the main collection), dbEST (Expressed Sequence Tags), and dbGSS (Genome Survey Sequences). Figure 1.2 and Figure 1.3 show how to retrieve FTO data through Entrez Nucleotide.

① We need to choose 'Nucleotide' in the drop down box, and enter the gene name in the text box (Figure 1.2), click on 'Search', and then we will get the list of results (Figure 1.3). Selecting an item and clicking it, the data record form of the sequence of FTO entry in GenBank is showed in Table 1.2.

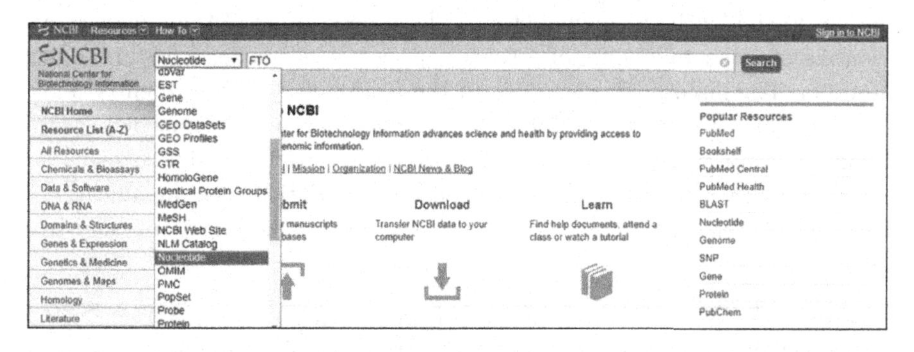

Fig. 1.2 Retrieve GenBank in the homepage of NCBI

② Search and align GenBank sequences to a query sequence using BLAST (Basic Local Alignment Search Tool). BLAST searches CoreNucleotide, dbEST, and dbGSS independently.

③ Search, link, and download sequences programatically using NCBI e-utilities.

④ The ASN.1 and flatfile formats are available at NCBI's anonymous FTP server: ftp://ftp.ncbi.nlm.nih.gov/ncbi-asn1 and ftp://ftp.ncbi.nlm.nih.gov/genbank.

Figure 1.4 is an example of the FTO gene, showing the data record form

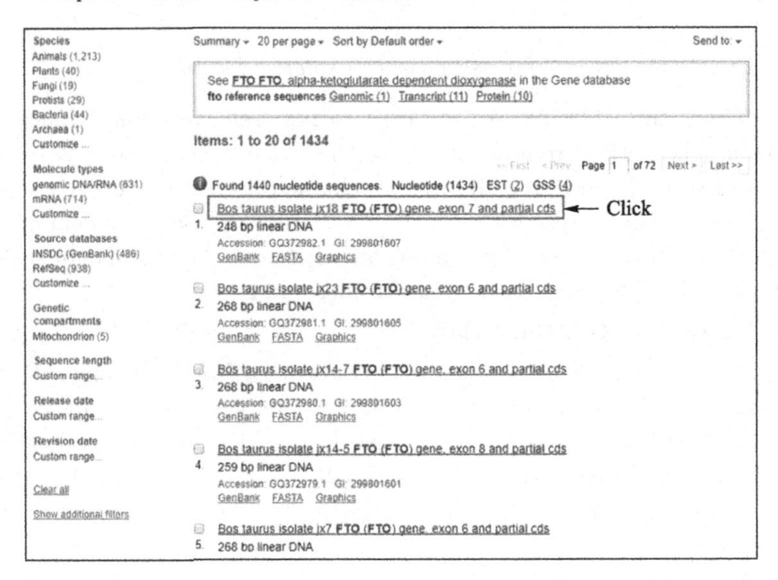

Fig. 1.3 The list of search results in Genbank

of the sequence entry in GenBank. The GenBank database not only gives the sequence information, but also contains comprehensive annotation information. Each GenBank entry contains a brief description of the sequence, a scientific name, a taxonomy of a source species, a reference, a feature, and an origin. The main fields contained in the GenBank data record entry and their meanings are shown in Table 1.2. Some sequence regions of biological significance are given in the characteristic data, such as coding sequences (CDS), transcription units, repeat regions, and mutations and modification sites that can encode proteins (Sites of mutations or modifications) and so on.

LOCUS	GQ372982 248 bp DNA linear MAM 06-OCT-2011
DEFINITION	Bos taurus isolate jx18 FTO (FTO) gene, exon 7 and partial cds.
ACCESSION	GQ372982
VERSION	GQ372982.1
KEYWORDS	.
SOURCE	Bos taurus (cattle)
ORGANISM	Bos taurus
	Eukaryota; Metazoa; Chordata; Craniata; Vertebrata; Euteleostomi;
	Mammalia; Eutheria; Laurasiatheria; Cetartiodactyla; Ruminantia;
	Pecora; Bovidae; Bovinae; Bos.
REFERENCE	1 (bases 1 to 248)
AUTHORS	Zhang,B., Zhang,Y., Zhang,L., Wang,J., Li,Z. and Chen,H.
TITLE	Allelic Polymorphism Detected in the Bovine FTO Gene

```
JOURNAL     Mol. Biotechnol. 49 (3), 257-262 (2011)
PUBMED      21479694
REFERENCE   2 (bases 1 to 248)
AUTHORS     Zhang,B. and Chen,H.
TITLE       Direct Submission
JOURNAL     Submitted (09-JUL-2009) College of Animal Science and Technology,
            Northwest A&F University, Yangling, Shaanxi 712100, China
FEATURES    Location/Qualifiers
source      1..248
            /organism="Bos taurus"
            /mol_type="genomic DNA"
            /isolate="jx18"
            /db_xref="taxon:9913"
            /chromosome="18"
            /PCR_primers="fwd_seq: tactggaggagaactgaat, rev_seq:
            gcacaacatcccaagaaa"
            /note="breed: Jiaxian"
gene        <1..>248
            /gene="FTO"
misc_difference 58
            /gene="FTO"
            /replace="t"
mRNA        <92..>211
            /gene="FTO"
            /product="FTO"
CDS         <92..>211
            /gene="FTO"
            /codon_start=1
            /product="FTO"
            /protein_id="ADJ51677.1"
/translation="VEFEWLRQFWFQGSRYKKCTDWWCQPMSQLEEMWRKMEWL"
exon        92..211
            /gene="FTO"
            /number=7
ORIGIN
1           tactggagga gaactgaatg acgaagttca ttgtccttga ttgtccatca tggggacatt
61          ataaccacgg tcttgtcttt atgatcccca ggtcgagttt gaatggctga gacagttttg
121         gtttcaaggc agtcgataca aaaagtgcac tgactggtgg tgtcagccca tgagtcagct
181         ggaagagatg tggagaaaga tggagtggtt ggtaagtatc caggagtgga tttcttggga
241         tgttgtgc
//
```

Fig. 1.4 The entry for human FTO at GenBank

Table 1.2 The main fields contained in the GenBank data record entry and their meanings

Field	Meaning
LOCUS	The locus name, length, molecule type, GenBank division, and modification date of the sequence
DEFINITION	Brief description of sequence; includes information such as source organism, gene name/protein name, or some description of the sequence's function (if the sequence is non-coding), such as "complete cds"
ACCESSION	The unique identifier for a sequence record
VERSION	The version information of the entry
KEYWORDS	Word or phrase describing the sequence
SOURCE	Free-format information including an abbreviated form of the organism name sometimes followed by a molecule type
ORGANISM	The formal scientific name for the source organism
REFERENCE	Publications by the authors of the sequence
AUTHORS	The authors of references
TITLE	The title of the published work
JOURNAL	MEDLINE abbreviation of the journal name
PUBMED	PubMed Identifier (PMID)
COMMENT	Information about genes and gene products, as well as regions of biological significance reported in the sequence
FEATURES	Feature data of the sequence
ORIGIN	Start sign of the original sequence
//	End sign of the original sequence

Different series of items vary in size; some have only a few bases, and some have hundreds of thousands of bases. The above feature data are for reference only and not all sequences are included.

1.1.2 EMBL

EMBL-DNA database was established in 1982 by the European Molecular Biology Laboratory (EMBL), and it was the earliest database of DNA sequence. EBI is EMBL's Hinxton department in the UK, and it is mainly responsible for the establishment of EMBL-DNA database, which provides nucleotide sequence searching and sequence similarity querying. The website of EBI is http://www.ebi.ac.uk (Figure 1.5).

The main sources of data of EMBL include: ① directly submitted by the discoverer of the sequence, most of the international authoritative biology magazines require authors to submit their sequencing sequences to GenBank,

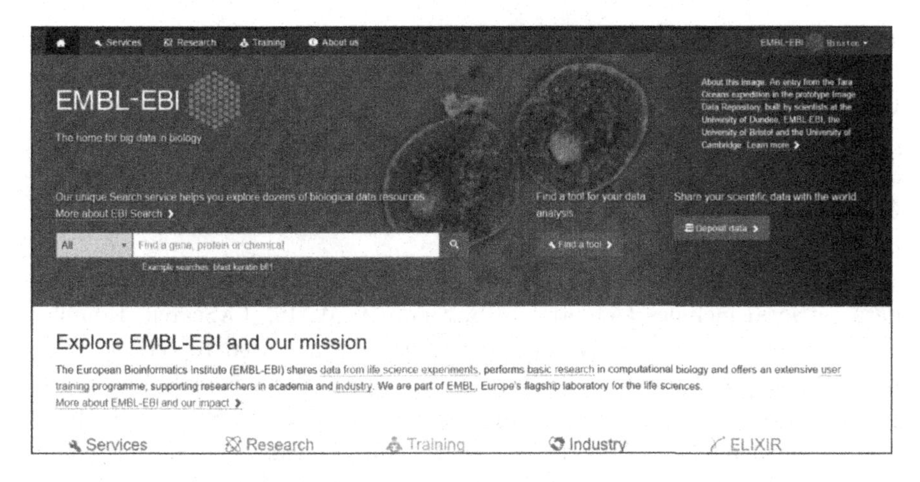

Fig. 1.5 The homepage of EMBL-EBI

EMBL or DDBJ, and get the sequence code issued by the database management system before the article be published; ② the published sequences from the *Journal of Biomedical Information.*

EMBL database is maintained by the ORACLE database management system. The identification mark and their meanings in EMBL is showed in Table 1.3.

Table 1.3 The major identification marks and their meanings in EMBL

Identification mark	Meaning
ID	Identification
AC	Accession number(s)
DE	Description
OS	Organism species
OC	Organism classification
KW	Keywords
RN	Reference number
RA	Reference author
RT	Reference title
RL	Reference location
DR	Database cross-references
DT	Date
SQ	Sequence header
//	Termination line

1.1.3 DDBJ

DDBJ nucleotide sequence database was established in 1984 and maintained by National Institute of Genetics (NIG). DDBJ collects nucleotide sequence data as a member of INSDC (International Nucleotide Sequence Database Collaboration) and provides freely available nucleotide sequence data to support research activities in life science.

Major activities of DDBJ are shown in Figure 1.6. Among them, the data retrieval includes Getentry, SRS, Sfgate & WAIS, TXSearch, Homology and several other ways. The first four ways are used to retrieve the raw data in the database. Homology is used to perform homologous analysis of user-submitted sequences or fragments through FASTA / BLAST. The retrieval methods provided by DDBJ are divided into sequence code retrieval, keyword retrieval and classification search. Getentry belongs to sequence code retrieval. SRS and Sfgate & WAIS belong to keyword retrieval, and TXSearch belongs to classification search. Users can submit their data through SAKURA, MSS and Sequin. DDBJ also provides multi-fragment sequence analysis and production of phylogenetic tree.

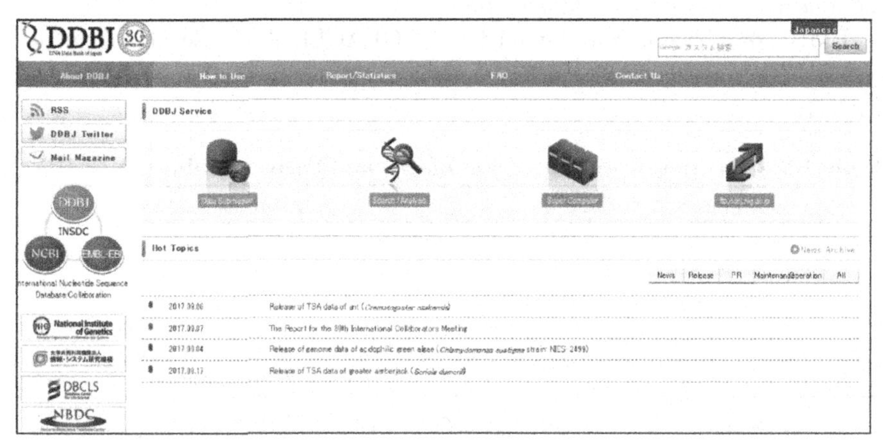

Fig. 1.6 Major activities of DDBJ

1.2 Protein Database

The amount of information available about proteins continues to increase at a rapid pace. Protein interactions, expression profiles and structures are being discovered on a large scale, while completely sequenced genomes cover the taxonomic tree with both breadth and depth. Biological and biochemical functions of individual proteins continue to be elucidated. Furthermore, improved analytical tools are available to make intelligent predictions about function, localization, secondary structure and other important protein properties.

The ability to store and interconnect this expanding universe of protein information is crucial to modern biological research. Accordingly, the Universal Protein Database plays an ever more important role by providing a central resource on protein sequences and functional annotation for biologists and scientists active in functional proteomics and genomics research.

Prior to 2002 there used to be three separate protein databases located around the globe each with its own data and rules for annotation. EMBL-EBI and SIB together used to produce Swiss-Prot and TrEMBL, while PIR produced the Protein Sequence Database (PIR-PSD). In 2002, three institutes pooled their resources and expertise and formed the UniProt consortium to provide a unified source with common annotation standards and datasets.

1.2.1 UniProt (Universal Protein Resource) Overview

UniProt strives to provide a centralized repository of protein sequences with comprehensive coverage and a systematic approach to protein annotation, incorporating, interpreting, integrating and standardizing data from large and disparate sources, and is the most comprehensive catalog of protein sequence and functional annotation. It has four components optimized for different uses. The UniProt Knowledgebase (UniProtKB) is an expertly curated database, a central access point for integrated protein information with cross-references to multiple sources. The UniProt Archive (UniParc) is a comprehensive sequence repository, reflecting the history of all protein sequences. UniProt Reference Clusters (UniRef) merge closely related sequences based on sequence identity to speed up searches, while the UniProt Metagenomic and Environmental Sequences database (UniMES) were created to respond to the expanding area of metagenomic data. UniProt is freely and easily accessible by researchers to conduct interactive and custom tailored analyses for proteins of interest to facilitate hypothesis generation and knowledge discovery.

The website of the UniProt homepage is http://www.uniprot.org/, see Figure 1.7.

1.2.2 The UniProt Knowledgebase (UniProtKB)

The centerpiece of UniProt database is the UniProtKB—a richly annotated protein sequence database with extensive cross references. Much of the annotation data are buried within the ever-increasing volume of scientific pub-

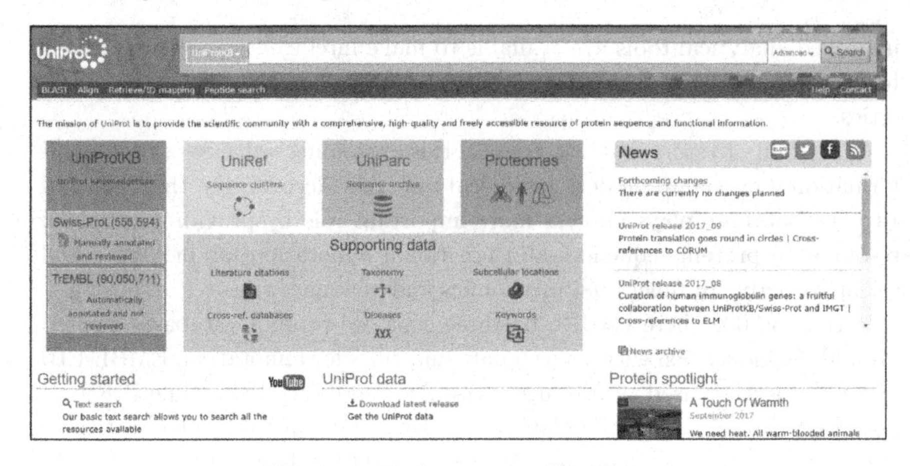

Fig. 1.7 The homepage of UniProt

lications or spread among individual databases stored at different locations with differing formats. The UniProtKB provides an integrated and uniform presentation of these disparate data, including annotations such as protein name and function, taxonomy, enzyme specific information (catalytic activity, cofactors, metabolic pathway, regulatory mechanisms), domains and sites, posttranslational modifications, subcellular locations, tissue specific or developmentally specific expression, interactions, splice isoforms, polymorphisms, diseases and sequence conflicts. Literature citations provide evidence for experimental data. Entries connect to various external data collections such as the underlying DNA sequence entries, protein structure databases, protein domain and family databases, and species and function-specific data collections. As a result, UniProtKB acts as a central hub connecting biomolecular information archived in 100 cross-referenced databases.

The UniProtKB contains two sections: UniProtKB/Swiss-Prot and UniProtKB/TrEMBL. UniProtKB/Swiss-Prot contains records with full manual annotation or computer-assisted, manually-verified annotation performed by biologists and based on published literature and sequence analysis. UniProtKB/TrEMBL contains records with computationally generated annotation and large-scale functional characterization. The computer-assisted annotation may employ some automatically generated rules, or manually curated rules based on protein families, including HAMAP family rules, PIRSF classification based name and site rules and RuleBase rules.

UniProtKB/TrEMBL is a supplement to the Swiss-Prot database of the Swiss Bioinformatics Institute. UniProtKB/TrEMBL contains the translations of all coding sequences (CDS) present in the EMBL/GenBank/DDBJ Nucleotide Sequence Databases and also protein sequences extracted from the literature or submitted to UniProtKB/Swiss-Prot. The database is divided into two parts: SP-TrEMBL and REM-TrEMBL. Items in SP-TrEMBL will be merged into the Swiss-Prot database, as a pre-release version of Swiss-Prot. REM-TrEMBL contains the remaining sequences, including immunoglobulins, T cell receptors, small peptides of less than 8 amino acid residues, synthetic sequences, patent sequences, etc. Since TrEMBL is translated by a computer program based on nucleic acid sequences, the sequence error rate may be large and there may be a greater redundancy.

1.2.3 UniProt Reference Clusters (UniRef)

The UniRef includes three separate datasets that compress sequence space at different resolutions, achieved by merging sequences and sub-sequences that are 100% (UniRef100), >90% (UniRef90) or >50% (UniRef50) identical, regardless of source organism. Reduction of sequence redundancy speeds sequence similarity searches while rendering such searches more informative.

1.2.4 Example of UniProtKB Search: HBB

Figure 1.8 shows the data record of the sequence entry of hemoglobin subunit beta (HBB) protein in UniProtKB. Table 1.4 gives a description of identification mark.

```
ID   HBB_HUMAN                Reviewed;       147 AA.
AC   P68871; A4GX73; B2ZUE0; P02023; Q13852; Q14481; Q14510; Q45KT0;
AC   Q549N7; Q6FI08; Q6R7N2; Q8IZI1; Q9BX96; Q9UCD6; Q9UCP8; Q9UCP9;
DT   21-JUL-1986, integrated into UniProtKB/Swiss-Prot.
DT   23-JAN-2007, sequence version 2.
DT   30-AUG-2017, entry version 163.
DE   RecName: Full=Hemoglobin subunit beta;
DE   AltName: Full=Beta-globin;
DE   AltName: Full=Hemoglobin beta chain;
DE   Contains:
DE     RecName: Full=LVV-hemorphin-7;
DE   Contains:
DE     RecName: Full=Spinorphin;
GN   Name=HBB;
OS   Homo sapiens (Human).
OC   Eukaryota; Metazoa; Chordata; Craniata; Vertebrata; Euteleostomi;
OC   Mammalia; Eutheria; Euarchontoglires; Primates; Haplorrhini;
```

```
OC   Catarrhini; Hominidae; Homo.
OX   NCBI_TaxID=9606;
RN   [1]
RP   NUCLEOTIDE SEQUENCE [GENOMIC DNA].
RX   PubMed=1019344;
RA   Marotta C., Forget B., Cohen-Solal M., Weissman S.M.;
RT   "Nucleotide sequence analysis of coding and noncoding regions of human
RT   beta-globin mRNA.";
RL   Prog. Nucleic Acid Res. Mol. Biol. 19:165-175(1976).
RN   [2]
RP   NUCLEOTIDE SEQUENCE [GENOMIC DNA].
RX   PubMed=6254664; DOI=10.1016/0092-8674(80)90428-6;
RA   Lawn R.M., Efstratiadis A., O'Connell C., Maniatis T.;
RT   "The nucleotide sequence of the human beta-globin gene.";
RL   Cell 21:647-651(1980).
RN   [3]
RP   NUCLEOTIDE SEQUENCE [GENOMIC DNA], AND VARIANT LYS-7.
...
RA   Ross C., Mandal A.K.;
RT   "Mass spectrometry based characterization of Hb Beckman variant in a
RT   falsely elevated HbA(1c) sample.";
RL   Anal. Biochem. 489:53-58(2015).
CC   -!- FUNCTION: Involved in oxygen transport from the lung to the
CC       various peripheral tissues.
...
CC       URL="https://en.wikipedia.org/wiki/Hemoglobin";
CC   ---------------------------------------------------------------
CC   Copyrighted by the UniProt Consortium, see http://www.uniprot.org/terms
CC   Distributed under the Creative Commons Attribution-NoDerivs License
CC   ---------------------------------------------------------------
DR   EMBL; M25079; AAA35597.1; -; mRNA.
DR   EMBL; V00499; CAA23758.1; -; Genomic_DNA.
DR   EMBL; DQ126270; AAZ39745.1; -; Genomic_DNA.
...
DR   GO; GO:0010942; P:positive regulation of cell death; IDA:BHF-UCL.
DR     GO; GO:0045429; P:positive regulation of nitric oxide biosynthetic process;
     NAS:UniProtKB.
...
PE   1: Evidence at protein level;
KW   3D-structure; Acetylation; Complete proteome;
KW   Congenital dyserythropoietic anemia; Direct protein sequencing;
KW   Disease mutation; Glycation; Glycoprotein; Heme;
KW   Hereditary hemolytic anemia; Hypotensive agent; Iron; Metal-binding;
```

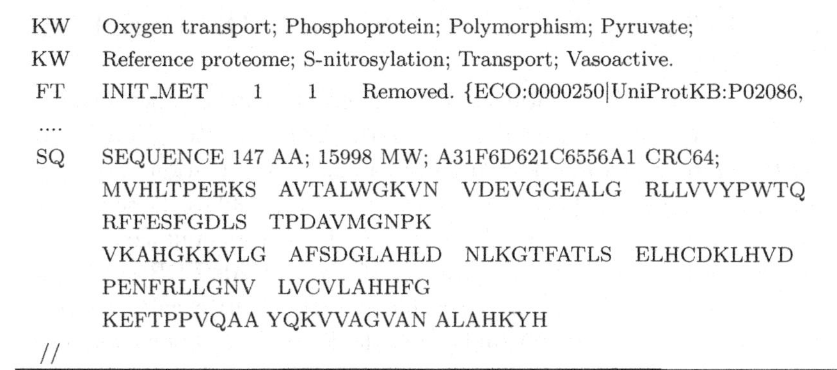

```
KW   Oxygen transport; Phosphoprotein; Polymorphism; Pyruvate;
KW   Reference proteome; S-nitrosylation; Transport; Vasoactive.
FT   INIT_MET   1   1   Removed. {ECO:0000250|UniProtKB:P02086,
....
SQ   SEQUENCE 147 AA; 15998 MW; A31F6D621C6556A1 CRC64;
     MVHLTPEEKS  AVTALWGKVN  VDEVGGEALG  RLLVVYPWTQ
     RFFESFGDLS  TPDAVMGNPK
     VKAHGKKVLG  AFSDGLAHLD  NLKGTFATLS  ELHCDKLHVD
     PENFRLLGNV  LVCVLAHHFG
     KEFTPPVQAA YQKVVAGVAN ALAHKYH
//
```

Fig. 1.8 The entry for HBB in UniProtKB

Table 1.4 The main identification marks contained in the UniProtKB data record entry and the meanings

Identification mark	Meaning	The number of occurrences in each record
ID	Identification	Once; starts the entry
AC	Acccession number(s)	Once or more
DT	Date	Tree times
DE	Description	Once or more
GN	Gene names(s)	Optional
OS	Organism species	Once
OG	Organelle	Optional
OC	Organism classification	Once or more
OX	Taxonomy cross-reference	Once
RN	Reference number	Once or more
RP	Reference position	Once or more
RC	Reference comment(s)	Optional
RX	Reference coss-reference(s)	Optional
RG	Reference group	Once or more
RA	Refrence authors	Once or more
RT	Reference title	Optional
RL	Reference location	Once or more
CC	Comments or notes	Optional
DR	Database cross-references	Optional
KW	Keywords	Optional
FT	Feature table data	Optional
SQ	Sequence header	Once
(blanks)	Sequence data	Once or more
//	Termination line	Once; ends the entry

1.3 Protein Three-Dimensional Structure Database PDB

Protein Data Bank (PDB) database is an international biological macromolecule structure database, established by the United States Brookhaven National Laboratory in 1971. This resource provides information about the 3D shapes of proteins, nucleic acids, and complex assemblies. It helps students and researchers understand all aspects of biomedicine and agriculture, from protein synthesis to health and disease. The Worldwide PDB (wwPDB) organization (http://www.wwpdb.org/) manages the PDB archive and ensures that the PDB is freely and publicly available to the global community.

Nowadays, structures and experimental data are deposited at and processed by the wwPDB partner sites in America (Research Collaboratory for Structural Bioinformatics Protein Data Bank, RCSB PDB, http://rcsb.org), Europe (Protein Data Bank in Europe, PDBe, http://pdbe.org), and Japan (Protein Data Bank Japan, PDBj, http://pdbj.org). Each site offers various tools for searching, visualizing, and analyzing PDB data.

The website of RCSB PDB is http://www.rcsb.org/pdb/home/home.do, and its homepage is showed in Figure 1.9.

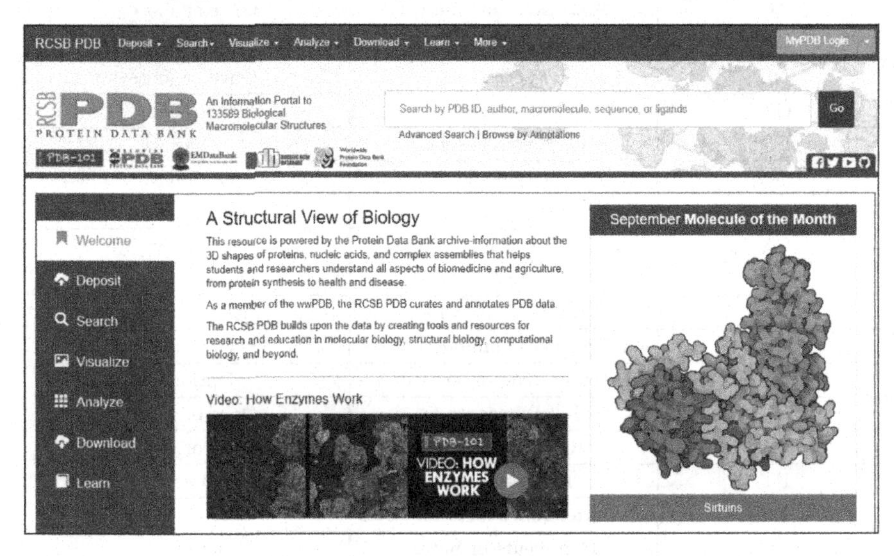

Fig. 1.9 The homepage of RCSB PDB

1.4 Genome Browser

The initial sequences, generated by the human genome project together with the draft genome sequences of several model organisms, including the house mouse (*Mus musculus*), fruit fly (*Drosophila melanogaster*), nematode (*Caenorhabditis elegan*), baker's yeast (*Saccharomyces cerevisiae*), Gram-negative bacterium (*Escherichia coli*) and thale cress (*Arabidopsis thaliana*), completed at the beginning of this millennium, create a paradigm shift within biological research, as predicted by Gilbert in the early 1990s. With the rapid development of next-generation sequencing technologies, hundreds of eukaryotic and thousands of prokaryotic genomes have been sequenced.

It is not helpful to have 3 billion letters of genomic DNA shown as plain text! By systematic integration of genome sequences together with annotations generated through much heterogeneous data, genome browser provides a unique platform for molecular biologists to browse, search, retrieve and analyze these genomic data efficiently and conveniently. With a graphical interface, genome browser helps users to extract and summarize information intuitively from huge amount of raw data. Furthermore, different types of annotations from multiple sources can be integrated into one genome browser, helping users to analyze data across different data providers effectively. With a uniform interface, users can navigate the whole genome using the same genomics coordinate system, and make comparative analysis across different lineages such as primates, mammalians, vertebrates and plants.

We will mainly introduce three human genome browsers: UCSC Genome Browser, Ensembl genome browser and NCBI Map Viewer.

1.4.1 UCSC Genome Browser

The University of California, Santa Cruz (UCSC) Genome Browser Database is an up to date source for genome sequence data integrated with a large collection of related annotations. The database is optimized to support fast interactive performance with the web-based UCSC Genome Browser, a tool built on top of the database for rapid visualization and querying of the data at many levels. The annotations for a given genome are displayed in the browser as a series of tracks aligned with the genomic sequence. Sequence data and annotations may also be viewed in a text-based tabular format or downloaded as tab-delimited flat files. The Genome Browser Database, browsing tools and downloadable data files can all be found on the UCSC Genome Bioinformatics web site (http://genome.ucsc.edu) (Figure 1.10), which also contains links to documentation and related technical information. The following is the homepage of the UCSC. The middle of the function menu bars show the main UCSC tools, including Genome Browser, BLAT, Table Browser, Gene Sorter, In-Silico PCR, VisiGene, Genome Graphs, etc. The bottom of the page is a brief introduction to UCSC and the recent information updates.

Fig. 1.10 The homepage of UCSC Genome Browser

1. Organization of the database

Sequence and annotation data for each genome assembly are stored in a MySQL relational database, which is quite efficient at retrieving data from indexed files. The database is loaded in large batches and is used primarily as a read-only database. To improve performance, each of the Genome Browser webservers has a copy of the database on its local disk.

The Genome Browser database contains both positional tables with data based on genomic start–stop coordinates and non-positional tables with data independent of position. The coordinates in positional tables are defined using half-open zero-based ranges, i.e. the first 100 bases of a chromosome are represented as (0,100), while the next 100 bases are represented as (100,200) and so on. Half-open coordinates allow the length of a feature to be obtained by simply subtracting the start from the end and tend to minimize $+/-1$ errors during software development.

The database is optimized to support the browser's range-based queries, with additional optimizations to accommodate the varying sizes of the positional tables in the database.

2. Viewing and downloading data

The UCSC Genome Bioinformatics website provides both graphical and text-based views of the sequence data and annotations.

The Genome Browser—accessed via the Genome Browser link on the home page—is a rapid interactive interface to the data at varying levels of detail. In response to a user's query, the Genome Browser displays a set of annotations tracks corresponding to the assembly and range dictated by the query and configured based on user preferences. Alternatively, the BLAT search tool can be used to quickly search for homologous regions to a DNA or protein sequence, which can be displayed in the browser. The user can navigate through the assembly and zoom the image in or out using navigation buttons. A set of track controls allows the user to display each annotation in single line (dense) or expanded (full) mode or hide the track completely. Some tracks have filters to fine-tune the data displayed. Individual entries within a track have associated details pages that provide information about the annotation, related links to outside sites and database, and in some cases links to the genomic, mRNA and protein sequence.

At the chromosome level (Figure 1.11), the Genome Browser display pro-

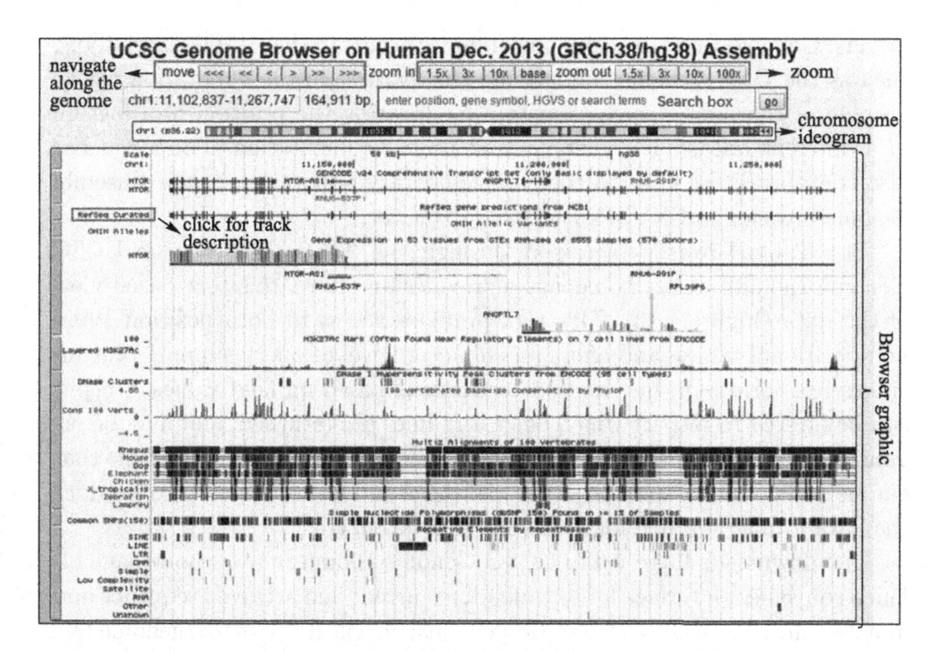

Fig. 1.11 The chromosome level viewing in UCSC

vides an overview of the coverage and completeness for the region. At a zoomed-in level, the display can focus on a specific area of research interest, for example the examination of alternative splicing patterns of a gene. Most annotation tracks are displayed in a horizontal linear fashion, with the exception of a few 'wiggle' tracks that plot the feature's scores on a vertical axis.

Users can configure the Track Controls section (Figure 1.12) to change the display mode (hide, dense, full, squish or pack) of each annotation tracks.

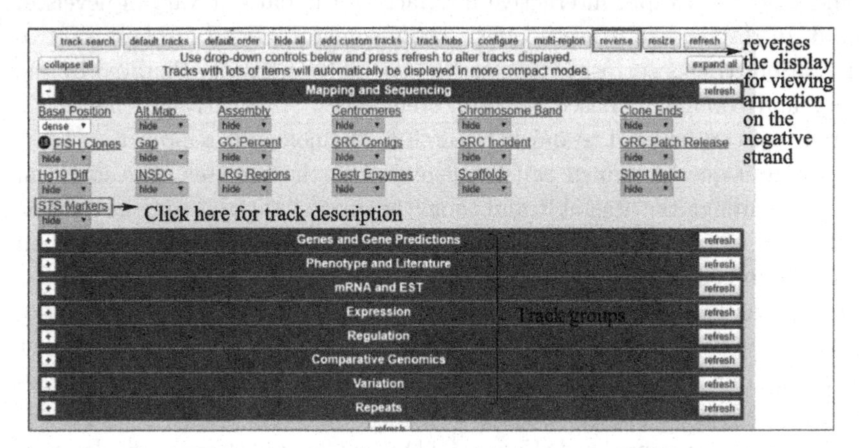

Fig. 1.12 The track display interface in UCSC

The Genome Browser menu bar provides access to the BLAT search tools, such as the DNA sequence underlying the features in the annotation tracks graphic, a coordinate conversion tool for locating the position of a feature in a different release of a genome, and user documentation. The menu bar also contains direct links to the complementary annotation in the Ensembl genome browser and NCBI's Entrez MapViewer.

The Table Browser—accessed through the Tables link on the UCSC Genome Bioinformatics homepage—provides an alternative text-based view of the data (Figure 1.13). This tool provides access to both positional and non-positional tables, and offers an enhanced level of query support that includes free-form SQL queries and restrictions based on field values. Output can be filtered to restrict which fields and lines are returned, and may be organized into one of several formats, including a simple tab-delimited file that can be loaded into a spreadsheet or database and then processed to produce the data format suitable for a custom annotation track.

The downloads link on the UCSC Genome Bioinformatics homepage offers a convenient interface for downloading current and archived sequence and annotation data. Data can also be downloaded via ftp at ftp://genome.cse.ucsc.edu/goldenPath/. Zipped data for each genome assembly are organized

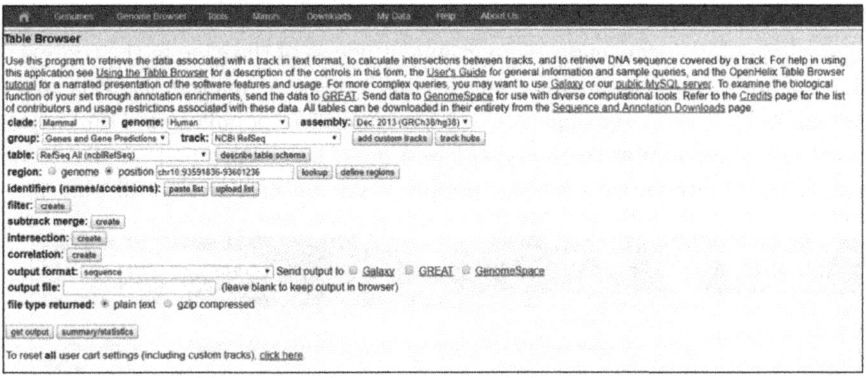

Fig. 1.13 The Table Browser of UCSC Genome Browser

into three folders. BigZips contains the sequence data, which is packaged by chromosome and by contig. Files ended with suffix .fa contain the sequence in Fasta format for the contig layout. Files with suffix .agp contain an index that shows how the corresponding .fa file is built, with each line representing an actual sequence record or a gap. The Chromosomes folder contains the assembled sequence in Fasta format divided up by chromosome. The Database folder contains the genome annotation database tables in tab-delimited format.

A large portion of the Genome Browser Database is accessible using the Distributed Annotation Service (DAS) protocol. The UCSC DAS server is located at http://genome.ucsc.edu/cgi-bin/das. To accommodate the large size of some of the annotation tables, it is best to enable compression on DAS clients when accessing the UCSC DAS server.

In addition to the browsing tools, the UCSC Genome Bioinformatics homepage provides links to the BLAT alignment tool, user and technical documentation, the Genome Browser mirror sites, archives, etc.

1.4.2 Ensembl Genome Browser

The Ensembl (http://www.ensembl.org/) database project provides a bioinformatics framework to organize the sequences of large genomes. It is a comprehensive source of stable automatic annotation of human, mouse and other genome sequences, available as either an interactive web site or as flat files.

Ensembl browser organizes different views into four classes: Location, Gene, Transcript and Variation, which can be easily navigated through tabs at the top of each web page (Figure 1.14). The location class includes views of the genome sequence at a range of resolutions and genome sequence based comparative views. Gene based views include textual information about the gene, views of its local genomic environment, views of the gene in the context of its orthologs and paralog relationships with other genomes in the Ensembl system and views of sequence variation within that population. Transcript based views are similar to the gene based ones, but focus on the individual transcript structures with more detail. Variation based views display in-

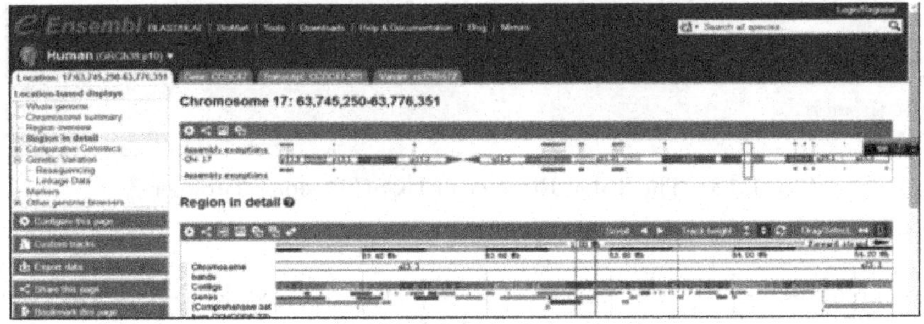

(a) location-based view

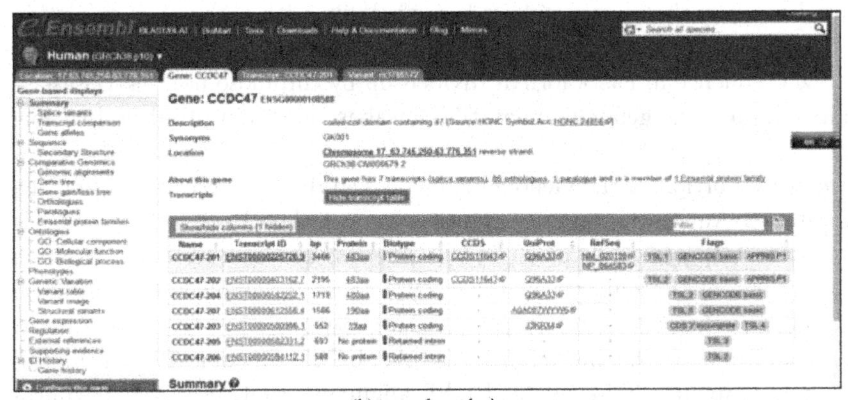

(b) gene-based view

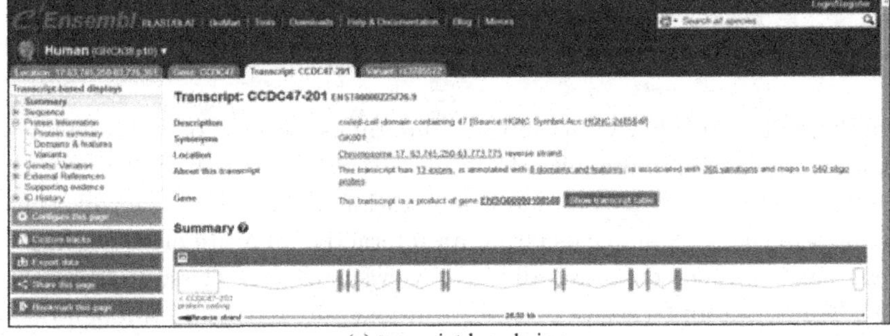

(c) transcript-based view

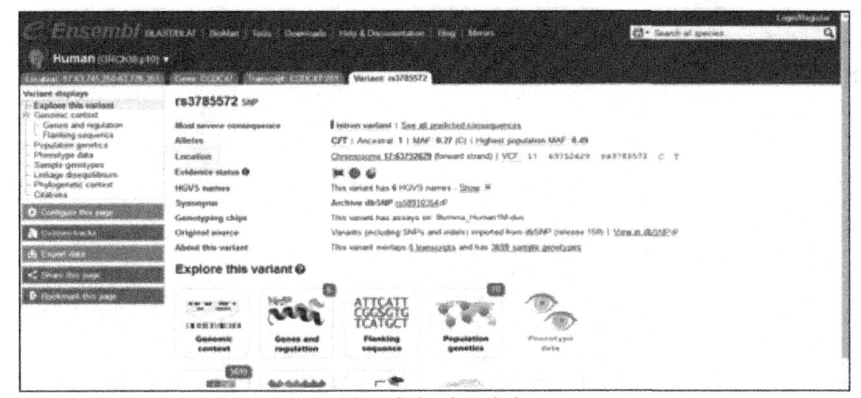

(d) variation-based view

Fig. 1.14 Ensembl genome browser

formation focused on the individual SNPs. The relationship between these new views is clearly shown by the left hand hierarchical menu which is context specific for each class. Each view within a class has a common header panel, summarizing the location or object. It is very easy to navigate among these views. Since only a specific chunk of information is shown in each view, this makes pages easier to read as well as improving the responsiveness of the servers. Configuration controls take the form of a context specific pop-up panel for most views, e.g. allowing tracks to be enabled and disabled in genome sequence based display elements.

1.4.3 NCBI Map Viewer

NCBI Map Viewer provides graphical display of features on NCBI's assembly of human genomic sequence data as well as cytogenetic, genetic, physical, and radiation hybrid maps.

Users can click on the 'Genome' on the homepage of the NCBI and click on 'Human Genome' to access the human genome map page (Figure 1.15). (https://www.ncbi.nlm.nih.gov/projects/genome/guide/human/index.shtml). Select a chromosome and click to enter a page similar to UCSC Genome Browser.

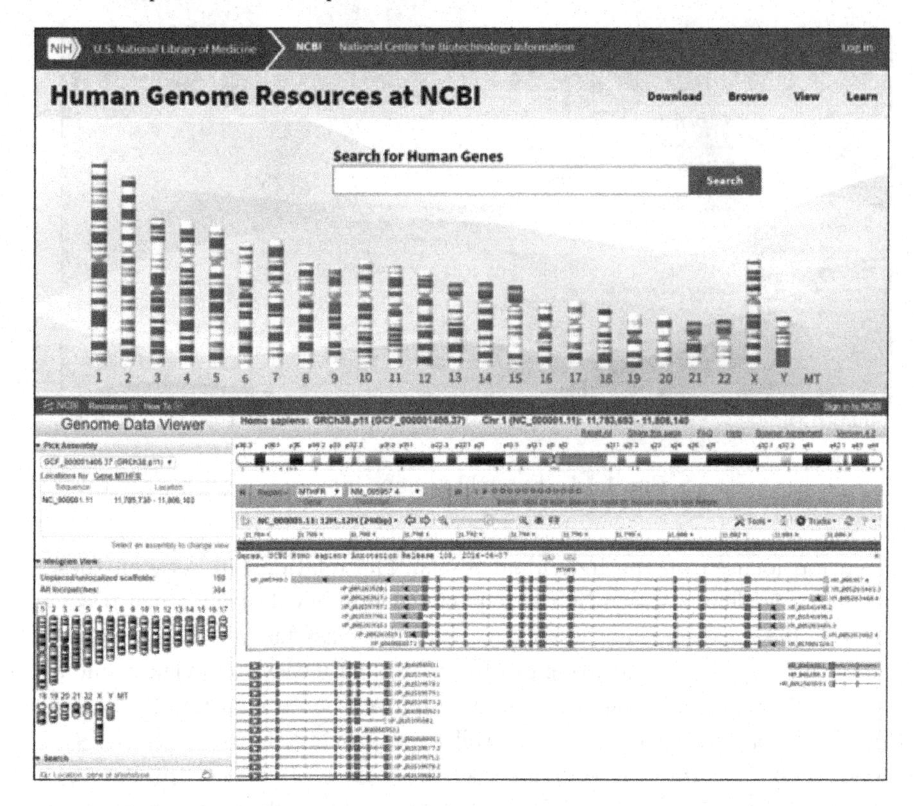

Fig. 1.15 Human Genome Resources in NCBI Map Viewer

References

[1] Anon. [2017-08-17]. https://www.ncbi.nlm.nih.gov/genbank/tbl2asn2/.

[2] Clamp M. Ensembl 2002: accommodating comparative genomics. *Nucleic Acids Research*, 2003, 31(1), 38-42.

[3] Clark K, Karsch-Mizrachi I, Lipman J D, et al. GenBank. *Nucleic Acids Research*, 2005(43): D30-D35.

[4] Cunningham F, Amode M, and Barrell D. Ensembl 2015. *Nucleic Acids Research*, 2015(43): D662–D669.

[5] Database resources of the National Center for Biotechnology Information. *Nucleic Acids Research*, 2004, 43(D1): D6-D17.

[6] Genome.ucsc.edu. Table Browser. (2017). [2017-10-22]. http://genome.ucsc.edu/goldenPath/help/hgTablesHelp.html#GettingStarted.

[7] Hubbard T, Aken B, Ayling S, and Ballester B. Ensembl 2009. *Nucleic Acids Research*, 2009(37): D690–D697.

[8] Karolchik D. The UCSC Genome Browser Database. *Nucleic Acids Research*, 2003, 31(1): 51-54.

[9] Pevsner J. *Bioinformatics and Functional Genomics.* 2nd ed. Hoboken, New Jersey: John Wiley & Sons, Inc, 2009: 13-47.

[10] Protein Data Bank (PDB): The Single Global Macromolecular Structure Archive. *Methods in Molecular Biology*, 2017: D627-D641.

[11] The Genome Database. [2017-08-30]. http://ftp://ftp.ncbi.nih.gov/pub/factsheets/Factsheet_Genome.pdf.

[12] The Universal Protein Resource (UniProt) in 2010. *Nucleic Acids Research*, 2009, 38(Database): D142-D148.

[13] Tyner C, Barber P G and Casper J. The UCSC Genome Browser database: 2017 update. *Nucleic Acids Research*, 2017(45): D626–D634.

[14] Uniprot.org. section: manual in Help. [2017-08-30]. http://www.uniprot.org/help/?query=*&fil=section:manual.

[15] Wang J, Kong L, Gao G and Luo J. A brief introduction to Web-based Genome Browsers. *Briefings in Bioinformatics*, 2012, 14(2): 131-143.

[16] Wu C. The Universal Protein Resource (UniProt): an expanding universe of protein information. *Nucleic Acids Research*, 2006, 34(90001): D187-D191.

[17] Zerbino D, Wilder S, Johnson N, Juettemann T and Flicek P. The Ensembl Regulatory Build. *Genome Biology*, 2015, 16(1).

Chapter 2 Sequence Alignment

Ning Zhao[1], Ziling Wang[2]

2.1 Pairwise Sequence Alignment

Sequence alignment is the most common task in bioinformatics, and also the basic composition and important basis of bioinformatics. Sequence alignment is not only the most common and most classical sequence analysis method for database search, gene comparison and new gene discovery, but also the results of sequence alignment as a secondary biological database provide valuable information for protein structure and functional prediction, phylogenetic tree establishment, gene disease treatment, new drug design, and many other biological studies. Through sequence alignment, we can discover the functional, structural and evolutionary information of biological sequences.

In this section, we will introduce the pairwise sequence alignment. The pairwise sequence alignment is very widely used in bioinformatics. Firstly, we will describe some important concepts of the pairwise sequence alignment, and then in section 2.3, we will describe the most common method (BLAST) of pairwise sequence alignment.

2.1.1 Choose of the Sequence Alignment

In the case of sequence alignment, we can analyze the similarity between the two sequences from the nucleotide and amino acid levels, and how should we choose?

In general, we choose amino acid sequences for sequence similarity comparison and analysis. There are several reasons for it. Firstly, many changes in a DNA sequence do not change the amino acid that is specified. For example,

1. Zhao Ning, College of Life Sciences and Bioengineering, School of Science, Beijing Jiao Tong University, Beijing, China, 100044.

2. Wang Ziling, College of Life Sciences and Bioengineering, School of Science, Beijing Jiao Tong University, Beijing, China, 100044.

when the coden changes from CGU to CGC, the corresponding amino acid remains arginine. Secondly, amino acid sequences are often more informative. For example, many amino acid sequences have similar biophysical properties. Important relationships of the related amino acids in an alignment can be accounted for using scoring systems. Finaly, protein sequence comparisons can identify homologous sequences while the corresponding DNA sequence comparisons cannot.

Of course not all cases using amino acid sequences alignment are appropriate. For example, DNA sequence alignment is used to confirm the identity of a DNA sequence in a database search.

2.1.2 Definitions: Homology, Similarity, Identity

Two sequences are homologous if they share a common evolutionary ancestry. We need to note that homology is a qualitative concept, there are no degrees of homology, that is, sequences are either homologous or not. The two concepts associated with homology are similarity and identity. While homology is a qualitative concept, similarity and identity are quantities that describe the relatedness of sequences. When two sequences are homologous, their amino acid or nucleotide sequences usually share significant identity. Of course, two sequences may be homologous without sharing statistically significant amino acid or nucleotide identity. It is worth noting that the similarity of the two sequences is high, but they may not be homologous sequences, and the similarity of the two sequences may be generated by stochastic factors, which are called convergent evolution.

Sequences that are homologous may be orthologous or paralogous. Orthologs are homologous sequences in different species that arose from a common ancestral gene during speciation. Orthologs are presumed to have similar biological functions. For example, human and rat myoglobins both transport oxygen in muscle cells. Paralogs are homologous sequences that arose by a mechanism such as gene duplication. For example, human alpha 1 globin is paralogous to alpha 2 globin, and human alpha 1 globin and beta globin are also paralogs. Notably, orthologs and paralogs do not necessarily have the same function.

2.1.3 Gaps

In the case of pairwise sequence alignment, the two sequences can't be exactly the same, so when the two sequences are aligned, the gaps are inserted so that the two sequences can be sufficiently aligned. For example, assuming that the two sequences are CTAGCTAATCTG and CTGTGCTAATCTG, we can see that the second alignment is better than the first one, because it leads to more character alignment.

CTAGCTAATCTG-- CTA--GCTAATCTG

CTGTGCTAATCTG CTGTGCTAATCTG

The difference in sequence is caused by mutations. Mutations often occur during evolution. The most common mutations are substitutions, insertions and deletions. Substitutions don't affect the length of the sequence, so there is no need to insert a gap. Insertions and deletions (indels) occur when residues are added or removed. In the sequence alignment, due to the difference of the length between two sequences, one or more sites need to be inserted to obtain the best alignment result, and the inserted site is a gap. Therefore, in the sequence alignment, gaps are inserted in order to reflect the insertion or deletion of nucleotides or amino acids.

The gap penalty is a scoring system used in bioinformatics for pairwise sequence alignment. The penalty is divided into the gap opening and gap extension components, where the gap opening penalty is the cost for adding a new gap and the gap extension penalty is the incremental cost incurred along the length of the gap. There are three main types of gap penalties: constant, linear and affine gap penalty. A constant gap penalty occurs when there is a cost associated with opening a gap, but no cost for the length of a gap. A linear gap penalty occurs when there is no cost associated with opening a gap, but there is a cost for the length of a gap. The most common gap penalty in sequence alignment is the affine gap penalty. The affine gap penalty combines the components in both the constant and linear gap penalty, taking the form $A + (B \cdot L)$. The letter A represents the gap opening penalty, the letter B represents the gap extension penalty, and the letter L represents the length of the gap.

2.1.4 Scoring Matrix

In calculating the score for the alignment, it is necessary to consider the difference between the substitution of the 'similar character' and the 'different character'. In fact, for different types of character substitution, the score is not the same, especially for the protein sequence. For example, if alanine is replaced with another smaller and hydrophobic amino acid, such as valine, the effect on protein function may be less; if it is replaced by a larger and charged residue, such as lysine, then the impact on protein function may be greater than the former. Thus, the score of the substitution between amino acid residues with similar physical and chemical properties is clearly higher than the amino acid residue substitution score, which is far from the physicochemical properties. Likewise, conservative amino acid substitutions should be higher than non-conservative amino acid substitutions. This is the reason for proposing a scoring matrix (or a substitution matrix).

Scoring matrix is the basis of sequence alignment. Choosing different scoring matrices will get different alignment results, and understanding the theoretical basis of the scoring matrix will help to select the appropriate scoring matrix in practical application. Here are some of the commonly used scoring matrices.

1. Scoring matrix for DNA sequence alignment

We know that DNA consists of four nucleotides, namely adenine nucleotide (A), guanine nucleotide (G), cytosine nucleotide (C) and thymidine nucleotide (T). Then the alphabet of the DNA sequence can be expressed as $A = \{A, T, C, G\}$. Here are a few common scoring matrices.

- Unitary matrix. The unitary matrix (Figure 2.1) is the simplest scoring matrix, where the match score between the same nucleotides is '1' and the substitution score between the different nucleotides is '0'. The matrix is less used in the actual sequence alignment because it does not contain any physical and chemical information about the base and does not discriminate against the different substitutions.

	A	T	C	G
A	1	0	0	0
T	0	1	0	0
C	0	0	1	0
G	0	0	0	1

Fig. 2.1 Unitary matrix

- BLAST matrix. BLAST is currently the most popular nucleic acid sequence alignment program. Figure 2.2 is its scoring matrix. This is also a very simple matrix, and if the two nucleotides are the same, the score is '+5', and vice versa '–4'.

	A	T	C	G
A	5	–4	–4	–4
T	–4	5	–4	–4
C	–4	–4	5	–4
G	–4	–4	–4	5

Fig. 2.2 BLAST matrix

- Transition-transversion matrix. The transition is the substitution between purine and purine or pyrimidine and pyrimidine, such as A → G,

C → T; transversion is the substitution between purine and pyrimidine, such as A → C, G → T, etc. In the evolutionary process, the frequency of transition occurs much higher than transversion, and the matrix shown in Figure 2.3 reflects this situation, where the transition score is '–1' and the transversion score is '–5'.

	A	T	C	G
A	1	–5	–5	–1
T	–5	1	–1	–5
C	–5	–1	1	–5
G	–1	–5	–5	1

Fig. 2.3 Transition-transversion matrix

2. Scoring matrix for protein sequence alignment

When the amino acid sequence alignment is performed, the amino acid is first encoded using a single character to obtain a sequence of Fasta format. Table 2.1 shows the alphabet of 20 amino acids.

Table 2.1 Common amino acids and their abbreviations

Amino acid	Three-letter code	One-letter code	Amino acid	Three-letter code	One-letter code
Alanine	Ala	A	Leucine	Leu	L
Arginine	Arg	R	Lysine	Lys	K
Asparagine	Asn	N	Methionine	Met	M
Aspartate	Asp	D	Phenylalanine	Phe	F
Cysteine	Cys	C	Proline	Pro	P
Glutamine	Gln	Q	Serine	Ser	S
Glutamate	Glu	E	Threonine	Thr	T
Glycine	Gly	G	Tryptophan	Trp	W
Histidine	His	H	Tyrosine	Tyr	Y
Isoleucine	Ile	I	Valine	Val	V

Here are a few common scoring matrices.

- Unitary matrix. The principle of protein unitary matrix is the same as the DNA unitary matrix, where the match score between the same nucleotides is '1', and the substitution score between the different nucleotides is '0'.
- Genetic code matrix (GCM). The genetic code matrix is obtained by calculating the number of codons required to convert one amino acid to another amino acid, usually '1' or '2'. All three codon positions are changed only from Met to Tyr.
- Hydrophobic matrix. The scoring matrix is obtained based on the change

in hydrophobicity before and after substitution of amino acid residues. If the hydrophobicity doesn't change much during amino acid substitution, the substitution score is high, otherwise, the substitution score is low.

- PAM matrix. PAM matrices are a common family of scoring matrices. It was introduced by Margaret Dayhoff in 1978. PAM matrix is a point mutation model based on the evolution of amino acids, that is, if the two amino acids substitute frequently, the nature is easy to accept the substitutions, and then the substitution score is high.

The most basic PAM matrix is the PAM1 matrix, which represents average occurrence of one mutation per 100 amino acids. All the data being used in PAM matrix comes from sequence alignments of closely related proteins with more than 85% amino acid identity.

In the process of computing PAM matrix, it is quite difficult to collect the statistics about amino acid substitution in distantly diverged sequences. Hence, it is easier to compute PAM matrix with closely related protein sequence. All the other PAM matrices are extrapolated from PAM1 by assuming that repeated mutations would follow the same pattern as those in the PAM1 matrix, and multiple substitutions can occur at the same site. Using this logic, Dayhoff derived matrices as high as PAM250. Dayhoff used matrix multiplication: they multiplied the PAM1 matrix by itself, up to hundreds of times, to obtain other PAM matrices. For example, PAM1 matrix multiplied by itself by 250 times to obtain the PAM250 matrix (Table 2.2).

As an example, if the original amino acid is an alanine, there is just a 13% chance that the second sequence will also have an alanine. Similarly, there is a 6% chance that the alanine will have been replaced by an arginine.

- BLOSUM matrix. In addition to the PAM matrix, another very common set of scoring matrices is the blocks substitution matrix (BLOSUM). The PAM matrix and the BLOSUM matrix result in the same scoring outcome, but use difference methodologies. BLOSUM directly looks at mutations in motifs of related sequences while PAM's extrapolate evolutionary information is based on closely related sequences. Since both PAM and BLOSUM use different methods for showing the same scoring information, the two can be compared but due to the very different method of obtaining this score, a PAM100 does not equal a BLOSUM100.

The BLOSUM62 matrix is the default scoring matrix for the BLAST protein search programs at NCBI. It merges all proteins in an alignment that has 62% amino acid identity or greater into one sequence. If a block of aligned globin orthologs includes several that have 62%, 80%, and 95% amino acid identity, these would all be grouped as one sequence. Substitution frequencies for the BLOSUM62 matrix are weighted more heavily by blocks of protein sequences having less than 62% identity.

Table 2.2 The PAM250 mutation probability matrix

	A	R	N	D	C	Q	E	G	H	I	L	K	M	F	P	S	T	W	Y	V
A	13	6	9	9	5	8	9	12	6	8	6	7	7	4	11	11	11	2	4	9
R	3	17	4	3	2	5	3	2	6	3	2	9	4	1	4	4	3	7	2	2
N	4	4	6	7	2	5	6	4	6	3	2	5	3	2	4	5	4	2	3	3
D	5	4	8	11	1	7	10	5	6	3	2	5	3	1	4	5	5	1	2	3
C	2	1	1	1	52	1	1	2	2	2	1	1	1	1	2	3	2	1	4	2
Q	3	5	5	6	1	10	7	3	7	2	3	5	3	1	4	3	3	1	2	3
E	5	4	7	11	1	9	12	5	6	3	2	5	3	1	4	5	5	1	2	3
G	12	5	10	10	4	7	9	27	5	5	4	6	5	3	8	11	9	2	3	7
H	2	5	5	4	2	7	4	2	15	2	2	3	2	2	3	3	2	2	3	2
I	3	2	2	2	2	2	2	2	2	10	6	2	6	5	2	3	4	1	3	9
L	6	4	4	3	2	6	4	3	5	15	34	4	20	13	5	4	6	6	7	13
K	6	18	10	8	2	10	8	5	8	5	4	24	9	2	6	8	8	4	3	5
M	1	1	1	1	0	1	1	1	1	2	3	2	6	2	1	1	1	1	1	2
F	2	1	2	1	1	1	1	1	3	5	6	1	4	32	1	2	2	4	20	3
P	7	5	5	4	3	5	4	5	5	3	3	4	3	2	20	6	5	1	2	4
S	9	6	8	7	7	6	7	9	6	5	4	7	5	3	9	10	9	4	4	6
T	8	5	6	6	4	5	5	6	4	6	4	6	5	3	6	8	11	2	3	6
W	0	2	0	0	0	0	0	0	1	0	1	0	0	1	0	1	0	55	1	0
Y	1	1	2	1	3	1	1	1	3	2	2	1	2	15	1	2	2	3	31	2
V	7	4	4	4	4	4	4	5	4	15	10	4	10	5	5	5	7	2	4	17

The BLOSUM62 matrix is shown as Figure 2.4.

3. The relationship of the PAM and BLOSUM matrices

PAM matrix is based on data from the alignment of closely related protein families, and it involves the assumption that substitution probabilities for highly related proteins can be extrapolated to probabilities for distantly related proteins. To compare the closely related sequences, PAM matrix with lower numbers is used. To compare the distantly related proteins, PAM matrix with higher numbers is used.

BLOSUM matrix is based on empirical observations of more distantly related protein alignments. To compare the closely related sequences, BLOSUM matrix with higher numbers is used. To compare the distantly related proteins, BLOSUM matrix with lower numbers is used.

The relationships of the PAM and BLOSUM matrices are depicted in Figure 2.5.

A	4																			
R	-1	5																		
N	-2	0	6																	
D	-2	-2	1	6																
C	0	-3	-3	-3	9															
Q	-1	1	0	0	-3	5														
E	-1	0	0	2	-4	2	5													
G	0	-2	0	-1	-3	-2	-2	6												
H	-2	0	1	-1	-3	0	0	-2	8											
I	-1	-3	-3	-3	-1	-3	-3	-4	-3	4										
L	-1	-2	-3	-4	-1	-2	-3	-4	-3	2	4									
K	-1	2	0	-1	-1	1	1	-2	-1	-3	-2	5								
M	-1	-2	-2	-3	-1	0	-2	-3	-2	-1	2	-1	5							
F	-2	-3	-3	-3	-2	-3	-3	-3	-1	0	0	-3	0	6						
P	-1	-2	-2	-1	-3	-1	-1	-2	-2	-3	-3	-1	-2	-4	7					
S	1	-1	1	0	-1	0	0	0	-1	-2	-2	0	-1	-2	-1	4				
T	0	-1	0	-1	-1	-1	-1	-2	-2	-1	-1	-1	-1	-2	-1	1	5			
W	-3	-3	-4	-4	-2	-2	-3	-2	-2	-3	-2	-3	-1	1	-4	-3	-2	11		
Y	-2	-2	-2	-3	-2	-1	-2	-3	2	-1	-1	-2	-1	3	-3	-2	-2	2	7	
V	0	-3	-3	-3	-1	-2	-2	-3	-3	3	1	-2	1	-1	-2	-2	0	-3	-1	4
	A	R	N	D	C	Q	E	G	H	I	L	K	M	F	P	S	T	W	Y	V

Fig. 2.4 The BLOSUM scoring matrix

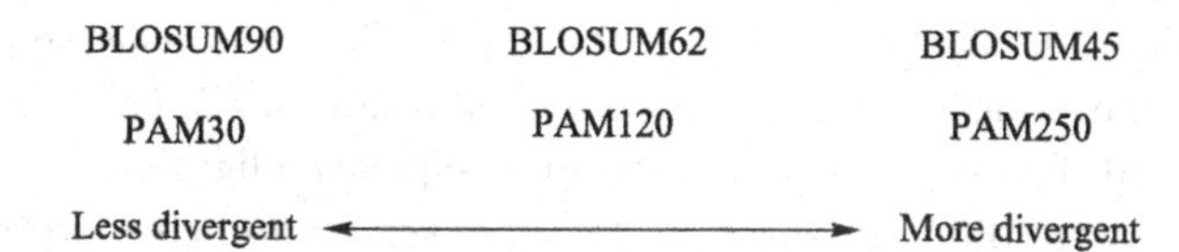

Fig. 2.5 The relationships of the PAM and BLOSUM matrices

2.2 Multiple Sequence Alignment

Multiple sequence alignment (MSA) has become one of the indispensable tools of bioinformatics, in fact of biology, as scientists try to make sense of the rapidly increasing flood of sequence information. A multiple sequence alignment is an alignment of more than two sequences obtained by inserting gaps into sequences such that the resulting sequences have the same length L and can be arranged in a matrix with N rows and L columns where each column represents a homologous position. Multiple sequence alignments are fundamental to tasks such as homology searches, genomic annotation, protein structure prediction, and the areas of computational evolutionary biology, gene regulation networks, and functional genomics.

In this section, we will discuss multiple sequence alignment. The principle is that multiple sequence alignments are achieved by successive application of pairwise methods. Unlike the pairwise sequence alignment, the goal of multiple sequence alignment is to discover the consensus of multiple sequences. If pairwise sequence alignment is mainly used to establish the homology of two sequences and to speculate on their structure and function, then multiple sequence alignment is more useful for studying molecular structure, function and evolutionary relationship. Likewise, conserved sequence fragments associated with a domain or function can only be found after multiple sequence alignments. For a series of homologous proteins, it is desirable to study the phylogenetic relationships implicit in the protein sequence in order to better understand the evolution of these proteins. In actual research, biologists do not just analyze individual proteins, but rather focus on the relationship between proteins, and then analyze the structure and function of proteins. Thus, multiple sequence alignment is one of the most common methods of proteome analysis.

2.2.1 Applications of Multiple Sequence Alignment

Compared with the pairwise sequence alignment, multiple sequence alignment has a wider range of applications.

- Determination of the consensus sequence. The consensus sequence is the calculated order of the most frequent residues, either nucleotide or amino acid, found at each position in a sequence alignment. Consensus sequences are commonly used in database search and chip probe design to identify sequences with high similarity.
- Mutation analysis. Multiple sequence alignment is used to reveal genomic variation in the same line of different individuals, and the most common of which is the single nucleotide polymorphism analysis. The purpose of mutation analysis is to identify the SNP that is the most likely to alter the function.
- Conserved segments analysis. In the genome, there are a large number of conserved segments that remain unchanged during evolution, such as exons, promoters, enhancers and non-coding RNAs. Multiple sequence alignment can be used to identify these conserved segments.
- Functional analysis of protein. With the development of sequencing technology, more and more protein sequences are determined, and the function of a large number of proteins can be revealed. By comparing the similarity of unknown proteins with known protein sequences to obtain the function of unknown proteins is also an application of multiple sequence alignment.

- Structural analysis of protein. The multiple sequence alignment method is used to compare the germline family of proteins, and the structure of the unknown protein is deduced by the known protein.
- Genomic structure analysis. Multiple sequence alignment of multiple genomes can reveal the structure and evolutionary characteristics of the genome. We can analyze genomic DNA via UCSC, Ensembl and Galaxy.
- Phylogenetic analysis. If we choose the similar sequence to be included in our multiple sequence alignment, we can reconstruct the history of these proteins.

2.2.2 Multiple Sequence Alignment Approaches

There are five approaches to multiple sequence alignment: ① dynamic programming approach; ② progressive alignment (e.g., Clustal Omega); ③ iterative approaches (e.g., PRALINE, IterAlign, MUSCle); ④ consistency-based approaches (e.g., ProbCons, T-COFFEE); ⑤ structure-based methods.

The most commonly used approach is a progressive method. It is called 'progressive' because the strategy entails calculating pairwise sequence alignment scores between all the proteins (or nucleic acid sequences) being aligned, then beginning the alignment with the two most similar sequences and progressively adding more sequences to the alignment. A benefit of this approach is that it permits the rapid alignment of hundreds of even thousands of sequences. The progressive alignment is an instance of a heuristic algorithm, so it isn't guaranteed to find the best possible alignment. In practice, however, it is efficient and produces biologically meaningful results.

Progressive multiple sequence alignment requires the following three steps (Figure 2.6):

- First perform all possible pairwise alignments between each pair of sequences. Then calculate the 'distance' between each pair of sequences based on these isolated pairwise alignments. Finally, generate a distance matrix.
- A guide tree is calculated from the distance matrix. There are two principal ways to construct a guide tree: the unweighted pair group method of arithmetic averages (UPGMA) and the neighbor-joining method. We will describe them in detail in the next chapter.
- The multiple sequence alignment is created in a series of steps based on the order presented in the guide tree. The algorithm first selects the two most closely related sequences from the guide tree and creates a pairwise alignment. The next sequence is added to the pairwise alignment or used in another pairwise alignment. The alignment continues progressively until all sequences have been aligned.

Recently the most popular web-based program for performing progressive multiple sequence alignment has been Clustal series. Clustal Omega is the latest addition to the Clustal family. It offers a significant increase in scalability over previous versions, allowing hundreds of thousands of sequences to be aligned in only a few hours. It will also make use of multiple processors. In addition, the quality of alignments is superior to previous versions, as measured by a range of popular benchmarks. Next we illustrate the procedure by aligning three distantly related globins (human neuroglobin, NP_067080.1; human beta globin, NP_000509.1; human myoglobin, NP_05359.1), selected from NCBI protein and pasted into a text document in the FASTA format. The results are shown in Figure 2.7.

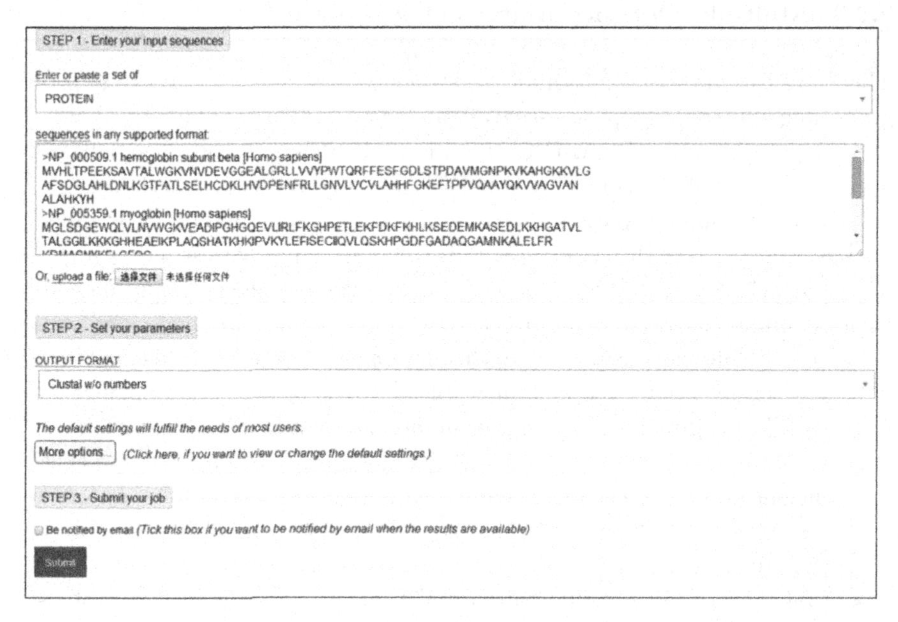

Fig. 2.6 Multiple sequence alignment of three distantly related proteins using Clustal Omega

The output is a screen capture from Clustal Omega using the progressive alignment algorithm. '*' (asterisk) indicates positions which have a single, fully conserved residue. ':' (colon) indicates conservation between groups of strongly similar properties - scoring > 0.5 in the Gonnet PAM250 matrix. '.' (period) indicates conservation between groups of weakly similar properties – scoring $\leqslant 0.5$ in the Gonnet PAM250 matrix.

```
CLUSTAL O(1.2.4) multiple sequence alignment


NP_067080.1      ---MERPEPELIRQSWRAVSRSPLEHGTVLFARLFALEPDLLPLFQYNCRQFSSPEDCLS
NP_000509.1      MVHLTPEEKSAVTALWGKVNVDEVG--GEALGRLLVVYPWTQRFFES-FGDLSTPDAVMG
NP_005359.1      -MGLSDGEWQLVLNVWGKVEADIPGHGQEVLIRLFKGHPETLEKFDK-FKHLKSEDEMKA
                    :   * . :   *  *. .        : **:   *     *:    .:. : :  .

NP_067080.1      SPEFLDHIRKVMLVIDAAVTNVEDLSSLEEYLASLGRKH-RAVGVKLSSFSTVGESLLYM
NP_000509.1      NPKVKAHGKKVLGAFSDGLAHLDNL---KGTFATLSELHCDKLHVDPENFRLLGNVLVCV
NP_005359.1      SEDLKKHGATVLTALGGILKKKGHH---EAEIKPLAQSHATKHKIPVKYLEFISECIIQV
                 . ..  *  .*: .:.  : :  .    :  :  *.. *     :  . :  : : :: :

NP_067080.1      LEKCLGPAFTPATRAAWSQLYGAVVQAMSRGWDGE---
NP_000509.1      LAHHFGKEFTPPVQAAYQKVVAGVANALAHKYH------
NP_005359.1      LQSKHPGDFGADAQGAMNKALELFRKDMASNYKELGFQG
                 *       *   .:.* .:    . : ::  ::
```

Fig. 2.7 Multiple sequence alignment of three distantly related proteins

2.2.3 Multiple Sequence Alignment Parameters

There are many parameters involved in multiple sequence alignment, including scoring matrices and gap penalties. They have the same meaning in multiple sequence alignment as they do in pairwise sequence alignment.

One of Clustal Omega's heuristics is that, in protein sequence alignment, different scoring matrices are used for each alignment based on expected evolutionary distance. If two sequences are close neighbors in the tree, a scoring matrix optimized for close relationships aligns them. Distant neighbors are aligned using matrices optimized for distant relationships. Thus, when prompted to choose a series of matrices in the multiple alignment parameters menu, it means just that: using BLOSUM62 for close relationships and BLOSUM45 for more distant relationships, rather than the same scoring matrix for all pairwise alignments.

Another heuristic that Clustal Omega uses is scalable gap penalties for protein alignments. A gap opening next to a conserved hydrophobic residue can be penalized more heavily than a gap opening next to a hydrophilic residue. A gap opening too close to another gap can be penalized more heavily than an isolated gap.

2.3 Basic Local Alignment Search Tool

Basic Local Alignment Search Tool (BLAST) is the main NCBI tool (http://www.ncbi.nlm.nih.gov/BLAST/) for comparing nucleotide or protein sequences to sequence databases. The BLAST algorithm was designed by Stephen Altschul. As the sizes of the sequence databases increase, the need for better and faster algorithm arises leading to the development of BLAST. BLAST searching is one of the fundamental ways of learning about a protein or gene, because it reveals what related sequences are present in the same organism and other organisms.

BLAST searching allows the user to perform pairwise sequence alignments between a sequence (termed the query) and a database (termed the target). This means that tens of millions of sequences are evaluated in a BLAST search, and only the most closely related matches are returned. BLAST provides a local alignment strategy having both speed and sensitivity. Meanwhile, it also provides convenient accessibility on the World Wide Web.

BLAST searching uses heuristic algorithm system, that is, local alignment algorithm, rather than global alignment algorithm (Figure 2.8). All the fragments of the two sequences in the global alignment algorithm are aligned and run through the entire sequence length. The local alignment algorithm is to find out the most similar fragments of the two sequences.

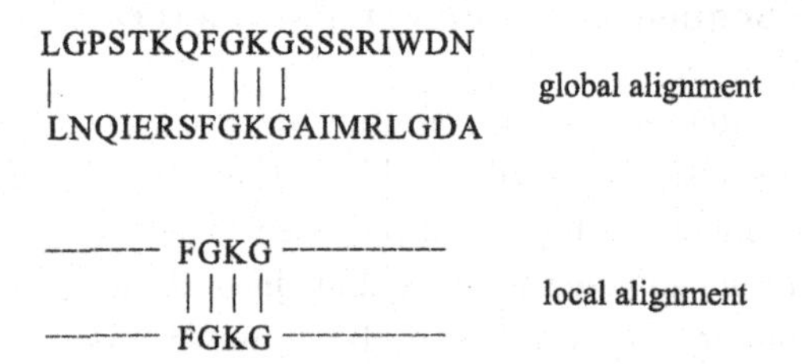

Fig. 2.8 Comparison of global alignment and local alignment

BLAST can be used for several purposes:

- Inferring and determining the function of the query sequence.
- Determining what orthologs and paralogs are known for a particular protein or gene.
- Discovering new genes or proteins.
- Inferring the evolutionary distance from sequence similarity.
- Determining what variants have been described for a particular protein or gene.
- Investigating expressed sequence tags (ESTs) that may exhibit alternative splicing.
- Exploring conserved regions that are important in the function and/or structure of a protein.

2.3.1 BLAST Search Steps

There are a few key steps when performing a web-based BLAST search, the homepage of a BLAST search at NCBI (Figure 2.9).

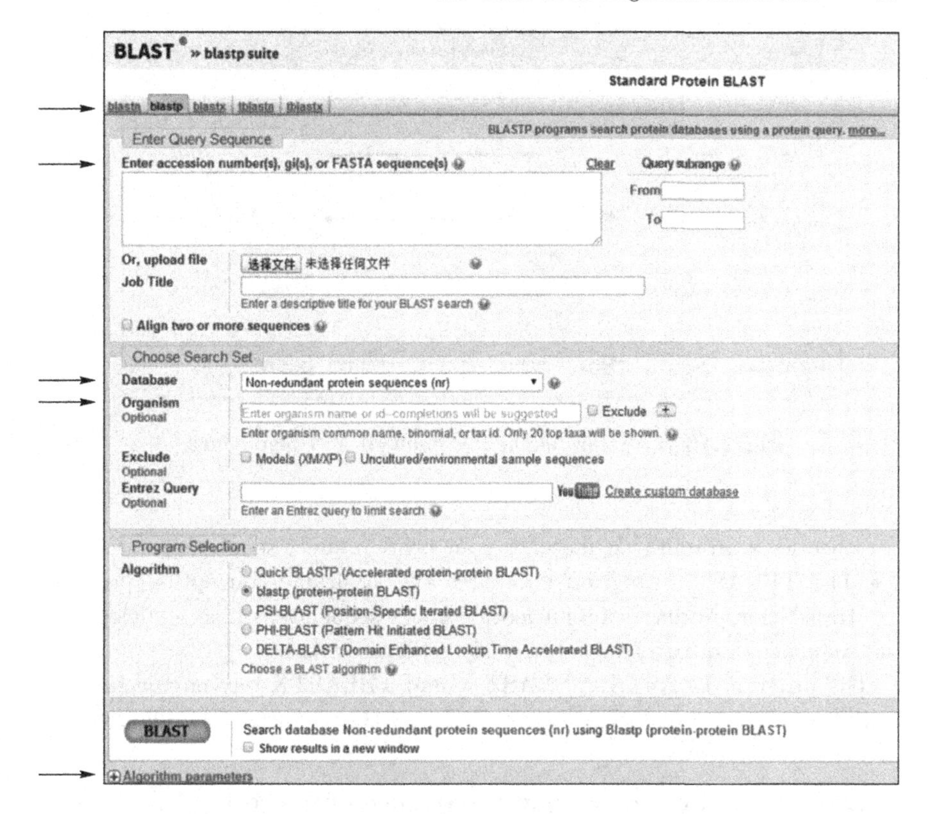

Fig. 2.9 Main page for a BLAST search at NCBI

1. **Step 1: selecting BLAST program**

BLAST is a fast heuristic homology search algorithm which comprises five basic routines to search with a query sequence against a sequence database, including all combinations of nucleotide and protein sequences, as summarized in Figure 2.10.

- The BLASTP program compares a protein query sequence against a protein sequence database.
- The BLASTN program compares a nucleotide query sequence against a nucleotide sequence database.
- The BLASTX program compares the six-frame conceptual protein translation products of a nucleotide query sequence against a protein sequence database. If you have a DNA sequence and you want to know what protein sequence it encodes, you can perform a BLASTX search.
- The TBLASTN program compares a protein query sequence against a nucleotide sequence database translated in six reading frames. You can use the program TBLASTN to determine whether a DNA database

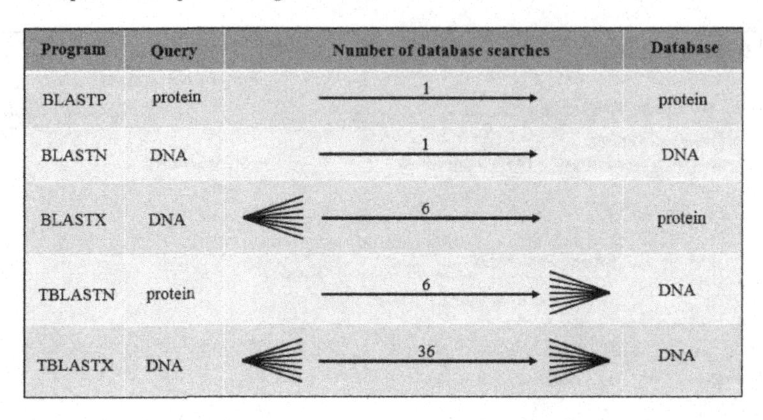

Fig. 2.10 Overview of the five main BLAST algorithms

encodes a protein that matches your protein query sequence.

- The TBLASTX program compares the six-frame conceptual protein translation products of a nucleotide query sequence against a nucleotide sequence database translated in six reading frames.

The program BLASTX, TBLASTN, and TBLASTX rely on the fundamental relationship of DNA to protein. In all three cases, these algorithms perform protein-protein alignments.

2. Step 2: entering the query sequence of interest

After selecting the BLAST program, the user is required to enter the query sequence into the query box. The protein or DNA query sequence can be input as an accession number, GI identifier or FASTA-formatted sequence. A sequence in FASTA format consists of a single-line description and lines of sequence data. The description line begins with a greater than ('>') symbol. Generally, we choose to input the accession number to a BLAST search.

3. Step 3: limiting search range

Sequence alignment allows the user to specify a database for BLAST searching. Databases that are available for BLAST searching are listed on each BLAST page. For protein database searches (BLASTP and BLASTX), the default option is the non-redundant (nr) database. Another commonly used database is Reference proteins. Table 2.3 lists the available protein databases for BLAST searching at NCBI.

For DNA database searches (BLASTN, TBLASTN and TBLASTX), the default option is the nucleotide collection (nr/nt) database. Other commonly used options include the human (or mouse) genomic plus transcript database or the EST database. Table 2.4 lists all the available nucleotide databases for BLAST searching at NCBI.

Table 2.3 Protein sequence databases that can be searched by BLAST searching

Program	Database
BLASTP BLASTX	Non-redundant protein sequences (nr)
	Reference proteins (refseq_protein)
	Model Organisms (landmark)
	UniProtKB/Swiss-Prot (swissprot)
	Patented protein sequences (pat)
	Protein Data Bank proteins (pdb)
	Metagenomic proteins (env_nr)
	Transcriptome Shotgun Assembly proteins (tsa_nr)

Table 2.4 Nucleotide sequence databases that can be searched by BLAST searching

Program	Database
BLASTN TBLASTN TBLASTX	Human genomic plus transcript (Human G+T) Mouse genomic plus transcript (Mouse G+T)
	Nucleotide collection (nr/nt) 16S ribosomal RNA sequences (Bacteria and Archaea)
	Reference RNA sequences (refseq_rna)
	RefSeq Representative genomes (refseq representative_genomes)
	RefSeq Genome Database (refseq_genomes) Whole-genome shotgun contigs (wgs)
	Expressed sequence tags (est)
	Sequence Read Archive (SRA)
	Transcriptome shotgun Assembly (TSA)
	High throughput genomic sequences (HTGS) Patent sequences (pat)
	Protein Data Bank (pdb) Reference genomic sequences (refseq_genomic)
	Genomic survey sequences (gss)
	Sequence tagged sites (dbsts)
	NCBI Genomes (chromosome)

BLAST can be restricted by an organism (Figure 2.9). Some popular groups are Archaea, Metazoa, Bacteria, Vertebrata, Eukaryota, Mammalia, Embryophyta, Rodentia, Fungi and Primates. BLAST searches can be restricted to any genus and species or other taxonomic groupings, and also can be limited using any terms that are used in an 'Entrez search'.

4. Step 4: selecting algorithm parameters

When the user runs the BLAST program, the desired result can be obtained by adjusting the algorithm parameters (Figure 2.11). The BLAST program contains three sets of algorithm parameters: 'General Parameters' 'Scoring Parameters' and 'Filters and Masking'. Next let us focus our attention on the BLASTP program.

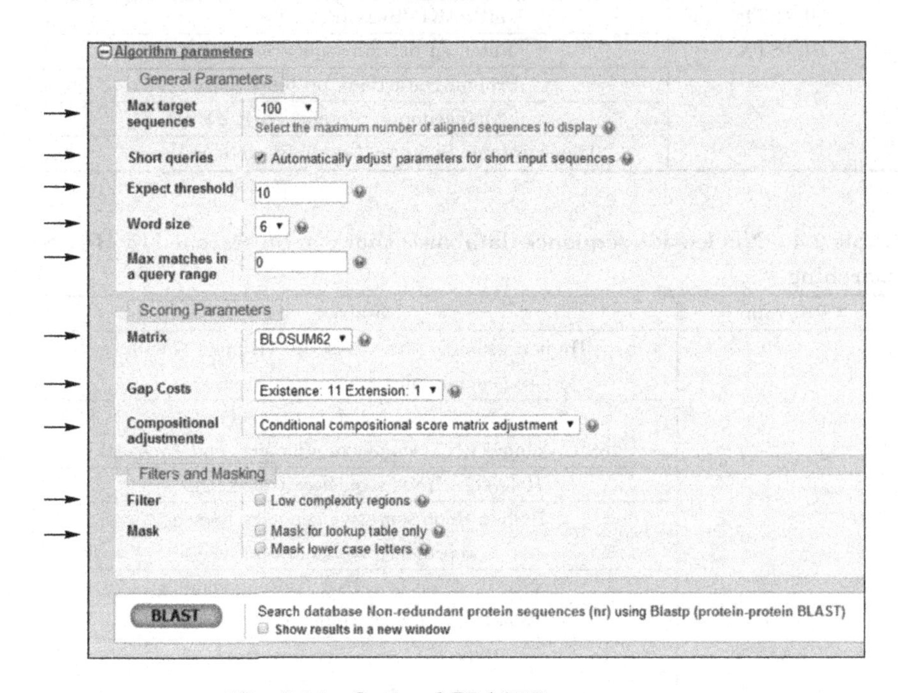

Fig. 2.11 Optional BLASTP parameters

(1) General Parameters

- Max target sequences. It is the maximum number of aligned sequence to display. The default value is 100. You can select fewer or more than the default value.
- Short queries. If the query sequence is short, you should select this option. The expect value and word size are automatically adjusted to get the best matches.
- Expect threshold. The expect value E is the number of different alignments with score equal to or greater than some score S that is expected to occur in a database search by chance. This parameter controls the sensitivity of the search. The default setting for the expect value is 10. By changing the expect option to a lower number, few database hits are returned; fewer chance matches are reported. Increasing E returns more hits.

- Word size. When a query sequence is used to search a database, the BLAST algorithm first divides the query into a series of smaller sequences (words) of a particular length (word size). For different BLAST programs, the word size is not the same.
- Max matches in a query range. Matches to one region of interest can be obscured by frequent matches to a different region of a protein. This feature offers a solution in which redundant database hits are discarded.

(2) Scoring Parameters

- Matrix. The default scoring matrix is BLOSUM62. For BLASTP, there are seven other alternative substitution matrices: PAM30, PAM70, PAM250, BLOSUM45, BLOSUM50, BLOSUM80, and BLOSUM90. It is sometimes advisable to try a BLAST search using several substitution matrices. For BLASTN, the default scoring system is '+2' for a match and '–3' for a mismatch. A variety of other scoring schemes are available.
- Gap costs. The gap cost is the penalty for the gap. A gap is a space to compensate for insertions and deletions in one sequence relative to another. The gap score consists of two parts: the gap existence penalty and the gap extension penalty. There are many combinations of the gap existence penalty and the gap extension penalty to choose. The choice of gap costs is typically 10—15 for gap existence penalty and 1—2 for gap extension penalty.
- Compositional adjustments. The default option is the 'conditional compositional score matrix adjustment'. Compositional adjustments generally increase the accuracy of BLAST searches considerably.

(3) Filters and Masking

- Filtering masks portions of the query sequence that have low complexity. Examples are dinucleotide repeats (e.g., the repeating nucleotides CACACACA...) or regions of a protein that are extremely rich in one or two amino acids. Filtering is applied to the query sequence, but not the entire database.
- The 'mask for lookup table only' option masks the matching of words above threshold to database hits. The 'mask lower case letters' option allows you to enter a query in the FASTA format using upper case characters for the search but filtering those residues you choose to filter by entering them in lower case.

2.3.2 Selecting Formatting Parameters

There are many options for formatting the output of a BLAST search. Next we will perform a web-based BLASTP search with human beta globin (NP_000509.1) as a query sequence and restrict the search to RefSeq proteins from the mouse (*Mus musculus*).

Figure 2.12 shows the details of the search results, including the description of the query sequence and the database, the type of BLAST search and the query length... By clicking 'Search Summary', more details can be viewed, including expect value, matrix, threshold and composition-based stats (Figure 2.13).

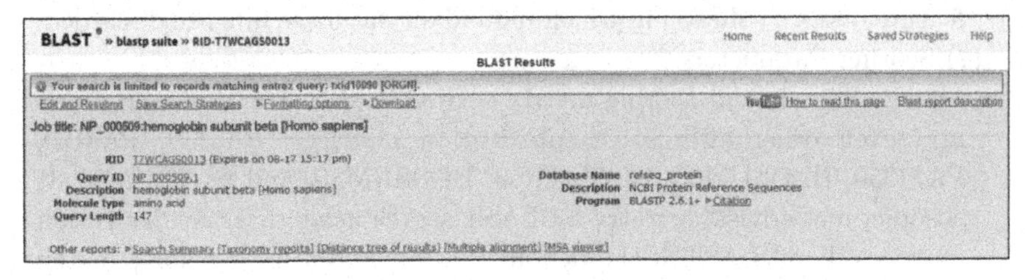

Fig. 2.12 The details of the search results

Search Parameters	
Program	blastp
Word size	6
Expect value	10
Hitlist size	100
Gapcosts	11,1
Matrix	BLOSUM62
Filter string	F
Genetic Code	1
Window Size	40
Threshold	21
Composition-based stats	2

Database	
Posted date	Aug 14, 2017 2:45 AM
Number of letters	34,421,658,504
Number of sequences	90,332,248
Entrez query	txid10090 [ORGN]

Karlin-Altschul statistics		
Lambda	0.320339	0.267
K	0.136843	0.041
H	0.422367	0.14
Alpha	0.7916	1.9
Alpha_v	4.96466	42.6028
Sigma		43.6362

Fig. 2.13 Search Summary

Next is the 'graphic summary' section of the results (Figure 2.14). This includes conserved domains followed by a color-coded summary, with the length of the query sequence represented across the x axis. Each bar drawn below the map represents a database protein (DNA) sequence that matches the query sequence. The most similar hits are shown at the top, and this section also contains specific hits and superfamilies of the query sequence.

Next is followed by a table of sequences producing significant alignments

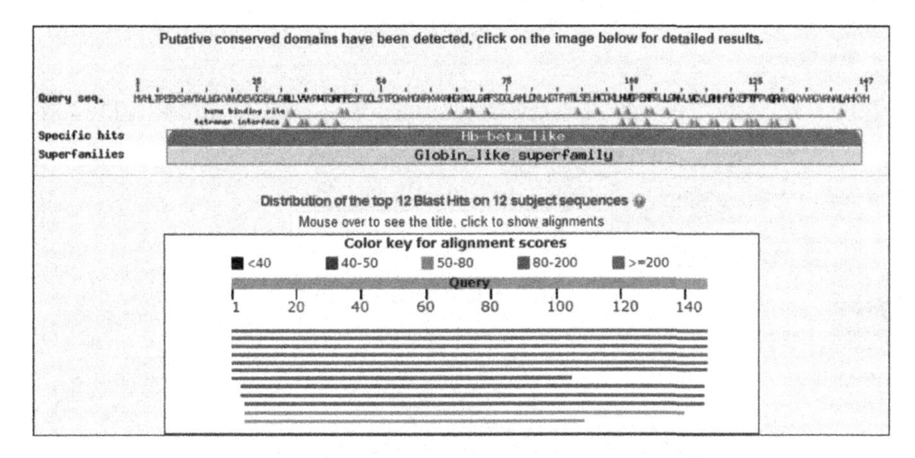

Fig. 2.14 The graphic summary of BLAST results

(Figure 2.15). These sequences are arranged according to the E value. The most significant alignment (the lowest E value) is at the top. The table includes several columns: description (name and species), score, query cover, E value, identity and accession number. Furthermore, you can click on the boxes on the left of each row for further analyses such as a distance tree or multiple alignment.

Sequences producing significant alignments:

Select: All None Selected 0

Alignments Download GenPept Graphics Distance tree of results Multiple alignment

Description	Max score	Total score	Query cover	E value	Ident	Accession
hemoglobin subunit beta-1 [Mus musculus]	251	251	100%	6e-87	80%	NP_001265090.1
hemoglobin subunit beta-2 [Mus musculus]	249	249	100%	4e-86	80%	NP_058652.1
hemoglobin, beta adult t chain [Mus musculus]	233	233	100%	6e-80	80%	NP_032246.2
hemoglobin subunit epsilon-Y2 [Mus musculus]	229	229	100%	2e-78	73%	NP_032247.1
hemoglobin subunit beta-H1 [Mus musculus]	226	226	100%	7e-77	69%	NP_032245.1
hemoglobin beta, bh2 [Mus musculus]	160	160	100%	9e-51	57%	NP_001121158.1
PREDICTED: hemoglobin beta, bh2 isoform X1 [Mus musculus]	125	125	71%	8e-38	58%	XP_006508068.1
hemoglobin alpha, adult chain 2 [Mus musculus]	111	111	97%	9e-32	42%	NP_001077424.1
hemoglobin subunit zeta [Mus musculus]	96.7	96.7	97%	7e-26	34%	NP_034535.1
hemoglobin subunit theta-1 [Mus musculus]	91.3	91.3	96%	1e-23	36%	NP_001029153.1
hemoglobin, theta T2 [Mus musculus]	75.5	75.5	92%	1e-17	34%	NP_778165.1
PREDICTED: hemoglobin subunit theta-1 isoform X1 [Mus musculus]	68.6	68.6	71%	3e-15	37%	XP_006514823.1

Fig. 2.15 Sequences producing significant alignments

The lower portion of a BLAST search output consists of a series of pairwise sequence alignments, such as those in Figure 2.16. Five scoring measures are provided: 'Score' 'Expect Method' 'Identities' 'Positives' and 'Gaps'.

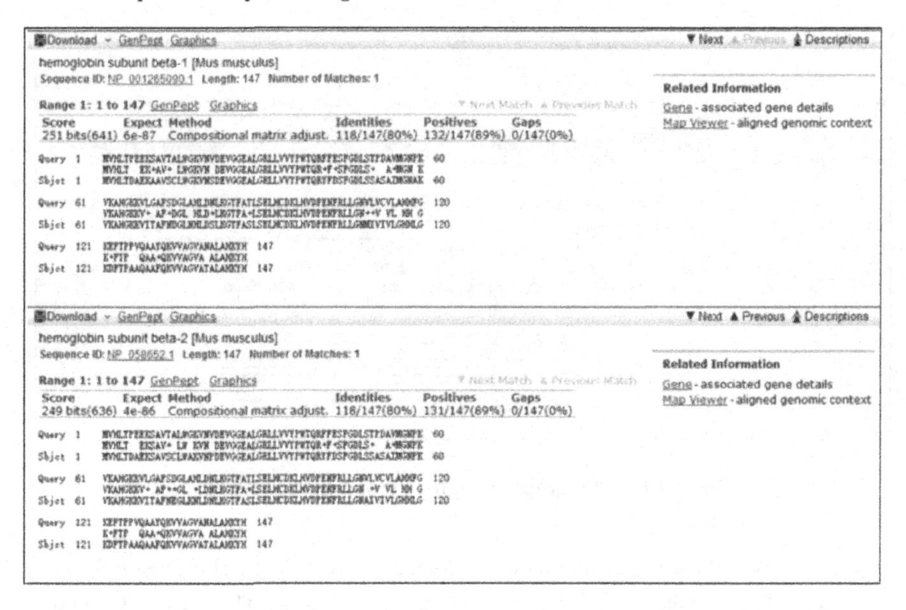

Fig. 2.16 The coding regions of the query and subject sequence

References

[1] Deng Y, Wang H, Hamamoto R, et al. Functional genomics, genetics, and bioinformatics. *Biomed Res.* Int, 2015: 184824.

[2] Agrawal A, Brendel V P, and Huang X. Pairwise statistical significance and empirical determination of effective gap opening penalties for protein local sequence alignment. *Int. J. Comput. Biol. Drug. Des*, 2008, 1(4): 347-367.

[3] Do C B and Katoh K. Protein multiple sequence alignment. *Methods Mol. Biol.*, 2008, 484: 379-413.

[4] Simossis V, Kleinjung J, and Heringa J. An overview of multiple sequence alignment. *Curr. Protoc. Bioinformatics*, 2003(3): Unit 3.7.

[5] Koehl P. Protein structure similarities. *Curr. Opin. Struct. Biol.*, 2001, 11(3): 348-353.

[6] Chatzou M, et al. Multiple sequence alignment modeling: methods and applications. *Brief Bioinform*, 2016, 17(6): 1009-1023.

[7] Diniz W J and Canduri F. REVIEW-ARTICLE Bioinformatics: an overview and its applications. *Genet. Mol. Res.*, 2017, 16(1).

[8] Mitrophanov A Y and Borodovsky M. Statistical significance in biological sequence analysis. *Brief Bioinform*, 2006, 7(1): 2-24.

[9] Chowdhury B, and Garai G. A review on multiple sequence alignment from the perspective of genetic algorithm. *Genomics*, 2017.

[10] Gaskell G J. Multiple sequence alignment tools on the Web. *Biotechniques*, 2000, 29(1): 60-62.

[11] Bystroff C, Shao Y, and Yuan X. Five hierarchical levels of sequence-structure correlation in proteins. *Appl. Bioinformatics*, 2004, 3(2-3): 97-104.

[12] Ladunga I. Finding homologs to nucleotide sequences using network BLAST searches. *Curr. Protoc. Bioinformatics*, 2002(3): Unit 3.3.

[13] Program Selection Tables of the Blast NCBI [2017-08-30]. https://www.ncbi.nlm.nih.gov/Class/BLAST/blast_course.short.html.

[14] Hancock J M, Zvelebil M J. Dictionary of Bioinformatics and Computational Biology// *Dictionary of bioinformatics and computational biology.* New York: Wiley-Liss, 2004: 641-661.

[15] Rohl R B C. Bioinformatics and functional genomics. *Quarterly Review of Biology*, 2005.

[16] Nguyen K, Guo X, Pan Y. Multiple Biological Sequence Alignment: Scoring Functions. *Algorithms and Applications*, 2016.

Chapter 3 Molecular Phylogeny and Evolution

Tian Tian[1]

3.1 Introduction to Molecular Evolution

In 1859, *The Origin of Species*, a landmark book was published by Charles Darwin. In this book, he wrote, 'As many more individuals of each species are born than can possibly survive; and as, consequently, there is a frequently recurring struggle for existence, it follows that any being, if it vary however slightly in any manner profitable to itself, under the complex and sometimes varying conditions of life, will have a better chance of surviving, and thus be naturally selected. From the strong principle of inheritance, any selected variety will tend to propagate its new and modified form.'

Evolution is the theory that groups of organisms change over time that is descendants inherit morphological, and physiological features from their ancestors, and meanwhile descendants alter in functions and structures with their ancestors. When we chase the relationship between the descendants and their ancestors, we are exploring the tree of life.

Evolution is a process of change. Heredity is normally conservative as offsprings resemble their parents. However, the structure and function of organisms alter through generations. Changes may occur through the following three mechanisms [1]:

- Development is affected by the conditions of growth. Environmental factors such as accidents and disease-causing infections are not hereditary in nature.
- Sexual reproduction ensures changes from one generation to the next. When an offspring inherits DNA from parents, the genes included are also shuffled in a different combination.
- Mutation as well as genetic drift is the source of producing changes in genes and chromosomes.

---

1. Tian Tian, College of Life Sciences and Bioengineering, School of Science, Beijing Jiao Tong University, Beijing, China, 100044.

Evolution is a process of mutation with selection at the molecular level. The definition of molecular evolution is that the study of alterations in genes and proteins throughout different branches of the tree of life. This principle also utilizes data from organisms in nowadays to rebuild the evolutionary history of certain species.

Phylogeny is the deduction process of evolutionary relationships between different organisms from a variety of species. This was conducted by comparing morphological characteristics between organisms traditionally [2]. Nowadays, however, molecular sequence data analysis may be the main part in phylogenetic study. The evolutionary relationships, which are usually demonstrated as a form of a tree, can offer hypotheses of past biological events.

3.1.1 Aims of Molecular Phylogeny

All living organisms share a common ancestor and are part of the tree of life. More than 99% of all living organisms that once lived in the earth have died out [3]. In the existing species, closely related organisms are descended from more recent common ancestors than distantly related organisms. One objective of phylogeny is to infer the correct trees for all species of life. Phylogenetic analyses were traditionally based on easily observable features, such as the presence or absence of wings or a spinal cord. Recently, phylogenetic analyses are more and more dependent on molecular sequence data defining families of genes and proteins rather than the obvious characteristics. Deducing the time divergence between organisms since the time when they last shared a common ancestor is another objective of phylogeny. While the tree of life provides an appealing metaphor, evolution is not that ideally a single tree. Instead, mutation and selection are common in the process of evolution. Actually, genes can be laterally transferred between species. Thus the tree of life has depicted more like a densely interconnected bush instead of a single tree with well-defined branches [4].

A true tree is essentially impossible to generate, because we don't know the actual evolution process. However, we generate inferred trees, which depict hypothesized historical evolution events. Based on some model from the available data, a series of evolutionary events are described by such trees.

Bacteria, archaea, and eukaryotes are the three major branches of the tree of life. In this chapter we will discuss the topic of phylogenetic trees that are utilized to assess the relationships of homologous proteins (or homologous nucleic acid sequences) in a family. A phylogenetic tree can depict any group of homologous proteins (or nucleic acid sequences). Homologous proteins are defined as they share a common ancestor. You may perform a BLAST search and get several homologous proteins with high scores, which also possibly have associated functions. However, in an evolutionary context, we also need to distinguish orthologs and paralogs. Various approaches can be utilized to study the proteins relations, such as pairwise alignment using Dayhoff 's scoring matrices, BLAST searching, and multiple sequence alignment, and we will address the identification of related protein folds. Observed similarities and differences between molecular sequences are necessary for evolutionary models:

- Dayhoff, et al. [5] introduced scoring matrices in explicit evolutionary terms, 'An accepted point mutation in a protein is a replacement of one amino acid by another, accepted by natural selection. It is the result of two distinct processes: the first is the occurrence of a mutation in the portion of the gene template producing one amino acid of a protein; the second is the acceptance of the mutation by the species as the new predominant form.

  'To be accepted, the new amino acid usually function in a way similar to the old one: chemical and physical similarities are found between the amino acids that are observed to interchange frequently.' Dayhoff, et al. compared the observed amino acid sequences from two proteins not with each other but with their inferred ancestor obtained from phylogenetic trees.
- Feng and Doolittle [6] performed the Needleman and Wunsch pairwise alignment progressively 'to achieve the multiple alignments of a set of protein sequences and to construct an evolutionary tree depicting their relationship. The sequences are assumed a priority to share a common ancestor, and the trees are constructed from different matrices derived directly from the multiple alignment. The thrust of the method involves putting more trust in the comparison of recently diverged sequences than in those evolved in the distant past.'
- In our description of protein families, we provided the example of the Pfam Jalview tool that allows distance information from the multiple sequence alignments of any Pfam family to be depicted as a tree.

In this chapter, to generate phylogenetic trees, multiple sequence alignments of protein (or DNA or RNA) are utilized. These trees provide an obvious demonstration of the evolutionary history of these molecular sequences.

3.1.2 Neutral Theory of Molecular Evolution

There is a tremendous amount of DNA polymorphism in all species that is difficult to account for by conventional natural selection. We examine single nucleotide polymorphisms (SNPs), an extremely common form of polymorphism that does not appear to be under selection in most instances. Similarly, many chromosomal copy number variants occur in apparently normal individuals. These involve multiple regions of up to millions of base pairs of DNA that are deleted or duplicated, and the majority of copy number variants appears to be sporadic, benign, and not under positive or negative selective pressure.

In the decades up to the 1960s, the prevailing model of molecular evolution was that most changes in genes are selected for or against in a Darwinian sense. Motoo Kimura [7, 8] proposed a different model to explain evolution at the DNA level. Kimura [7] noted that the rate of amino acid substitution averages approximately one change per 28~106 years for proteins of 100 residues. He further estimated that the corresponding rate of nucleotide substitution must be extremely high (one base pair of DNA replaced in the genome of a population every 2 years on average).

Kimura's conclusion was that most observed DNA substitutions must be neutral or nearly neutral, and that the main cause of evolutionary changes (or variability) at the molecular level is random drift of mutant alleles. Most nonsynonymous mutations are deleterious, and thus are not observed as substitutions in the population. Under this model, called the neutral theory of evolution, positive Darwinian selection plays an extremely limited role. Indeed, the existence of a molecular clock makes sense in the context of the neutral hypothesis because most amino acid substitutions are neutral. (Thus, substitutions are tolerated by natural selection to change in a manner that has clock-like properties. If substitutions occurred primarily in the context of positive or negative selection, it is unlikely that they could account for clock-like evolution.) In the decades since his 1983 publication, the neutral theory continues to be tested in a variety of organisms. We will explore some of these studies when we consider the eukaryotic chromosome.

3.2 Models of DNA and Amino Acid Substitution

Phylogenetic analyses rely on models of DNA or amino acid substitution. These models may be implicit or explicit. For distance-based methods, statistical models are employed to estimate the number of DNA or amino acid changes that occurred in a series of pairwise comparisons of sequences. For maximum likelihood and Bayesian approaches, statistical models are applied to individual characters (residues) in order to assess the most likely topology as well as other features such as substitution rates along individual branches. For maximum parsimony, the criterion for finding the best tree is based on the shortest branch lengths, and while individual characters are also evaluated, many of these statistical models are not applicable.

The simplest approach to define the relatedness of a group of nucleotide (or amino acid) sequences is to align pairs of sequences and count the number of differences. The degree of divergence is sometimes called the Hamming distance.

For an alignment of length N with n sites at which there are differences, the degree of divergence d is defined as

$$d = n/N \times 100 \tag{3.1}$$

Earlier in this chapter we discussed an example of this type of calculation by Zuckerkandl and Pauling [9], who counted the number of amino acid differences between human beta globin and delta, gamma, and alpha globin. The Hamming distance is simple to calculate, but it ignores a large amount of information about the evolutionary relationships among the sequences. The main reason is that character differences are not the same as distances: the differences between two sequences are easy to measure, but the genetic distance involves many mutations that cannot be observed directly. We discuss a correction implemented by Dickerson [10] that was proposed by Margoliash and Smith [11] and by Zuckerkandl and Pauling [12]; In MEGA software this is referred to as the Poisson correction [13]. The Poisson correction for distance d assumes equal substitution rates across sites and equal amino acid frequencies. It uses the following formula to correct for multiple substitutions at a single site:

$$d = -\ln(1 - p) \tag{3.2}$$

where d is the distance, and p is the proportion of residues. We make the following assumptions [14]. Firstly, the probability of observing a change is small but nearly identical across the genome. This probability is proportional to the length of the time interval $\lambda\Delta t$ for some constant λ. The probability of observing no changes is thus $1 - \lambda\Delta t$. Secondly, we assume the number of nucleotide or amino acid changes is constant over the time interval t. When a mutation does occur, this does not alter the probability of another mutation occurring at this same position. Thirdly, we assume that changes occur independently. Equation 3.2 is derived from the Poisson distribution, which describes the random occurrence of events when that probability of occurrence is small. The Poisson distribution is used to model a variety of phenomena, such as the decay of radioactivity over time. It is given by the formula:

$$P(X) = \frac{e^{-\mu}\mu^X}{X!} \tag{3.3}$$

where $P(X)$ is the probability of X occurrences per unit of time, μ represents the population mean number of changes over time, and e is $\sim$2.71828 [15].

Let us consider a practical example of how different substitution models affect the distances that are measured in a set of homologous proteins. We enter the proteins into MEGA and select the Distances pull-down menu to compute pairwise distances between these proteins. We can view the number of amino acid differences per sequence, highlighting several pairwise comparisons that are relatively closely or distantly related. Next we estimate the differences based on the Hamming distance (Equation 3.1, called the p-distance in MEGA). When we next use the Poisson correction, the distance values are comparable (relative to the Hamming distance) for closely related sequences such as globins from two lampreys. However, the estimated evolutionary divergence for distantly related sequences is dramatically different using the Poisson correction. The consequence of these differences is that entirely different phylogentic trees may be constructed depending on the particular model you choose. We can use this data set of proteins to construct a neighbor-joining tree (defined below) using either the p-distance (Figure 3.1 a) or the Poisson correction (Figure 3.1 b). Note that the topologies of the two trees differ in this example (inspect soybean and insect globin), and the branch lengths differ.

For the optimal tree using the p-distance correction, the sum of the branch lengths is 2.81, while for the tree made with the Poisson correction, the sum of the branch lengths is 4.93. Such differences can have large effects on the interpretation of a phylogenetic tree. Thus, to choose an appropriate model is the key point.

In order to model substitutions that occur in DNA sequences, Jukes and Cantor [16] proposed another fundamentally useful corrective formula:

$$D = -\frac{3}{4} \ln \left(1 - \frac{4}{3}p\right) \tag{3.4}$$

As an example of how to use Equation 3.4, consider an alignment where 3 nucleotides out of 60 aligned residues differ. The normalized Hamming distance is $3/60 = 0.05$. The Jukes-Cantor correction $D = 0.052$. In this case, applying the correction causes only a small effect. When $30/60$ nucleotides differ, the Jukes–Cantor correction is 0.82, a far more substantial adjustment.

The Jukes–Cantor one-parameter model describes the probability that each nucleotide will mutate to another (Figure 3.2 a). It makes the simplifying assumption that each residue is equally likely to change to any of the other three residues and that the four bases are present in equal frequencies. Thus, this model assumes that the rate of transitions equals the rate of transversions. The corrections are minimal for very closely related sequences but can be substantial for more distantly related sequences. Beyond about 70% differences, the corrected distances are difficult to estimate. This approaches the percent differences found in randomly aligned sequences.

Dozens of models have been developed that are more sophisticated than Jukes–Cantor. Usually, the transition rate is greater than the transversion

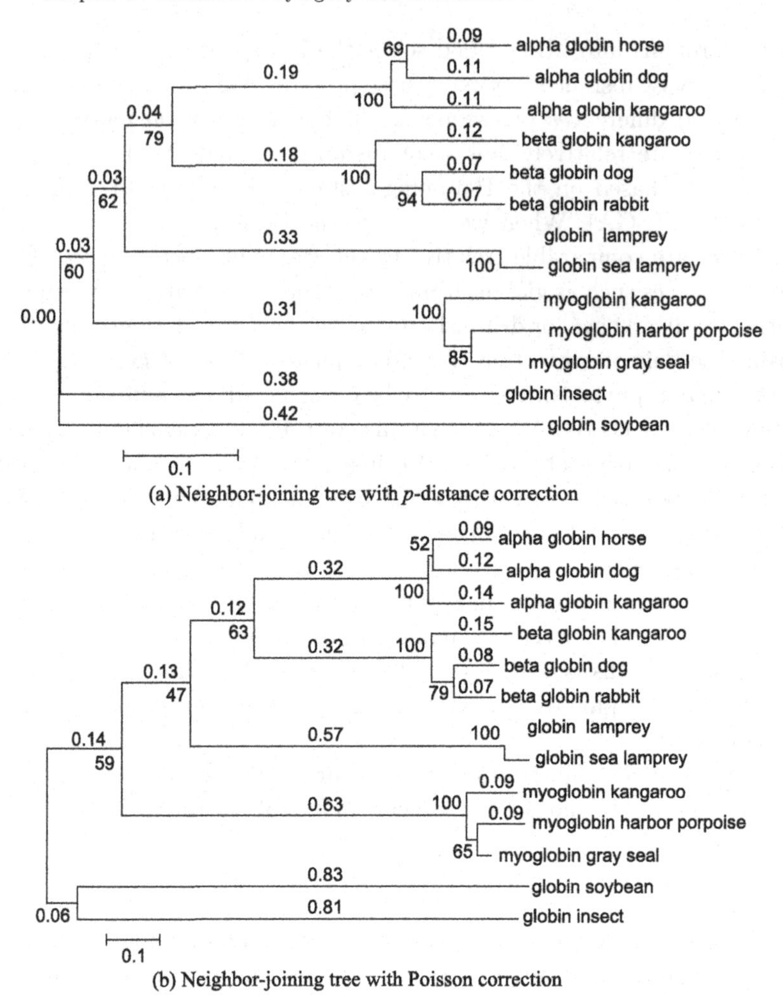

Fig. 3.1 The effect of differing models of amino acid substitution on phylogenic trees

rate; for eukaryotic nuclear DNA it is typically twofold higher. The Kimura's [17] two-parameter model adjusts the transition and transversion ratios by giving more weight to transversions to account for their likelihood of causing nonsynonymous changes in protein-coding regions (Figure 3.2 b). In any region of DNA (including noncoding sequence), the transition/transversion ratio corrects for the biophysical threshold for creating a purine-purine or pyrimidine-pyrimidine pair in the double helix. For example, Tamura [18] extended the two-parameter model to adjust for the guanine and cytosine (GC) content of the DNA sequences (Figure 3.2 c). We will see in Part III of this book that the GC content varies greatly among different organisms and different chromosomal regions within an organism's genome.

Changes in nucleotide substitution at a given position of an alignment represent one kind of DNA variation, and we have been discussing several ways to correct for changes that occur. Substitution rates are often variable across the length of a group of sequences. This represents a second distinct category of DNA variation, and we can also model these changes. Some sites (columns of aligned residues) are invariant, while others do undergo substitutions.

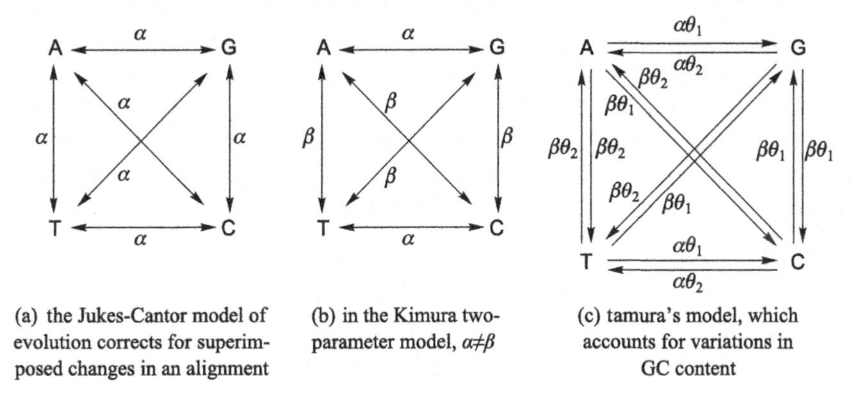

(a) the Jukes-Cantor model of evolution corrects for superimposed changes in an alignment

(b) in the Kimura two-parameter model, $\alpha \neq \beta$

(c) tamura's model, which accounts for variations in GC content

Fig. 3.2 Models of nucleotide substitution

- Because of the degeneracy of the genetic code, the third position of a codon almost always has a higher substitution rate than the first and second codon position.
- Some regions of a protein have conserved domains. Viruses or immunoglobulins often display hypervariable regions of mutation.
- Noncoding RNAs often have functional constraints such as stem and loop structures that include highly conserved positions with low substitution rates.

Zhang and Gu [19] measured α for protein sequences from 51 vertebrate nuclear genes and 13 mammalian mitochondrial genes. They reported a range of values from 0.17 to 3.45 (median value 0.71) for the 51 nuclear genes. There was a negative correlation between the extent of among-site rate variation and the mean substitution rate. Genes with a high level of rate variation among sites (A) have a low mean substitution rate and thus are slowly evolving. Rapidly evolving proteins have a low level of rate variation among sites.

When we create a phylogenetic tree using 13 globin protein sequences in PAUP or MEGA software, we can specify that there is a uniform rate of variation among sites (thus not invoking the gamma distribution), or we can set the shape parameter of α to any positive value. For a group of globin proteins, there are dramatic differences in the branch lengths and the topologies of trees created using the same neighbor-joining method and the Poisson correction with varying gamma distributions and shape parameters $\alpha = 0.25$, $\alpha = 1$, or $\alpha = 5$.

There are many choices for nucleotide or amino acid substitution models in programs such as PAUP, Phylip, and MEGA (Figure 3.3). In addition to and independent of the substitution model, there are many choices for the shape parameter α of the gamma distribution. Several groups have developed strategies to estimate the appropriate models to apply to a data set for phylogenetic analysis. For example, the ModelTest program implements a log likelihood ratio test to compare models [20, 21]. The log likelihood ratio test is a statistical test of the goodness-of-fit between two models. After a DNA data set is executed in PAUP software, ModelTest systematically tests up to 56 models of variation. The likelihood scores of a null model (L_0) and an alternative model (L_1) are calculated for comparisons of a relatively simple model and a relatively complex model. A likelihood ratio test statistic is obtained:

$$\delta = -2 \log \Lambda \tag{3.5}$$

where

$$\Lambda = \frac{\max[L_0(\text{NullModel}|\text{Data})]}{\max[L_1(\text{AlternativeModel}|\text{Data})]} \tag{3.6}$$

Λ is the Greek letter corresponding to L.

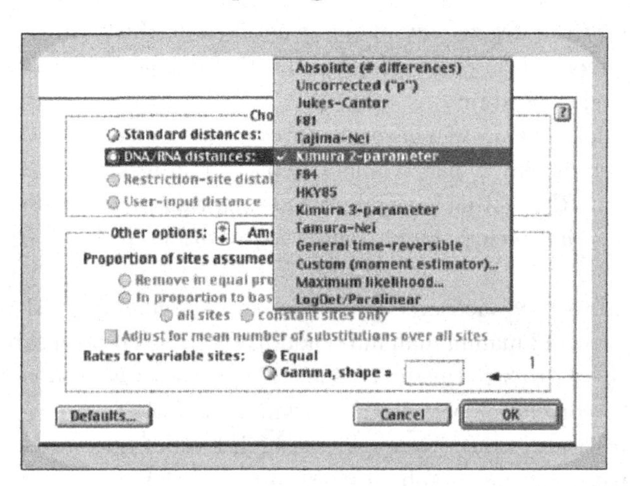

Fig. 3.3 Models of nucleotide and amino acid substitution. Software packages for phylogenetic analysis include many models

This test statistic follows a χ^2 distribution, and so given the number of degrees of freedom (equal to the number of additional parameters in the more complex model), a probability value is obtained. As an alternative to log likelihood ratio tests, ModelTest also uses the Akaike information criterion (AIC) [22]. This measures the best fitting model as that having the smallest AIC value:

$$\mathrm{AIC} = -2 \ln L + 2N \tag{3.7}$$

where L is the maximum likelihood for a model using N independently adjusted parameters for that model. In this way, good maxium likelihood scores are rewarded, while using too many parameters is penalized.

For the example of 11 myoglobin coding sequences, ModelTest selected the Kimura [23] two-parameter model with a G distribution shape parameter of $\alpha = 0.45$. You can perform a similar analysis using the web-based Findmodel tool of the HCV database.

3.3 Tree-Building Method

There are many ways to build a phylogenetic tree, reviewed in books [24-32] and articles [33-38]. We will consider four principal methods of making trees: distance-based methods, maximum parsimony, maximum likelihood, and Bayesian inference. Distance-based methods begin by analyzing pairwise alignments of the sequences and using those distances to infer the relatedness between all the taxa. Maximum parsimony is a character-based method in which columns of residues are analyzed in a multiple sequence alignment to identify the tree with the shortest overall branch lengths that can account for the observed charcter differences. Maximum likelihood and Bayesian inference are model-based statistical approaches in which the best tree is inferred that can account for the observed data.

Molecular phylogeny captures and visualizes the sequence variation that occurs in homologous DNA, RNA, or protein molecules. As we learn how to make trees we will also use some of the most popular software tools for phylogeny. All are extremely versatile and offer a broad range of approaches to making trees.

- PAUP (Phylogenetic Analysis Using Parsimony) was developed by David Swofford.
- MEGA (Molecular Genetic Evolutionary Analysis) was written by Sudhir Kumar, Koichiro Tamura, and Masatoshi Nei. Many of its concepts are explained in an excellent textbook by Nei and Kumar [39], *Molecular Evolution and Phylogenetics*.
- PHYLIP (the PHYLogeny Inference Package) was developed by Joseph Felsenstein. It is perhaps the most widely used phylogeny program, and together with PAUP is the most-used software for published phylogenetic trees. Felsenstein has written an outstanding book, *Inferring Phylogenies*

[32].

- TREE-PUZZLE was developed by Korbinian Strimmer, Arndt von Haeseler, and Heiko Schmidt. It implements a maximum likelihood method which is a model-based approach to phylogeny.
- MrBayes was developed by John Huelsenbeck and Fredrik Ronquist. It implements Bayesian estimation of phylogeny, another model-based approach. MrBayes estimates a quantity called the posterior probability distribution which is the probability of a tree conditioned on the observed data.

All of these programs are useful. PAUP has a particularly user-friendly interface for the Macintosh platform, and although the programs discussed in this chapter it is the only one that is a commercial package. The others are freely available by download. MEGA is particularly inviting for the PC platform. PHYLIP is perhaps the most popular program, and is command-line driven without an accessible graphical user interface.

3.3.1 Phylogenetic Methods

Distance-based methods begin the construction of a tree by calculating the pairwise distances between molecular sequences [40, 41]. A matrix of pairwise scores for all the aligned proteins (or nucleic acid sequences) is used to generate a tree. The goal is to find a tree in which the branch lengths correspond as closely as possible to the observed distances. The main distance-based methods include the unweighted pair group method with arithmetic mean (UPGMA) and neighbor joining (NJ). Distance-based methods of phylogeny are computationally fast, and thus they are particularly useful for analyses of a larger number of sequences (e.g., >50 or 100).

These methods use some distance metric, such as the number of amino acid changes between the sequences, or a distance score. A distance metric is distinguished by three properties: ① the distance from a point to itself must be zero, that is, $D(x, x) = 0$; ② the distance from point x to y must equal the distance from y to x, that is, $D(x, y) = D(y, x)$; and ③ the triangle inequality must apply in that $D(x, y) \leqslant D(x, z) + D(z, y)$. While similarities are also useful, distances (which differ from differences when they obey the above properties) offer appealing properties for describing the relationships between objects [42].

The observed distances between any two sequences i, j can be denoted d_{ij}. The sum of the branch lengths of the tree from taxa i and j can be denoted d'_{ij}. Ideally, these two distance measures are the same, but phenomena such as the occurrence of multiple substitutions at a single position typically cause d_{ij} and d'_{ij} to differ. The goodness of fit of the distances based on the observed data and the branch lengths can be estimated as follows [40]:

$$\sum_i \sum_j w_{ij}(d_{ij} - d'_{ij})^2 \tag{3.8}$$

The goal is to minimize this value; it is zero when the branch lengths of a tree match the distance matrix exactly.

We can inspect the multiple sequence alignment as well as the tree to think about the essence of distance-based molecular phylogeny. In this approach, one can calculate the percentage of amino acid similarity between each pair of proteins in the multiple sequence alignment. Some pairs, such as dog and rabbit beta globins, are very closely related and will be placed close together in the tree. Others, such as insect globin and soybean globin, are more distant than the other sequences and will be placed farther away on the tree. In a sense, we can look at the sequences horizontally, calculating distance measurements between the entire sequences. This approach discards a large amount of information about the characters (i.e., the aligned columns of residues), instead summarizing information about the overall relatedness of sequences. In contrast, character information is evaluated in maximum parsimony, maximum likelihood, and Bayesian approaches. All strategies for inferring phylogenies must make some simplifying assumptions, but nonetheless the simpler approaches of distance-based methods very often produce phylogenetic trees that closely resemble those derived by character-based methods.

3.3.2 The UPGMA Distance-Based Method

We introduce UPGMA here because the tree-building process is relatively intuitive and UPGMA trees are broadly used in the field of bioinformatics. However, the algorithm most phylogeny experts employ to build distance-based trees is neighbor-joining (described below). We can make a distance-based tree in PAUP by selecting the distance criterion from the analysis menu. A dialog box allows you to choose either the UPGMA or neighbor-joining algorithm. MEGA4 similarly offers a pull-down menu for these options. UPGMA clusters sequences based on a distance matrix. As the clusters grow, a tree is assembled. As we would expect, different homologous proteins are clustered in distinct clades. The most closely related proteins are clustered most closely together.

The UPGMA algorithm was introduced by Sokal and Michener [43] and works as follows. Consider five sequences whose distances can be represented as points in a plane. We also represent them in a distance matrix. Some protein sequences, such as 1 and 2, are closely similar, while others (such as

1 and 3) are far less related. UPGMA clusters the sequences as follows, and you can draw a draft while we are discussing sequentially [42].

① We begin with a distance matrix. We identify the least dissimilar groups (i.e. the two OTUs i and j that are most closely related). All OTUs are given equal weights. If there are several equidistant minimal pairs, one is picked randomly.

② Combine i and j to form a new group ij. In our example, groups 1 and 2 have the smallest distance (0.1) and are combined to form cluster (1, 2). This results in the formation of a new, clustered distance matrix having one fewer row and column than the initial matrix. Dissimilarities that are not involved in the formation of the new cluster remain unchanged; for example, in the distance matrix of taxa 3 and 4 still maintain a distance of 0.3. The values for the clustered taxa (1, 2) reflect the average of OTUs 1 and 2 to each of the other OTUs. The distance from OTU1 to OTU4 was initially 0.8, from OTU2 to OTU4 was 1.0, and then the distance from OTU (1, 2) to OTU 4 becomes 0.9.

③ Connect i and j through a new node on the nascent tree. This node corresponds to group ij. The branches connecting i to ij and j to ij each has a length $D_{ij}/2$. In our example, OTUs 1 and 2 are connected through node 6, and the distance between OTU1 and node 6 is 0.05. We label the internal node 6 because we reserve the numbers 1 to 5 on the x axis as the terminal nodes of the tree.

④ Identify the next smallest dissimilarity (between OTUs 4 and 5), and combine those taxa to generate a second clustered dissimilarity matrix. In this step it is possible that two OTUs will be joined (if they share the least dissimilarity), or a single OTU (denoted i) will be joined with a cluster (denoted jk), or two clusters will be joined (ij, kl). The dissimilarity of a single OTU i with a cluster jk is computed simply by taking the average dissimilarity of ij and ik. In this process a new distance matrix is formed, and the tree continues to be constructed. The smallest distance in the matrix is 0.3 corresponding to the relation of OTU3 to the combined OTU4,5. These are joined in the graphic representation, in the distance matrix, and in the tree.

⑤ Continue until there are only two remaining groups, and join these. We demonstrate how to perform UPGMA calculations on available data sets in a series of 12 tables on the supplementary website.

From the tree, we can deduce the proteins which have closest OTUs of the distance matrix have the shortest branch lengths. The second closest group has the next shortest branch lengths. These relationships are visualized in the phylogenetic tree.

A critical assumption of the UPGMA approach is that the rate of nucleotide or amino acid substitution is constant for all the branches in the tree, that is, the molecular clock applies to all evolutionary lineages. If this assumption is true, branch lengths can be used to estimate the dates of divergence, and the sequence-based tree mimics a species tree. An UPGMA tree is rooted because of its assumption of a molecular clock. If it is violated and there are unequal substitution rates along different branches of the tree, the method can produce an incorrect tree. Note that other methods (including neighbor-joining) do not automatically produce a root, but a root can be placed by choosing an outgroup or by applying midpoint rooting.

The UPGMA method is a commonly used distance method in a variety of applications including microarray data analysis. In phylogenetic analyses using molecular sequence data its simplifying assumptions tend to make it significantly less accurate than other distance-based methods such as neighbor-joining.

3.3.3 Making Trees by Distance-Based Methods: Neighbor-Joining

The neighbor-joining (NJ) method is used for building trees by distance methods [44]. It produces both a topology and branch lengths. We begin by defining a neighbor as a pair of OTUs connected through a single interior node X in an unrooted, bifurcating tree. In general, the number of neighbor pairs in a tree depends on the particular topology. For a bifurcating tree with N OTUs, $N-2$ pairs of neighbors can potentially occur. The neighbor-joining method first generates a full tree with all the OTUs in a starlike structure with no hierarchical structure (Figure 3.4 a). All $N(N-1)/2$ pairwise comparisons are made to identify the two most closely related sequences. These OTUs give the smallest sum of branch lengths (see taxa 1 and 2 in Figure 3.4 b).

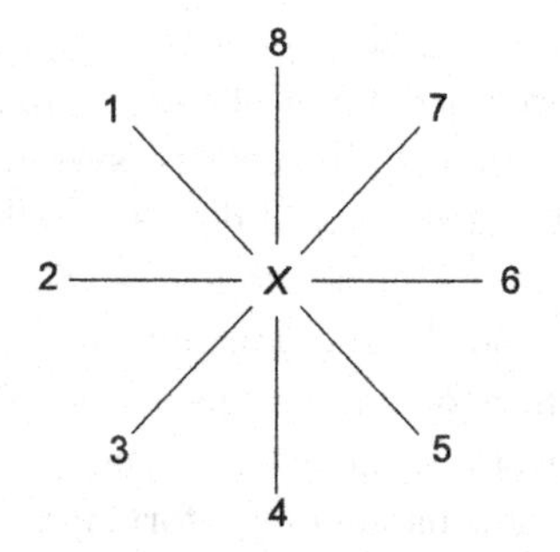

(a) the OTUs are first clustered in a starlike tree. "Neighbors" are defined as OTUS that are connected by a single, interior node in an unrooted, bifurcating tree

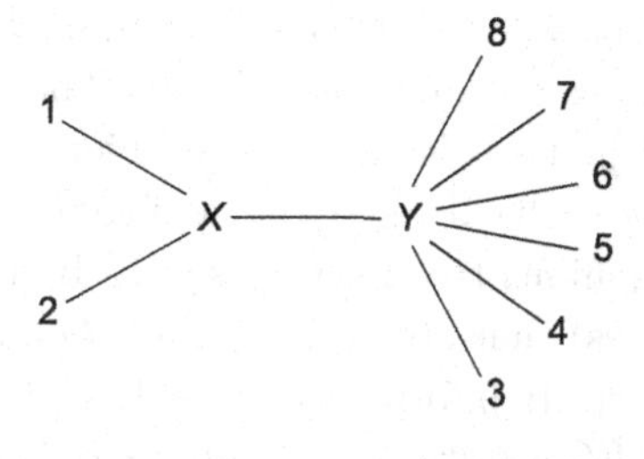

(b) the two closest OTUs are identified, such as OTUs 1 and 2. These neighbors are connected to the other OTUs via the internal branch XY

Fig. 3.4 The NJ method is a distance-based algorithm

OTUs 1 and 2 are now treated as a single OTU, and the method identifies the next pair of OTUs that gives the smallest sum of branch lengths. This could be two OTUs such as 4 and 6, or a single OUT such as 4 paired with the newly formed clade that includes OTUs 1 and 2. The tree has $N-3$ interior branches, and the neighbor-joining method continues to successively identify nearest neighbors until all $N-3$ branches are identified.

The process of starting with a starlike tree and finding and joining neighbors is continued until the topology of the tree is completed. The neighbor-joining algorithm minimizes the sum of branch lengths at each stage of clustering OTUs; although the final tree is not necessarily the one with the shortest overall branch lengths. Thus, its results may differ from minimum evolution strategies or maximum parsimony. Neighbor-joining produces an unrooted tree topology (because it does not assume a constant rate of evolution), unless an outgroup is specified or midpoint rooting is applied.

We have shown several examples of neighbor-joining trees. This algorithm is especially useful when studying large numbers of taxa. There are many recent examples of its use in the literature, such as studies of the 1918 influenza virus [45]. There are many alternative distance-based approaches, some of which have been systematically compared [46, 47].

3.3.4 Phylogenetic Inference: Maximum Parsimony

The main idea behind maximum parsimony is that the best tree is that with the shortest branch lengths possible [48]. Parsimony-based phylogeny based on morphological characters was described by Hennig [49], and Eck and Dayhoff [50] used a parsimony-based approach to generating phylogenetic trees. According to maximum parsimony theory, having fewer changes to account for the way a group of sequences evolved is preferable to more complicated explanations of molecular evolution. Thus we seek the most parsimonius explanations for the observed data. The assumption of phylogenetic systematics is that genes exist in a nested hierarchy of relatedness, and this is reflected in a hierarchical distribution of shared characters in the sequences. The most parsimonious tree is supposed to best describe the relationships of proteins (or genes) that are derived from common ancestors. The steps are as follows:

- Identify informative sites. If a site is constant, then it is not informative. MEGA software includes an option to view parsimony-informative sites. Noninformative sites include constant sites and positions in which there are not at least two states (e.g. two different amino acid residues) with at least two taxa having each state.
- Construct trees. Every tree is assigned a cost, and the tree with the lowest cost is sought. When a reasonable number of taxa are evaluated,

such as about a dozen or fewer, all possible trees are evaluated and the one with the shortest branch length is chosen. When necessary, a heuristic search is performed to reduce the complexity of the search by ignoring large families of trees that are unlikely to contain the shortest tree.

- Count the number of changes and select the shortest tree (or trees). Parsimony analysis assumes that characters are independent of each other. The length L of a full tree is computed as the sum of the lengths l_j of the individual characters:

$$L = \sum_{j=1}^{C} w_j l_j \tag{3.9}$$

where C is the total number of characters, and the weight w_j assigned to each character is typically 1. A different weight might be assigned if, for example, nucleotide transversions are more penalized than transitions.

As an example of how maximum parsimony works, consider five aligned amino acid sequences. Two possible trees describe these sequences; each tree has hypothetical sequences assigned to ancestral nodes. One of the trees requires fewer changes to explain how the observed sequences evolved from a hypothetical common ancestor. In this example, each site is treated independently.

In PAUP, you can set the tree-making criterion to parsimony. It is preferable to perform an exhaustive search of all possible trees to find the one with the shortest total branch lengths. In practice, this is not possible for more than 12 taxa, so it is often necessary to perform a heuristic search. Both heuristic and exhaustive searches often result in the identification of several trees having the same minimal value for total branch length of the tree. Trees can be visualized as a phylogram or a cladogram. An artifact called long-branch attraction sometimes occurs in phylogenetic inference, and parsimony approaches may be particularly susceptible. In a phylogenetic reconstruction of protein or DNA sequences, a branch length indicates the number of substitutions that occur between two taxa. Parsimony algorithms assume that all taxa evolve at the same rate and that all characters contribute the same amount of information. Long-branch attraction is a phenomenon in which rapidly evolving taxa are placed together on a tree, not because they are closely related, but artifactually because they both have many mutations. Consider the true tree in Figure 3.5, in which taxon 2 represents a DNA or protein that changes rapidly relative to taxa 1 and 3. The outgroup is (by definition) more distantly related than taxa 1, 2, and 3 are related to each other, respectively. A maximum parsimony algorithm may generate an inferred tree (Figure 3.5) in which taxon 2 is 'attracted' toward another long branch (the outgroup) because these two taxa have a large number of substitutions.

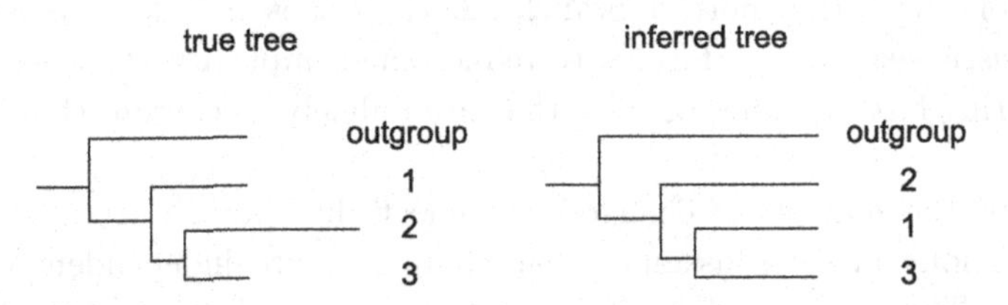

Fig. 3.5 Long branch chain attraction. The true tree includes a taxon (labeled 2) that evoloves more quickly than the other taxa. It shares a common ancestor with taxon 3. Whereas, in the inferred tree taxon 2 is placed separately from the other taxa because it is attracted by the long branch of the outgroup. From Philippe and Laurent (1998)

3.3.5 Model-Based Phylogenetic Inference: Maximum Likelihood

Maximum likelihood is an approach that is designed to determine the tree topology and branch lengths that have the greatest likelihood of producing the observed data sets. A likelihood is calculated for each residue in an alignment, including some model of the nucleotide or amino acid substitution process. It is among the most computationally intensive but most flexible methods available [51]. Maximum parsimony methods sometimes fail when there are large amounts of evolutionary change in different branches of a tree. Maximum likelihood, in contrast, provides a statistical model for evolutionary change that varies across branches. Thus, for example, maximum likelihood can be used to estimate positive and negative selection across individuals branches of a tree. The relative merits of maximum parsimony and maximum likelihood continue to be explored. For example, Kolaczkowski and Thornton [52] reported that when sequences evolve in a heterogeneous fashion over time maximum parsimony can outperform maximum likelihood.

A computationally tractable maximum likelihood method is implemented in the Tree-Puzzle program [53, 54]. The program allows you to specify various models of nucleotide or amino acid substitution and rate heterogeneity (e.g., the G distribution). There are three steps.

First, Tree-Puzzle reduces the problem of tree reconstruction to a series of quartets of sequences. For quartet A, B, C, D there are three possible topologies. In the maximum likelihood step the program reconstructs all quartet trees. For N sequences there are

$$\binom{N}{4}$$

possible quartets; for example, for 12 myoglobin DNA sequences there are

$$\binom{12}{4}$$

or 495 possible quartets. The three quartet topologies are weighted by their posterior probabilities.

$$\binom{n}{k}$$

is a binomial coefficient that is read as "n choose k." It describes the number of combinations, that is, how many ways there are to choose k things out of n possible choices. Given the factorial functions $n!$ and $k!$ we can write the binomial coefficient

$$\binom{n}{k} = \frac{n!}{k! \cdot (n-k)!}$$

for

$$\binom{12}{4}$$

this corresponds to

$$\frac{12!}{4!(8)!} \quad \text{or} \quad \frac{12 \cdot 11 \cdot 10 \cdot 9}{4 \cdot 3 \cdot 2 \cdot 1}$$

which is 495.

In the second step, called the quartet puzzling step, a large group of intermediate trees is obtained. The program begins with one quartet tree. Since that tree has four sequences, $N - 4$ sequences remain. These are added systematically to the branches that are most likely based on the quartet results from the first step. Puzzling allows estimates of the support to each internal branch of the tree that is constructed; such estimates are not available for distance- or parsimony-based trees.

In the third step, the program generates a majority consensus tree. The branch lengths and maximum likelihood value are estimated. An example of a consensus tree is shown in Figure 3.6 a.

The Tree-Puzzle program also allows an option called likelihood mapping which describes the support of an internal branch as well as a way to visualize the phylogenetic content of a multiple sequence alignment [53, 55]. The quartet topology weights sum to 1, and likelihood mapping plots them on a triangular surface. In this plot, each dot corresponds to a quartet that is positioned spatially according to its three posterior weights (Figure 3.6 b). For 13 globin protein sequences, 9.7% of the quartets were unresolved (as indicated in the center of the triangle). An additional 0.3% + 0.4% + 0.1% of the quartets were partially resolved. For 12 myoglobin DNA coding sequences, only 3% of the quartets were unresolved (not shown). Likelihood mapping summarizes the strength (or conversely the ambiguity) inherent in a data set for which you perform tree puzzling.

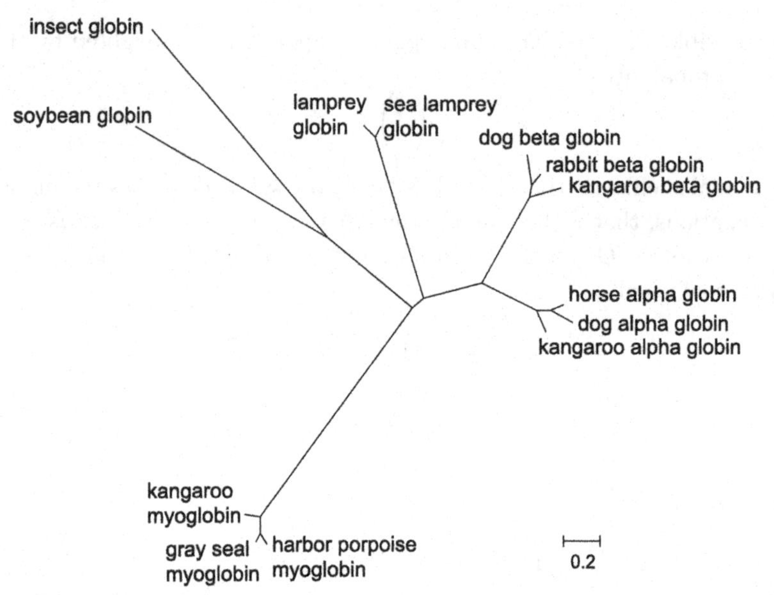

(a) the taxa in any tree with four or more sequences can be represented as quartets of sequences. This tree of 13 globin proteins was constructed using the Tree-Puzzle program

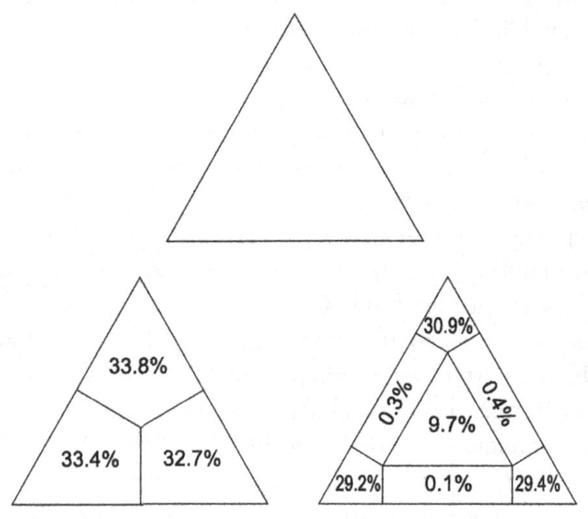

(b) likelihood mapping (in Tree-Puzzle) indicates the frequency with which quartets are successfully resolved

Fig. 3.6 Maximum likelihood inference of phylogenetic trees using quartet puzzling

3.3.6 Tree Inference: Bayesian Methods

Bayesian inference is a statistical approach to modeling uncertainty in complex models. Conventionally we calculate the probability of observing some data (such as the result of a coin toss) given some probability model. This probability is denoted Pr (data|model), that is, the probability of the data given the model (This is also read as 'the probability of the data conditional upon the model.') Bayesian inference instead seeks the probability of a tree conditional on the data (that is, based on the observations such as a given multiple sequence alignment). This assumes the form Pr (model|data), Pr (hypothesis|data), or in our case Pr (tree|data). According to Bayes's theorem [56],

$$\Pr[\text{Tree}|\text{Data}] = \frac{\Pr[\text{Data}|\text{Tree}] \times \Pr[\text{Tree}]}{\Pr[\text{Data}]} \tag{3.10}$$

Bayesian estimation of phylogeny is focused on a quantity called the posterior probability distribution of trees, Pr[Tree|Data] (This is read as the probability of observing a tree given the data). For a given tree, the posterior probability is the probability that the tree is correct, and our goal is to identify the tree with the maximum probability. On the right side of Equation 3.10, the denominator Pr[Data] is a normalizing constant over all possible trees. The numerator consists of the prior probability of a phylogeny (Pr[Tree]) and the likelihood Pr[Data|Tree]. These terms represent a distinctive feature of Bayesian inference of phylogeny: the user specifies a prior probability distribution of trees (although it is allowable for all possible trees to be given equal weight).

Practically, we can apply a Bayesian inference approach using the MrBayes software program. There are four steps. First, read in a Nexus data file. This can be accomplished by performing a multiple sequence alignment of interest, then converting it into the Nexus format with a tool such as ReadSeq. We will use an example of 13 globin protein-coding DNA sequences.

Second, specify the evolutionary model. This includes options for data that are DNA (whether coding or not), ribosomal DNA (for the analysis of paired stem regions), and protein. Before performing the analysis, one specifies a prior probability distribution for the parameters of the likelihood model. There are six types of parameters that are set as the priors for the model in the case of the analysis of nucleotide sequences: ① the topology of the trees (e.g. some nodes can be constrained to always be present), ② the branch lengths, ③ the stationary frequencies of the four nucleotides, ④ the six nucleotide substitution rates (for A → C, A → G, A → T, C → G, C → T, and G → T), ⑤ the proportion of invariant sites, and ⑥ the shape parameter of the gamma distribution of rate variation. For protein sequences, both fixed-rate and variable-rate models are offered. Your decisions on how to specify these parameters may be subjective. This can be considered either a strength of the Bayesian approach (because your judgment may help you to select optimal parameters) or a weakness (because there is a subjective element to the procedure). All priors do not have to be informative; one can select conservative settings.

Third, run the analysis. This is invoked with the MCMC (Monte Carlo Markov Chain) command. The posterior probability of the possible phylogenetic trees is ideally calculated as a summation over all possible trees, and for each tree, all combinations of branch lengths and substitution model parameters are evaluated. In practice this probability cannot be determined analytically, but it can be approximated using MCMC. This is done by drawing many samples from the posterior distribution [57]. MrBayes runs two simultaneous, independent analyses beginning with distinct, randomly initiated trees. This helps to assure that your analysis includes a good sampling from the posterior probability distribution.

Eventually the two runs should reach convergence. An MCMC analysis is performed in three steps: first, a Markov chain is started with a tree that may be randomly chosen. Second, a new tree is proposed. Third, the new tree is accepted with some probability. Typically tens to hundreds of thousands of MCMC iterations are performed. The proportion of time that the Markov chain visits a particular tree is an approximation of the posterior probability of that tree. Some authors have cautioned that MCMC algorithms can give misleading results, especially when data have conflicting phylogenetic signals [58].

Fourth, summarize the samples. MrBayes provides a variety of summary statistics including a phylogram, branch lengths (in units of the number of expected substitutions per site), and clade credibility values. The summary statistics for a Bayesian analysis are provided. They include a list of all trees sorted by their probabilities; this is used to create a 'credible' list of trees. A consensus tree showing branch lengths and support values for interior nodes can be generated.

Bayesian inference of phylogeny resembles maximum likelihood because each method seeks to identify a quantity called the likelihood which is proportional to observing the data conditional on a tree. The methods differ in that Bayesian inference includes the specification of prior information and uses MCMC to estimate the posterior probability distribution. Although they were introduced relatively recently, Bayesian approaches to phylogeny are becoming increasingly commonplace.

3.4 Evaluating Tree

After you have constructed a phylogenetic tree, how can you assess its accuracy? The main criteria by which accuracy may be assessed are consistency, efficiency, and robustness [59, 60]. One may study the accuracy of a tree-building approach or the accuracy of a particular tree. The most common approach is bootstrap analysis [61, 62]. Bootstrapping is not a technique to assess the accuracy of a tree. Instead, it describes the robustness of the tree topology. Given a particular branching order, how consistently does a tree-building algorithm find that branching order using a randomly permuted version of the original data set? Bootstrapping allows an inference of the variability in an unknown distribution from which the data were drawn [61].

Nonparametric bootstrapping is performed as follows. A multiple sequence alignment is used as the input data to generate a tree using some tree-building method. The program then makes an artificial data set of the same size as the original data set by randomly picking columns from the multiple sequence alignment. This is usually performed with replacement, meaning that any individual column may appear multiple times (or not at all). A tree is generated from the randomized data set. A large number of bootstrap replicates are then generated; typically, between 100 and 1000 new trees are made by this process. The bootstrap trees are compared to the original, inferred tree(s). The information you get from bootstrapping is the frequency with which each clade in the original tree is observed.

An example of the bootstrap procedure using MEGA4 is shown in Figure 3.1. The percentage of times that a given clade is supported in the original tree is provided based on how often the bootstraps supported the original tree topology. Bootstrap values above 70% are sometimes considered to provide support for the clade designations. Hillis and Bull [62] have estimated that such values provide statistical significance at the $p < 0.05$ level. This approach measures the effect of random weighting of characters in the original data matrix, giving insight into how strongly the phylogenetic signal that produces a tree is distributed through the multiple sequence alignment. In Figure 3.1 a and b, the clade containing three alpha globins has 100% bootstrap support, indicating that in all 500 bootstrap replicates that clade maintained its integrity (with none of the three alpha globins assigned to a different clade, and no non-alpha globin joining that clade). However, the clade containing horse and dog alpha globin received only 52% bootstrap support (Figure 3.1 b), suggesting that about half of the time kangaroo alpha globin was in a clade with the dog or horse orthologs. This example shows how viewing the bootstrap percentages can be useful to estimate the robustness of each clade in a tree. Note that bootstrapping supports a model in which alpha globins, beta globins, myoglobins, and lamprey globins each are assigned to a unique clade.

Maximum likelihood approaches report the tree with the greatest likelihood, and they also report the likelihood for internal branches. For Bayesian inference of phylogeny, the result is typically the most probable tree (called a maxiumum a posteriori probability estimate). The results are often summarized using a majority rule consensus tree in which the values represent the posterior probability that each clade is true. The confidence estimates may sometimes be too liberal [63]. For example, Mar, et al. [64] found that Bayesian posterior probabilities reached 100% at bootstrap percentages of 80%.

3.5 Perspectives

Molecular phylogeny is a fundamental tool for understanding the evolution and relationships of protein (and nucleic acid) sequences. The main output of this analysis is a phylogenetic tree, which is a graphical representation of a multiple sequence alignment. The recent rapid growth of DNA and protein sequence data, along with the visual impact of phylogenetic trees, has made phylogeny increasingly important and widely applied. We will show examples of trees in following chapters as we explore genomes across the tree of life.

The field of molecular phylogeny includes conceptually distinct approaches, including those outlined in this chapter (distance, maximum parsimony, maximum likelihood, and Bayesian methods). For each of these approaches software tools continue to evolve, so it is reasonable for you to obtain a multiple sequence alignment and perform phylogenetic analyses with all four tree-making approaches and with a variety of substitution models. The relative merits of these maximum-parsimomy versus model-based approaches continue to be debated [52].

References

[1] Simpson G G. *The Meaning of Evolution: A Study of the History of Life and of Its Significance for Man.* New Haven: Yale University Press, 1952.

[2] Mayr E. *The Growth of Biological Thought: Diversity, Evolution, and Inheritance.* Cambridge: Belknap Press of Harvard Univ., 1982.

[3] Wilson E O. *The Diversity of Life.* New York: W.W. Norton, 1992.

[4] Doolittle W F. Phylogenetic classification and the universal tree. *Science,* 1999(284): 2124–2129.

[5] Dayhoff M O. Atlas of Protein Sequence and Structure. *National Biomedical Research Foundation*, Silver Spring, 1978.

[6] Feng D F, and Doolittle R F. Progressive sequence alignment as a prerequisite to correct phylogenetic trees. *J. Mol.*, 1987(Evol. 25): 351–360.

[7] Kimura M. Evolutionary rate at the molecular level. *Nature*, 1968(217): 624–626.

[8] Kimura M. *The Neutral Theory of Molecular Evolution.* Cambridge: Cambridge University Press, 1983.

[9] Zuckerkandl E, and Pauling L. Molecular disease, evolution, and genic heterogeneity. In: Kasha M and Pullman B (eds.). *Horizons in Biochemistry.* Albert Szent-Gyorgyi Dedicatory Volume. New York: Academic Press, 1962.

[10] Dickerson R E. Sequence and structure homologies in bacterial and mammalian-type cytochromes. *J. Mol. Biol.*, 1971(57): 1–15.

[11] Margoliash E and Smith E L. Structural and functional aspects of cytochrome c in relation to evolution. In: Bryson V and Vogel H J (eds.). *Evolving Genes and Proteins.* New York: Academic Press, 1965, 221–242.

[12] Zuckerkandl E, and Pauling L. Evolutionary divergence and convergence in proteins. In: Bryson V and Vogel H J (eds.). *Evolving Genes and Proteins.* New York: Academic Press, 1965, 97–166.

[13] Nei M and Kumar S. *Molecular Evolution and Phylogenetics.* New York: Oxford University Press, 2000.

[14] Uzzell T, and Corbin K W. Fitting discrete probability distributions to evolutionary events. *Science*, 1971(172): 1089–1096.

[15] Zar J H. *Biostatistical Analysis*, 4th edition. Upper Saddle River: Prentice Hall, 1999.

[16] Jukes T H, and Cantor C. Evolution of protein molecules. In: Munro H N and Allison J B (eds.). *Mammalian Protein Metabolism.* New York: Academic Press, 1969.

[17] Kimura M. A simple method for estimating evolutionary rates of base substitutions through comparative studies of nucleotide sequences. *J. Mol. Evol.*, 1980(16): 111–120.

[18] Tamura K. Estimation of the number of nucleotide substitutions when there are strong transition-transversion and G + C content biases. *Mol. Biol. Evol.*, 1992(9): 678–687.

[19] Zhang J, and Gu X. Correlation between the substitution rate and rate variation among sites in protein evolution. *Genetics*, 1998(149): 1615–1625.

[20] Posada D, and Crandall K A. MODELTEST: Testing the model of DNA substitution. *Bioinformatics*, 1998(14): 817–818.

[21] Posada D. ModelTest Server: A web-based tool for the statistical selection of models of nucleotide substitution online. *Nucleic Acids Res.*, 2006, 34(Web Server issue), W700–W703.

[22] Posada D, and Buckley T R. Model selection and model averaging in phylogenetics: Advantages of Akaike information. criterion and Bayesian approaches over likelihood ratio tests. *Syst. Biol.*, 2004(53): 793–808, PMID: 15545256.

[23] Kimura M. A simple method for estimating evolutionary rates of base substitutions through comparative studies of nucleotide sequences. *J. Mol. Evol.*, 1980(16): 111–120.

[24] Nei M. *Molecular Evolutionary Genetics.* New York: Columbia University Press, 1987.

[25] Graur D, and Li W-H. *Fundamentals of Molecular Evolution*, 2nd ed. Sunderland: Sinauer Associates, 2000.

[26] Li W-H. *Molecular Evolution.* Sunderland: Sinauer Associates, 1997.

[27] Maddison D, and Maddison W. *MacClade 4: Analysis of Phylogeny and Character Evolution.* Sunderland: Sinauer Associates, 2000.

[28] Durbin R, Eddy S, Krogh A, and Mitchison G. *Biological Sequence Analysis.* Cambridge: Cambridge University Press, 1998.

[29] Baxevanis A D, and Ouellette B F. *Bioinformatics*, 2nd ed. New York: Wiley-Interscience, 2001.

[30] Clote P, and Backofen R. *Computational Molecular Biology: An Introduction.* New York: Wiley, 2000.

[31] Hall B G. *Phylogenetic Trees Made Easy: A How-To for Molecular Biologists.* Sunderland: Sinauer Associates, 2001.

[32] Felsenstein J. *Inferring Phylogenies.* Sunderland: Sinauer Associates, 2004.

[33] Bos D H, and Posada D. Using models of nucleotide evolution to build phylogenetic trees. *Dev. Comp. Immunol.*, 2005(29): 211–227.

[34] Felsenstein J. Inferring phylogenies from protein sequences by parsimony, distance, and likelihood methods. *Methods Enzymol*, 1996(266): 418–427.

[35] Felsenstein J. Phylogenies from molecular sequences: Inference and reliability. *Annu. Rev. Genet.*, 1988(22): 521–565.

[36] Hein J. Unified approach to alignment and phylogenies. *Methods Enzymol*, 1990(183): 626–645.

[37] Nei M. Phylogenetic analysis in molecular evolutionary genetics. *Annu. Rev. Genet.*, 1996(30): 371–403.

[38] Thornton J W, and DeSalle R. Gene family evolution and homology: Genomics meets phylogenetics. *Annu. Rev. Genomics Hum. Genet.*, 2000(1): 41–73.

[39] Nei M and Kumar S. *Molecular Evolution and Phylogenetics.* New York: Oxford University Press, 2000.

[40] Felsenstein J. Distance methods for inferring phylogenies: A justification. *Evolution*, 1984(38): 16–24.

[41] Desper R, and Gascuel O. Getting a tree fast: Neighbor Joining, FastME, and distance-based methods. *Curr. Protoc. Bioinformatics*, 2006(6): Unit 6.3.

[42] Sneath P H A and Sokal R R. *Numerical Taxonomy.* San Francisco: W. H. Freeman & Co., 1973.

[43] Sokal R R and Michener C D. A statistical method for evaluating systematic relationships. *Univ. Kansas Science Bull.*, 1958(38): 1409–1437.

[44] Saitou N, and Nei M. The neighbor-joining method: A new method for reconstructing phylogenetic trees. *Mol. Biol.*, 1987(Evol. 4): 406–425.

[45] Taubenberger J K, Reid A H, Lourens R M, Wang R, Jin G, and Fanning T G. Characterization of the 1918 influenza virus polymerase genes. *Nature*, 2005(437): 889–893.

[46] Hollich V, Milchert L, Arvestad L, and Sonnhammer E L. Assessment of protein distance measures and tree-building methods for phylogenetic tree reconstruction. *Mol. Biol.*, 2005(Evol. 22): 2257–6422.

[47] Desper R, and Gascuel O. Getting a tree fast: Neighbor Joining, FastME, and distance-based methods. *Curr. Protoc. Bioinformatics*, 2006(6): Unit 6.3.

[48] Czelusniak J, Goodman M, Moncrief N D, and Kehoe S M. Maximum parsimony approach to construction of evolutionary trees from aligned homologous sequences. *Methods Enzymol*, 1990(183): 601–615.

[49] Hennig W. *Phylogenetic systematics.* Urbana: University of Illinois Press, 1966.

[50] Eck R V and Dayhoff M O. *Atlas of Protein Sequence and Structure.* Maryland: Silver Spring, 1966.

[51] Felsenstein J. Evolutionary trees from DNA sequences: A maximum likelihood approach. *J. Mol.*, 1981(Evol. 17): 368–376.

[52] Kolaczkowski B, and Thornton J W. Performance of maximum parsimony and likelihood phylogenetics when evolution is heterogeneous. *Nature*, 2004(431): 980–984.

[53] Strimmer K, and von Haeseler A. Quartet puzzling: A quartet maximun likelihood method for reconstructing tree topologies. *Mol. Biol.*, 1996(Evol. 13): 964–969.

[54] Schmidt H A, Strimmer K, Vingron M, and von Haeseler A. TREE-PUZZLE: Maximum likelihood phylogenetic analysis using quartets and parallel computing. *Bioinformatics*, 2002(18): 502–504.

[55] Strimmer K, and von Haeseler A. Likelihood-mapping: a simple method to visualize phylogenetic content of a sequence alignment. Proc. *Natl. Acad. Sci.*, 1997(94): 6815–6819.

[56] Huelsenbeck J P, Ronquist F, Nielsen R, and Bollback J P. Bayesian inference of phylogeny and its impact on evolutionary biology. *Science*, 2001(294): 2310–2314.

[57] Huelsenbeck J P, Larget B, Miller R E, and Ronquist F. Potential applications and pitfalls of Bayesian inference of phylogeny. *Syst. Biol.*, 2002(51): 673–688.

[58] Mossel E, and Vigoda E. Phylogenetic MCMC algorithms aremisleading on mixtures of trees. *Science*, 2005(309): 2207–2209.

[59] Hillis D M. Approaches for assessing phylogenetic accuracy. *Syst. Biol.*, 1995(44): 3–16.

[60] Hillis D M, and Huelsenbeck J P. Signal, noise, and reliability in molecular phylogenetic analyses. *J. Hered.*, 1992(83): 189–195.

[61] Felsenstein J. Confidence limits on phylogenies: An approach using the bootstrap. *Evolution*, 1985(39): 783–791.

[62] Hillis D M, and Bull J J. An empirical test of bootstrapping as a method for assessing confidence in phylogenetic analysis. *Systematic Biol.*, 1993(42): 182–192.

[63] Suzuki Y, Glazko G V, and Nei M. Overcredibility of molecular phylogenies obtained by Bayesian phylogenetics. Proc. *Natl. Acad. Sci.*, 2002(99): 16138–16143.

[64] Mar J C, Harlow T J, and Ragan M A. Bayesian and maximum likelihood phylogenetic analyses of protein sequence data under relative branch-length differences and model violation. *BMC Evol. Biol.*, 2005(5): 1–20.

[65] Steel M. Should phylogenetic models be trying to "fit an elephant"?. *Trends Genet.*, 2005(21): 307–309.

Chapter 4 Predicting DNA and Protein Function from Sequence

Zhennan Wang[1], Haibin Wang[2]

With the rapid accumulation of genomic sequence information, there is a growing need to use computational approaches to accurately predict gene structure. Computational gene prediction is a prerequisite for detailed functional annotation of genes and genomes. The process includes a detection of the location of open reading frames (ORFs) and a delineation of the structures of introns as well as exons if the genes of interest are of eukaryotic origin. The ultimate goal is to describe all the genes computationally with high accuracy, which can significantly reduce the amount of experimental verification work required. However, this may still be a distant goal, particularly for eukaryotes, because many problems in computational gene prediction are still largely unsolved. Gene prediction, in fact, represents one of the most difficult problems in the field of pattern recognition. This is because coding regions normally do not have conserved motifs. Detecting coding potential of a genomic region has to rely on subtle features associated with genes that may be very difficult to detect.

The proteome corresponds to the genome, but they are fundamentally different. An organism has only one identified genome, and all the different cells that make up the organism share a genome. However, the conditions, time and location of the genes in the genome are different, and their expression products (proteins) are also different depending on the conditions, time and location. Therefore, the proteome is a dynamic concept. For the above reasons, coupled with gene splicing, protein post-translational modification and protein splicing. The expression of genetic information become more complex. The study of protein structure and function has a long history, however, because of its complexity, the structural and functional prediction

1. Wang Zhennan, College of Life Sciences and Bioengineering, School of Science, Beijing Jiao Tong University, Beijing, China, 100044.

2. Wang Haibin, College of Life Sciences and Bioengineering, School of Science, Beijing Jiao Tong University, Beijing, China, 100044.

of protein are more complex.

This chapter will describe a number of commonly used gene prediction algorithms, their theoretical basis, and their limitations. In addition, this chapter will also briefly introduce the bioinformatics pathways for the functional prediction of protein.

4.1 DNA Sequence Analysis

4.1.1 Repetitive Sequence

Repetitive sequence (also known as repetitive elements, or repeats) are patterns of nucleic acids that occur in multiple copies throughout the genome. The repetitive sequences are widely present in the genomes of eukaryotes and prokaryotes. These repetitive sequence, mostly collected and formed a repetitive sequence database. The presence of these repeats lead to serious errors in sequence analysis. For example, it leads to erroneous functional annotations in sequence alignment, especially those involving database search programs, making SNP analysis severely deviated. So, recognizing and blocking repetitive sequence is a crucial step in the analysis.

Eukaryotic DNA sequences can be divided into non repetitive sequence (unique sequence), low repetitive sequence, moderately repetitive sequence and highly repetitive sequence. Specific to repetitive element, it can be divided into SINE, ALU, MIR, LINE, LTR, MALR, ERVL, interspersed elements, small RNA, satellite DNA, Simple Sequence Repeat, etc.

Commonly used repetitive sequence analysis program is RepeatMasker (ftp.genome.washington.edu/cgi-bin/RepeatMasker) and XBLAST (bioweb. pasteur.fr/seqanal/interfaces/xblast.html # _data). Need to be reminded, most of these tools are designed for certain species. So, we need to figure out the scope of the species of these tools before using them. For example, RepeatMasker is mainly designed for primates, rodents, arabidopsis, herbs, fruit flies, as well as other mammals and vertebrates. The XBLAST can be applied to any species.

Now, let's use the RepeatMasker to shield the repeat sequence of the following sequence (Figure 4.1).

```
TGTAGGGAAA AGAAAGAGAG ATCAGACTGT TACTGTGTCT ATGTAGAAAG
GGAAGACATA AGAAACTCCA TTTTGACCTG TACCCTGAAC AATTGCCTTT
```

Fig. 4.1 The input sequence

The analysis results are shown in Figure 4.2

```
Repeat sequence:
    SW perc perc perc query      position in query matching repeat      position in repeat
score div . del . ins . sequence begin end (left) repeat class/ family begin end (left) ID
770    6.1    6.1 0.0 UnnamedSeq1    1    99 (1) + LTR5B    LTR/ ERVK    1    105 (928)    1

Alignments:
    Assumed background GC level in scoring matrices is 43 %

    770   6.06   6.06   0.00   UnnamedSeq1   1   99 (1) LTR5B # LTR/ ERVK   1   105 (928) 5

    UnnamedSeq1                                                               1
TGTAGGGAAAAGAAAGAGAGATCAGACTGTTACTGTGTCTATGTAGAAAG 50
                                  i
    LTR5B # LTR/ ERVK                                                         1
TGTAGGGAAAAGAAAGAGAGATCAGACTGTCACTGTGTCTATGTAGAAAG 50
    ......
    Transitions / transversions = 1.00 (5 / 0)
    Gap_init rate = 0.01 (1/ 105), avg . gap size = 6.00 (6/ 1)

Masked Sequence:
    > UnnamedSeq1
    NNNNNNNNNNNNNNNNNNNNNNNNNNNNNNNNNNNNNNNNNNNNNNNNNN
    NNNNNNNNNNNNNNNNNNNNNNNNNNNNNNNNNNNNNNNNNNNNNNNNNT
```

Fig. 4.2 The analysis results

The first part of the result is the attribute table generated by the RepeatMasker. The second part is the result of comparing it to the Repbase database. The third part is the output sequence after blocking the repetitive sequence, and the shielded part is represented by 'N'. The last part gives the composition of the repetitive elements in the input sequence.

RepeatView (http://l25.itba.mi.cnr.it/genebin/wwwrepeat.pl) can not only mask repetitive sequences, but also locate repeat sequences in the orig-

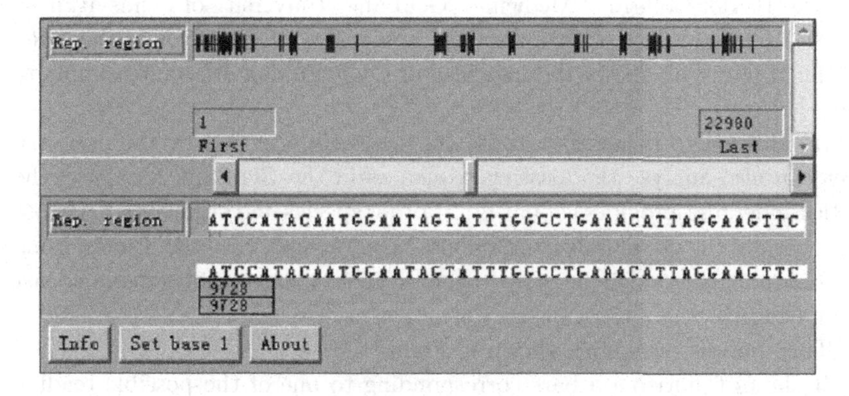

Fig. 4.3 The results from RepeatView

inal sequence. Figure 4.3 shows the results from RepeatView.

The general approach of repetitive sequence masking method is to determine the repetitive element by experiment first, then build the database, then compare the unknown sequence with the sequence in the database to find the repetitive sequence in the unknown sequence. All data analysis method has its own limitations. So, for the location and masking of repeating sequences in complex sequences, it is best to use a variety of methods for comparison.

4.1.2 The Identification of ORF

Regions of DNA that encode proteins are first transcribed into messenger RNA and then translated into protein. By examining the DNA sequence, we can determine the sequence of amino acids that will appear in the final protein. In translation codons of three nucleotides determine which amino acid will be added next in the growing protein chain. It is important to decide which nucleotide to start translation, and when to stop, this is called an open reading frame.

Once a gene has been sequenced, it is important to determine the correct open reading frame (ORF). Every region of DNA has six possible reading frames, three in each direction. The reading frame that is used to determine which amino acids will be encoded by a gene. Typically only one reading frame is used in translating a gene (in eukaryotes), and this is often the longest open reading frame. Once the open reading frame is known the DNA sequence can be translated into its corresponding amino acid sequence. An open reading frame starts with an atg (Met) in most species and ends with a stop codon (taa, tag or tga).

Many tools can be used for identifying the ORF, such as GENSCAN, Genie, GENEBUILDER, GLIMMERM, FGENES, GRAIL, ORF Finder, Gene feature (Baylor College of Medicine), GenLang (University of Pennsylvania).

NCBI provides us with a tool for analyzing ORF, which is the ORF Finder. Figure 4.4 shows the interface of ORF Finder. It's quite simple to use.

In addition to being able to use the sequence provided by the user, the user can also analyze the existing sequences in the GenBank database, the ORF Finder also provides codons for 21 other species and organelles to choose from besides the use of universal codons. Now we study the ORF Finder using the phosphoglyceraldehyde dehydrogenase gene sequence of the fission yeast as an example.

The analysis results are shown in Figure 4.5.

It shows 6 horizontal bars corresponding to one of the possible reading frame. In each direction of the DNA there would be 3 possible reading frames.

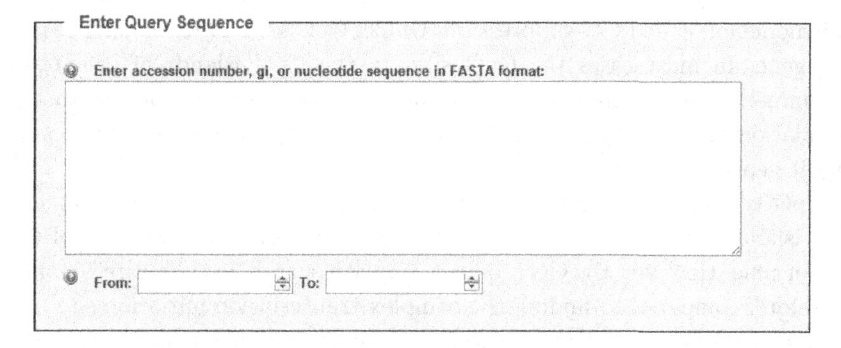

Fig. 4.4 The interface of ORF Finder

View | 1 GenBank | Redraw | 100 | SixFrames

Frame	from	to	Length
-2	26..	1366	1341
+2	32..	1351	1320
-3	64..	432	369
+3	633..	959	327
+3	234..	494	261
-3	952..	1113	162

Fig. 4.5 The analysis results of ORF Finder

So total 6 possible reading frames (6 horizontal bars) would be there for every DNA sequence. The 6 possible reading frames are +1, +2, +3 and –1, –2 and –3 in the reverse strand. The resultant amino acids can be saved and search against various protein databases using blast for finding similar sequences or amino acids. The result displays the possible protein sequence and the length of the open reading frame.

As can be seen from the results of the calculation, a DNA sequence could have many ORF, but only a few are true coding regions. In general, a long ORF may be a coding sequence, and a very short ORF is difficult to encode a protein. However, it is reported that some very short ORF may independently encode a short peptide which have important biological functions. So, it is also necessary to combine the sequence itself and other analytical methods to determine whether an ORF encodes a biologically significant protein.

4.1.3 CpG Island

CpG islands are short stretches of DNA sequences with an unusually high GC content and a higher frequency of CpG dinucleotides as compared to the rest of the genome. CpG islands are characterized by CpG dinucleotide content of at least 60% of that which would be statistically expected (~4%–6%), whereas the rest of the genome has much lower CpG frequency (~1%), a phenomenon called CG suppression. Unlike CpG sites in the coding region of a gene, in most cases the CpG sites in the CpG islands of promoters are unmethylated if the genes are expressed. This observation led to the speculation that methylation of CpG sites in the promoter of a gene may inhibit gene expression.

CpG islands are often found in the 5' region of the vertebrate gene. The CpG island is present at the 80% of the transcription initiation sites of the human gene. However, the CG sequence is relatively rare in the entire genome. Therefore, compared to finding the complex transcription initiation site and the 5' ends of the gene, CpG island is an important clue for the discovery of genes.

There are many computational tools for CpG island. We focus on the tools provided by EMBL, which is CpGPlot/CpGReport/Isochore(www.ebi.ac.uk/emboss/cpgplot/).

Figure 4.6 shows the interface of CpGPlot/CpGReport/Isochore.

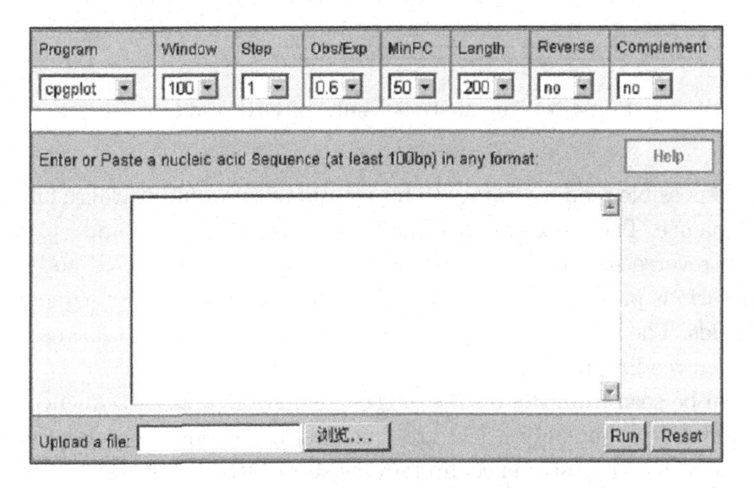

Fig. 4.6 The interface of CpGPlot/CpGReport/Isochore

The first drop-down box contains three programs for analyzing the CpG islands, which are CpGPlot, CpGReport, and Isochore.

CpGPlot graphically shows the Obs/Exp value, GC percentage and the estimated CpG island position after the sequence was analyzed.

CpGReport gives the form of the report on the CpG island location, size, C + G total and CG percentages found in the sequence.

Many genomes are chimeric by isochore, which are homogeneous on the base structure and are similar to the positions of the genes they contain on the physical map. Isochore are isolated by the sequences of specific GC contents. The role of Isochore is to graphically identify the GC content in different Isochore on a large fragment DNA sequence.

The second option is 'Window', which is 100, 200, 300, and 400, respectively. It means in what sequence length range to calculate GC percentages and apparent frequencies. The position of the window movement is controlled by the 'Step' parameter, which determines the number of bases that are traversed after each window has been calculated.

'Obs/Exp' means the ratio of average apparent and expected. 'MinPC' is used to specify the minimum GC content percentage. They are used to adjust the signal-to-noise ratio of the calculated results.

'Length' parameter specifies the minimum length of the sequence program identified CpG islands.

The 'Reverse' option is used to select whether or not to calculate the reverse sequence of the input sequence, and the 'Complement' option is used to select whether to compute the complementary sequence of the input sequence.

To identify the CpG island with an example, we select the mouse Mtch1 gene (gi: 20899759) as the input sequence, select CpGPlot, the other parameters using the default value. The results are shown in Figure 4.7.

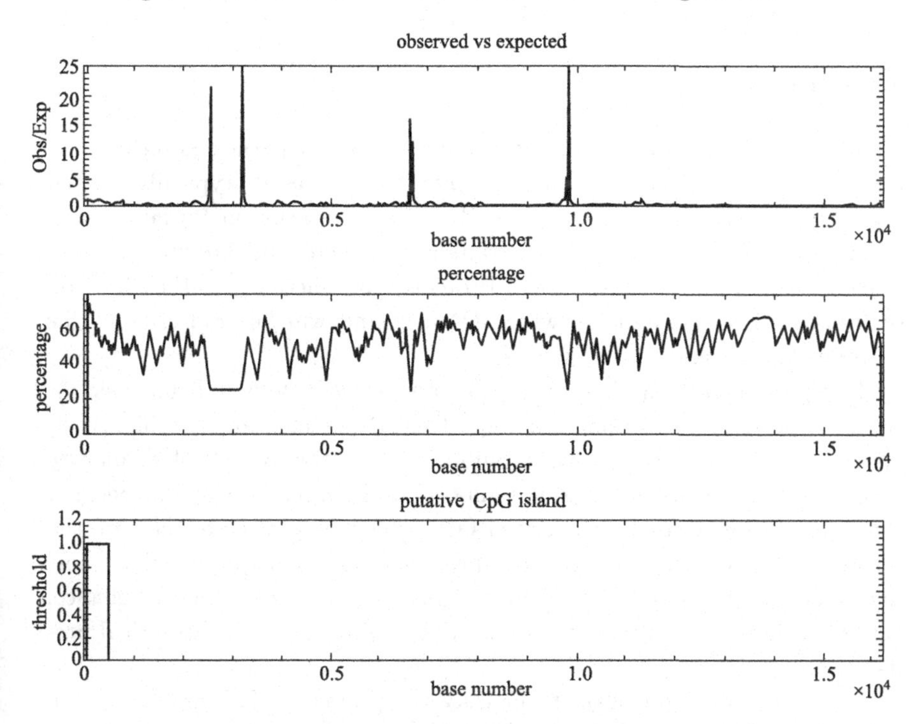

Fig. 4.7 The identification of CpG island in mouse Mtch1 gene

From the results of the calculation, it is clear that this sequence has one

CpG island, indicating that this sequence contains only one gene. In addition, the 5' end of the gene can be clearly analyzed. Figure 4.8 shows the calculation of CpGReport.

```
ID    16235 BP .
XX
DE    CpG island report .
XX
CC    Obs/ Exp ratio > 0 .60 .
CC    % C + % G > 50 .00 .
CC    Length > 200 .
XX
FH    Key              Location/ Qualifiers
FT    CpG island       47 . 499
FT                     / size = 453
FT                     / Sum C + G = 272
FT                     / Percent CG = 60 .04
FT                     / ObsExp = 0 .95
FT    numislands       1
//
```

Fig. 4.8 The calculation of CpGReport

4.1.4 Promoter

Promoter sequences are DNA sequences that define where the transcription of a gene by RNA polymerase begins. Promoter sequences are typically located directly at upstream or at the 5' end of the transcription initiation site. RNA polymerase and the necessary transcription factors bind to the promoter sequence and initiate transcription. Promoter sequences define the direction of transcription and indicate which DNA strand will be transcribed; this strand is known as the sense strand.

Computational identification of promoters is also a very difficult task, for several reasons. First, promoters are not clearly defined and are highly diverse. Each gene seems to have a unique combination of sets of regulatory motifs that determine its unique temporal and spatial expression. Second, the promoters can not be translated into protein sequences to increase the sensitivity for their detection. Third, promoter sites to be predicted are normally short and can be essentially found in any sequence by random chance, thus resulting in high rates of false positives associated with theoretical predictions.

There are two main ways to identify the promoter. The first way is to calculate the density of the transcription binding sites contained in the known promoter and non-promoter sequences first, and then generating a density ratio for each binding site on the promoter and non-promoter sequences, combining each individual density ratio to form a scoring blueprint. The second way is to analyze the six nucleotide frequency in the promoter region, non-promoter region, and the coding region sequence.

There are many tools that can be used for promoter recognition. Due to the heterogeneity and complexity of the functional region, it is difficult to correctly predict the promoter region. The results of the different methods and tools should be compared with each other.

TRES is the tool to analyze the promoter and transcription factor binding sites by using the first method. The full name of TRES transcription regulatory element search. The website is bioportal.bic.nus.edu.sg/tres/. TRES could detect the presence of known transcription factor binding sites, cis-acting elements, palindromic sequences, and phylogenetic-related promoters.

Neural network promoter prediction is another method of analyzing promoters. The basis of the Neural network promoter prediction program is a time-delay neural network. The time-delay network consists mainly of two feature layers, one for recognizing the TATA-box and the other for recognizing the 'Initiator', which is the region spanning the transcription start site. Both feature layers are combined into one output unit, which gives output scores between 0 and 1. The website is www.fruitfly.org/seq_tools/promoter.html.

Now, we use the human alpha-A lens globulin gene 5' side sequence (S79457) as an example to study its promoter region analysis. The analysis result of 'Neural Network Promoter Prediction' is shown in Figure 4.9.

Start	End	Score	Promoter Sequence
711	761	0.96	TCGCGCCACTATATGATCTGGGCGCCACTCTGGGTGACACAGCAAGACTC
1388	1438	0.87	AACCTCTGTGTCTAACGGGGGTGTGTGCTCTCCCTCCTCTGGCGACCATG
1755	1805	1.00	GCTGGTGGCATATATAGGGAGGGCTCGGCCTTGGCTCCACACTGGCTGCC

Fig. 4.9 The analysis result of 'Neural Network Promoter Prediction'

A total of three possible promoters were found and their positions and starting bases in the sequence were given.

The analysis of the promoter is complex and the results of the various analytical tools are also biased. Therefore, in order to build a complete promoter structure, when obtaining a candidate promoter sequence, there must be a relevant experimental evidence, such as RNA polymerase polymerase protection experiments in addition to the corresponding bioinformatics analysis.

4.1.5 Terminator

The end termination of an mRNA product generally has three main features: the poly A signal, the poly A tail, and an area of sequence that can form a stem loop structure. The poly A signal is a highly conserved site, six nucleotide long sequence. In eukaryotes the sequence is AATAAA and is located about 10~30 nucleotides from the poly A site. The AATAAA sequence is a highly conserved, eukaryotic poly A signal that signals for polyadenylation of the mRNA product 10~30 base pairs after the signal sequence.

The analysis of polyadenylation signals is simpler than the analysis of transcription initiation signals. Due to the presence of the AATAAA signal, the prediction of the poly A site is rarely dependent on the context of the sequence, but the upstream elements of other polyadenylation sites may affect the efficiency of the prediction.

Figure 4.10 is a schematic diagram of the analysis by the polyadenylation signal analysis tool 'Hcpolya'. The website is http://125.itba.mi.cnr.it/~webgene/wwwHC_polya.html. The analysis results show that the analyzed sequence has four possible transcription termination sites.

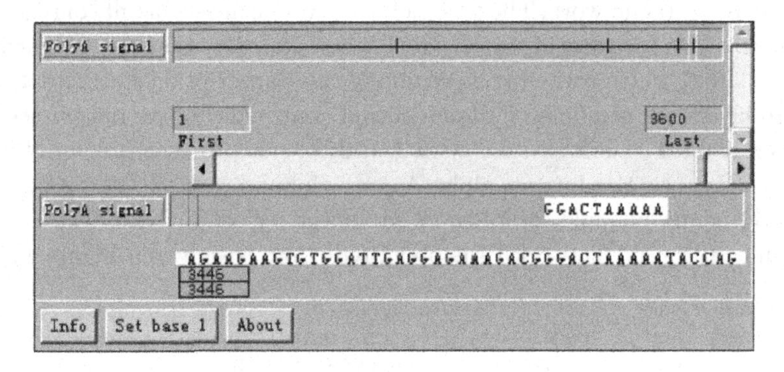

Fig. 4.10 The analysis result of 'Hcpolya'

4.2 Protein Sequence Analysis

4.2.1 The General Process of Function Prediction

If the sequence contig contains a protein coding region, the next analysis task is to determine the function of expression products, that is the protein. Many of the properties of the protein can be obtained directly from the sequence analysis, such as hydrophobicity, which can be used to predict whether the sequence is transmembrane helix or leader sequence. However, in general, the only way we predict the function of a protein based on the sequence is to search through the database to compare whether the protein is similar to the known function of the protein. There are two ways for the above comparative analysis.

① Compare the similarity of unknown protein sequences to known protein sequences.

② Search for if the unknown proteins contains sub-sequences or conserved segments associated with a particular protein family or domain.

Figure 4.11 shows the general process of predicting protein function according to the sequence. Because of the many technical routes involved, the resulting analysis is not always consistent.

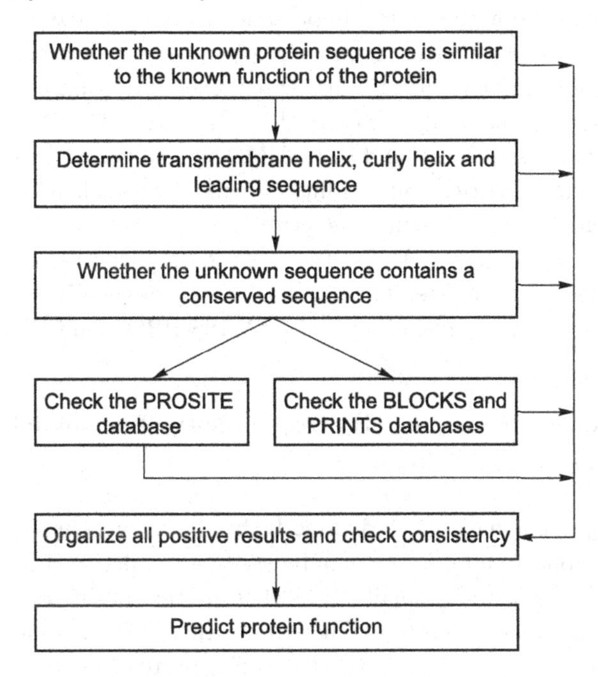

Fig. 4.11 The technical route of predicting protein function according to sequence

4.2.2 The Function Determination by Comparing Similar Sequences in Database

Proteins with similar sequences have similar functions. So, the most reliable way to determine protein function is to search the database for similarity. A significant match should have at least 25% of the same sequence and more than 80 amino acid segments.

There are many kinds of database search tools. Many of them are slow but sensitive, or fast but not sensitive. Quick search tools (such as BLASTP) are easy to find well-matched sequences, so there is no need to run more time-consuming tools (such as FASTA, BLITZ). These tools are only used if such a BLASTP can not find a significant match sequence. The general strategy is to start the BLAST search first. If the results can not be provided, run FASTA. If FASTA can not get clues about protein function, run the search program completely based on Smith-Waterman algorithm designed search program, such as BLITZ (www.ebi.ac.uk/searches/blitz.html). BLITZ does not make an approximate estimate (BLAST and FASTA could make approximate estimate by Smith-Waterman algorithm), so it is time-consuming but very sensitive.

It should also be aware of the importance of scoring matrix. There are important reasons for choosing different scoring matrices. First, the selected matrix must be consistent with the matching level. For example, PAM250 is used for long-range matching (<25% of the same rate), PAM40 is applied to non-similar protein sequences, and BLOSUM62 is a universal matrix. Second, by using different matrices, you can find the matching sequence that always appears, which is a way to reduce the error.

In addition to the use of different scoring matrix, you can also consider the use of different databases. The database that can usually be used is the non-redundant protein sequence database SWISS-PROT and PDB.

4.2.3 Sequence Properties: Hydrophobicity, Transmembrane Helix and Leader Sequences

Many functions can be predicted directly from the protein sequence. For example, hydrophobic information can be used for predictive transmembrane helices. There are also many small motifs that are the cells used to target the specific cell compartment protein. There are plenty of data resources on the web that help us use these properties to predict protein function.

The hydrophobic information can be created and presented with the ProtScale program of ExPASy (http://expasy.hcuge.ch/egibin/protscal.pl). This is a very useful tool that can calculate the properties of more than 50 proteins. The input of the program can be pasted through the input box, or you can enter the record number of SWISS-PROT. Only one parameter that requires additional settings is the width of the input box, which will indicate the number of residues calculated and displayed by the system each time. The default value is 9. If you want to consider the transmembrane helix, this parameter should be set to 20 because a transmembrane helix usually has a length of 20 amino acids.

There are several ways to predict the sequence of transmembrane helices. The simplest way is to find a segment containing 20 hydrophobic residues, and some complex and accurate algorithms can not only predict the position of the transmembrane helix but also determine its orientation on the membrane.

These methods are dependent on study results from a series of known transmembrane helical properties. TMbase is a transmembrane spiral database (http://ulrec3.unil.ch/tmbase/TMBASE_doc.html). Some of the related programs include TMPRED (http://ulrec3.unil.ch/software/TMPRED-form.html), PHDhtm (www.embl_heidelberg.de/services/sander/predictprotein/predictprotein.html), TMAP (http://www.embl-heidelberg.de/tmap/tmap/tmap_sin.html) and MEMSAT (ftp.biochem.ucl.ac.uk). These programs use different statistical models. In general, the prediction accuracy rate is 80%–95%. The transmembrane helix is one of the protein properties that can be accurately predicted from the sequence data.

The programs for predicting the preamble or special compartment target protein signals are SignalP (http://www.cbs.dtu.dk/services/SignalP) and PSORT (http://psort.nibbac.jp/form.html). The other functional sequence that can be determined from the sequence is a coil spiral. In this structure, the two spirals are entangled together by hydrophobic action to form a very stable structure. There are two related 2 programs, COILS (http://ulrec3.unil.ch/software/COILS_form.html) and Paircoil (http://ostrich.lcs.mit.edu/cgi-bin/score).

4.2.4 The Function Determination by Comparing Motif Database

The evolution rates of different sections of the protein are different. Some parts of the protein must maintain a certain residual pattern to maintain the function of the protein, and by determining these conserved regions, it is possible to provide clues for protein function. For example, there are many short sequences that can identify a protein active site or a binding region. The integrin receptor recognizes the RGD or LDV ligand motif, and if the unknown sequence contains an RGD pattern, it is presumed that one function of the unknown sequence may be a combination of integrin. This speculation is not to say that the protein sequence will certainly bind to integrins (many proteins that contain RGD do not bind to integrins), but it does provide us with an experimental hypothesis. Some examples are conserved sequences located at the site of enzyme activity, post-transcriptional modification sites, co-factor binding sites, or protein-classified signals, and many bioinformatics resources for these conserved patterns have been established and used in the sequence search and comparison.

There are two main methods that can be used for searching motif sequence. One way is to find a matching consensus sequence or motif. The advantages of this technology are fast, and the motif database is large and constantly expanding. Disadvantages are sometimes insensitive, because only the exact match with the consensus sequence or motif will be listed, and the relatively match will be ignored. In that case, the second method, a more elaborate sequence profile method, will work. In principle, the profile search is used to find the conserved sequence (not just a consistent sequence), which makes it more sensitive to finding those sequences that are not very relevant. The creation of profile and its databases is not easy, it requires a lot of computing and manpower. In the practical application, both of these types of databases should be searched at the same time. Where a significant match in one database may be completely missed in another database.

The most well-known motif database is PROSITE (http://expasy.hcuge.ch/sprot/prosite.html). A typical form of PROSITE recording (taking the consistent sequence of casein kinase II phosphorylation sites as an example): [ST] -x (2) - [DE], that is, a serine (S) or tyrosine (T) immediately followed by any two residues, and then a D or E. In addition, the record contains some other important information, such as the role of the site, where to be found and so on.

The profile database is mainly BLOCKS (http://www.blocks.fhcrc.org/blocks/), PRINTS (http://www.biochem.ucl.ac.uk/bsm/dbbrowers/PRINTS/) and ProDom (http://protein.toulouse.inra.fr/prodom/prodom.html). In general, the analysis should search all relevant databases to ensure that there is no omission. The details of the database listed above and the results of the output can refer to the database instructions.

References

[1] Jin Xiong. *Essential Bioinformatics*. Cambridge: Cambridge University Press. 2012.

[2] Yan Jiang. *Basic bioinformatics and application*. Beijing: Tsinghua University Press, 2003.

[3] Fan Longjiang. Bioinformatics. [2017-08-30]. http://ibi.zju.edu.cn/bioinplant/.

[4] Ashurst J L, and Collins J E. Gene annotation: Prediction and testing. *Annu. Rev. Genomics Hum. Genet.*, 2003(4): 69–88.

[5] Azad R K, and Borodovsky M. 2004. Probabilistic methods of identifying genes in prokaryotic genomes: Connections to the HMM theory. *Brief. Bioinform.*, 2004(5): 118–30.

[6] Wang Z, Chen Y, and Li Y. A brief review of computational gene prediction methods. *Geno. Prot. Bioinfo.*, 2004(4): 216–21.

[7] Zhang M Q. Computational prediction of eukaryotic protein coding genes. *Nat. Rev. Genetics.*, 2002(3): 698–709.

[8] Werner T. The state of the art of mammalian promoter recognition. *Brief. Bioinform.*, 2003(4): 22–30.

[9] Hershberg R1 Petrov DA. Selection on codon bias. *Annu Rev Genet.* 2008(42): 287-99.

[10] Liu J, and Rost B. Domains, motifs and clusters in the protein universe. *Curr. Opin. Chem. Biol.*, 2003(7): 5–11.

Chapter 5 Protein Structure

Yong Zeng[1], Lishu Zhang[2]

5.1 Overview of Protein Structure

Protein structure refers to the three-dimensional arrangement of atoms in a protein. Amino acid is the basic unit of protein. A single amino acid monomer may also be called a residue indicating a repeating unit of a polymer. Amino acids form to proteins undergoing a series of biochemical reactions. The primary chain formed under the mechanism: amino acids lose one water molecule per reaction and attach to one another with a peptide bond. A chain under 30 amino acids is often identified as a peptide, rather than a protein [1]. Usually, a mature protein with specific biological function was composed form several peptides via multiple non-covalent interactions such as hydrogen bonding, ionic interactions, van der Waals forces, and hydrophobic packing. There is a close relationship between protein structure and function. Usually, the three-dimensional structure determines the capacity to function. Analyzing the three-dimensional structure may help us to better understanding the functions of a protein at molecular level. However, a protein may suffer reversible structural changes in performing its biological function. Protein size range from ten to thousand amino acids. According to molecular weight, conformations and interactions, proteins are classified as structural proteins (such as tubulin and actin), membrane proteins (such as photoreceptors and ion channels), and globular proteins (such as globins). There are several methods to determine protein structure such as X-ray crystallography, NMR spectroscopy, and dual polarisation interferometry. Meanwhile, for protein structure data storage, retrieval and exchange, scientists established a comprehensive database: Protein Data Bank (PDB) which contains more

1. Zeng Yong, College of Life Sciences and Bioengineering, School of Science, Beijing Jiao Tong University, Beijing, China, 100044.

2. Zhang Lishu, College of Life Sciences and Bioengineering, School of Science, Beijing Jiao Tong University, Beijing, China, 100044.

than 50,000 structures. In this chapter, we will first introduce the structure of individual protein from the principles of primary, secondary, tertiary, and quaternary structure (Figure 5.1). Then, we will introduce some common tools and databases for protein structure prediction. At last, we will list and discuss some methods for protein structure determining and analysis.

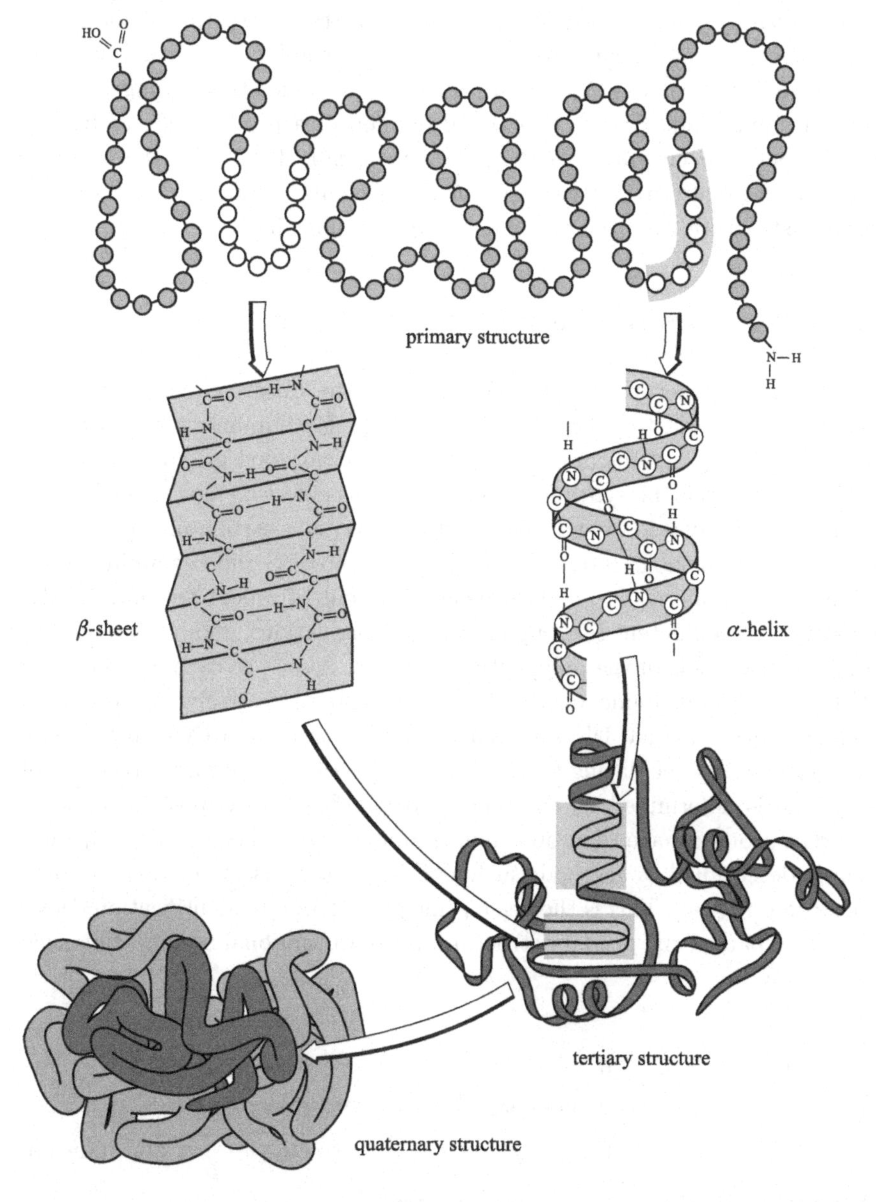

Fig. 5.1 Protein structure in multiple levels

5.2 Principles of Protein Structure

Protein structure is defined at several levels. Primary structure refers to the linear sequence of amino acid residues in a polypeptide chain. Secondary structure refers to the arrangements of the primary amino acid sequence into motifs such as α helices, β sheets, and coils. Tertiary structure is the three-dimensional arrangement formed by packing secondary structure elements into globular domains. Quaternary structure involves this arrangement of several polypeptide chains. Functional domain of a protein such as ligand-binding sites or enzymatic active sites are usually formed at the levels of tertiary and quaternary structure. In the following sections, we will discuss protein structure from the above four levels specifically.

5.2.1 Primary Structure

The primary structure of a protein refers to the linear sequence of amino acids which is held together by peptide bonds. Each amino acid consists of an amino group, a central carbon atom and a carboxyl group. During the process of protein biosynthesis, one water molecule is removed during the formation of a peptide bond. The peptide bond is a carbon-nitrogen amide linkage between the carboxyl group of one amino acid and the amino group of the next amino acid. The basic repeating unit of a peptide chain is NH-CH-CO with a different R group extending from various amino acids (Figure 5.2). The two ends of the polypeptide chain are named as carboxyl terminus (C-terminus) and amino terminus (N-terminus) respectively. According to genetic central dogma, DNA sequence is transcribed to mRNA, mRNA can be read by ribosome during translation. Essentially, the primary structure of a protein is determined by the gene corresponding to the protein. Usually, different proteins have their own unique amino acid sequences. The sequence can be determined by methods such as Edman degradation or tandem mass spectrometry [2-4]. This is the basic principle for protein identification. However, as a part of the primary structure, post-translational modification such

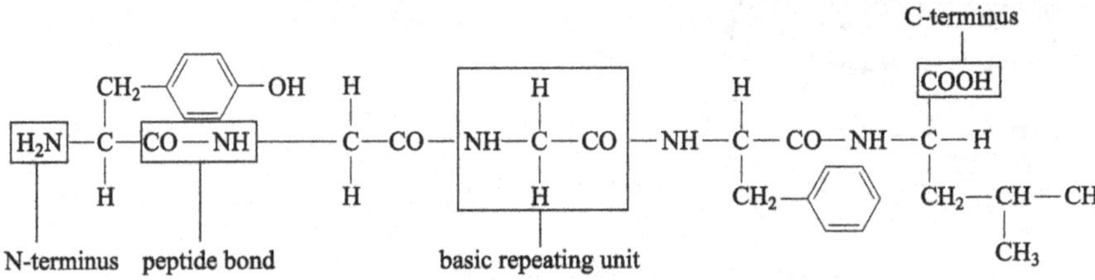

Fig. 5.2 Basic elements of protein primary structure

as disulfide bond formation, phosphorylations and glycosylations cannot be read and predicted from the gene [5, 6]. There are several tools for protein primary structure analysis and visualization based on well-established databases. DeepView is a popular software program for protein structures analysis and visualization based on PDB (Protein Data Bank). It can be downloaded from ExPASy website [7].

5.2.2 Secondary Structure

Protein secondary structure refers to highly regular local sub-structures on the actual polypeptide backbone chain. In general, proteins tend to be arranged with hydrophobic amino acids in the interior and hydrophilic residues exposed to the surface. This is the original driving force of protein secondary

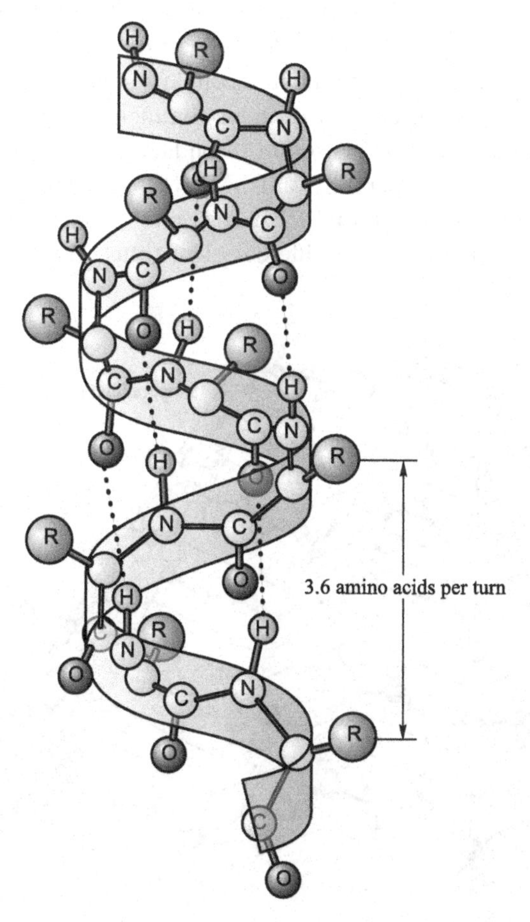

Fig. 5.3 Sketch of α-helix

structure formation [8, 9]. Two main types of secondary structure, the α-helix and the β-strand or β-sheets, were suggested in 1951 by Linus Pauling and coworkers [10]. Their hypothesis models were later confirmed by X-ray crystallography. These secondary structures are defined by patterns of hydrogen bonds between the main-chain peptide groups. They have a regular geometry, being constrained to specific values of the dihedral angles ψ and φ on the Ramachandran plot. Both the α-helix and the β-sheet represent a way of saturating all the hydrogen bond donors and acceptors in the peptide backbone [8, 11]. There are three types of helices: ① α-helix have 3.6 amino acids per turn, and represent 97% of all helices (Figure 5.3); ② 3.10 helices have 3.0 amino acids per turn (and thus are more tightly packed), and account for 3% of all helices; ③ π-helices, which occur only rarely, have 4.4 amino acids per turn. The β-sheets are formed from adjacent β strands composed of 2 to 15 residues (typically 5 to 10 residues). They are arranged in either parallel or antiparallel orientations that have distinct hydrogen bonding patterns (Figure 5.4). β-sheets have higher order properties, including the formation of barrels and sandwiches and 'super secondary structure motifs' such as $\beta - \alpha - \beta$ loops and $\alpha - \beta$ barrels. Proteins commonly contain combinations of both α helices and β sheets. Protein secondary structure can also be determined by X-ray crystallography and NMR [12, 13]. A variety of software and web servers such as DSSP and PBIL are applied for secondary-structure analysis and prediction based on different algorithms [11, 14, 15].

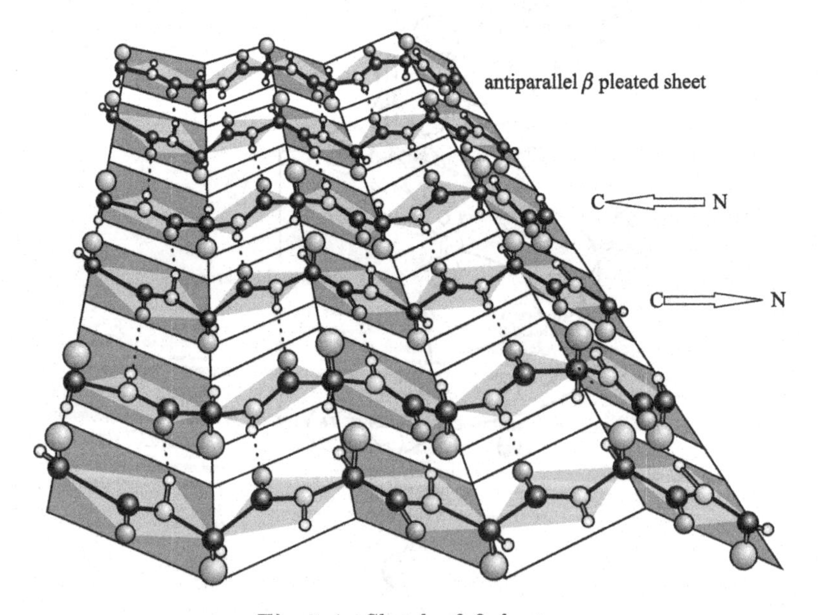

Fig. 5.4 Sketch of β-sheets

5.2.3 Tertiary Structure

It a core question that how does a protein fold into a three-dimensional structure in structural biology. There are too many possible conformations for a linear sequence of amino acids through random sampling of energy landscape. However, to maintain the maximum stability of protein structures and corresponding functions, protein secondary subunits must fold in specific ways [16, 17]. Protein tertiary structure refers to the three-dimensional structure of monomeric and multimeric protein molecules. The folding of α-helixes and β-pleated-sheets into a compact globular structure is driven by the non-specific hydrophobic interactions. Specific tertiary interactions, such as salt bridges, hydrogen bonds, and the tight packing of side chains and disulfide bonds may help to keep the proteins in most thermodynamic structures. There are several key words in stable maintenances of proteins tertiary structure such as chaperone proteins, kinetic traps and meta-stability [18].

5.2.4 Quaternary Structure

Quaternary structure is the three-dimensional structure consisting of the aggregation of two or more individual polypeptide chains. As subunits, polypeptides aggregate to more complex polymer by the same non-covalent interactions and disulfide bonds. Complexes of two or more polypeptides are called multimers [19]. Specifically, it would be called a dimer if it contains two subunits, a trimer if it contains three subunits and so on in the similar manner. The subunits are frequently related to one another by symmetry operations, such as a 2-fold axis in a dimer. Multimers consisting of identical subunits are referred to with a prefix of 'homo-' such as a homotetramer, consisting of different subunits are referred to with a prefix of 'hetero-' such as a heterotetramer.

We will set hemoglobin as an example to decipher the protein tertiary and quaternary structure specifically. Hemoglobin is the iron-containing oxygen-transport metalloprotein in the red blood cells of all vertebrates as well as the tissues of some invertebrates. The main function of hemoglobin in the blood is oxygen transportation from the respiratory organs to other organisms in metabolism processes. Hemoglobin is made up of many multi-subunit globular proteins, α-helixes are the main units of hemoglobin which are connected by non-helical segments. As Figure 5.5 showed, human hemoglobin contains two α chains, two β chains and some intro-containing heme-groups. Hydrogen bonds stabilize the helical sections and drive the peptide chains to fold into specific shape [20, 21].

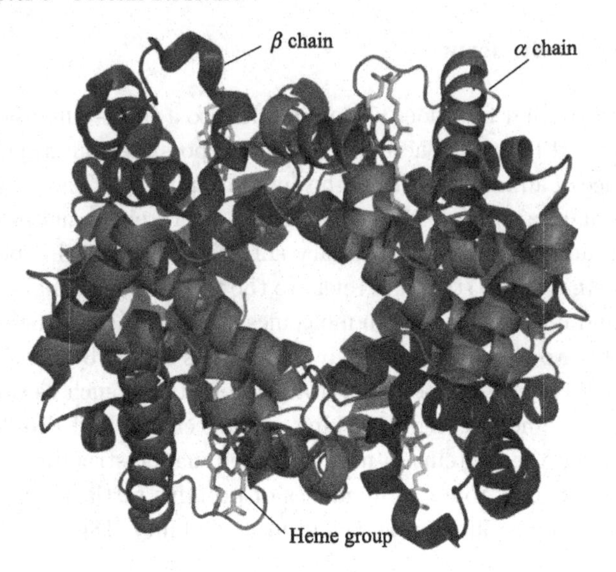

Fig. 5.5 Structure of human hemoglobin

5.3 Protein Structure Prediction

Protein structure prediction is the inference of the three-dimension structure from its amino acid sequence. It is a major goal of structure biology which is pursed by bioinformatics and theoretical chemistry. Since protein function is largely depends on the characterization and dynamic of protein structure, protein structure prediction is widely used in medicine and biotechnology such as drug design and enzymes design [22, 23]. There are three principal ways to predict the structure: first, for a protein target that shares substantial similarity to other proteins of known structure, homology modeling is applied. Second, for proteins that share folds but are not necessarily homologous, threading is a major approach. Third, for targets lacking identifiable homology (or analogy) to proteins of known structure, ab initio approaches are applied. In the following several sections, detailed description regarding the above three methods and corresponding software are presented.

5.3.1 Homology Modeling (Comparative Modeling)

Homology modeling, also known as comparative modeling of protein, refers to constructing a target protein from its amino acid sequence and a related homologous protein (the 'template') [24]. It is time-consuming to obtain structures information for every novel protein from experimental methods such as X-ray crystallography and protein NMR, computation based homology modeling can provide useful structural models for functional hypotheses and further experimental verification. For most proteins included PDB or SwissProt/TrEMBL databases, the assignment of structural models relies on computational biology approaches rather than experimental determination [25, 26]. Since evolutionarily related proteins have similar sequences and protein three-dimensional structures, the most reliable method of modeling and evaluating new structures is by comparison to previously known structures [27, 28]. As a fundamental method in structure genomics, homology modeling is widely used for protein structure prediction [28]. Homology modeling consists of four sequential steps [29].

① Template selection and sequence alignment. This can be accomplished by searching for homologous protein sequences and/or structures in databases such as BLAST and FAST. As a crucial task, structurally conserved and variable regions should be identified.

② Target-template alignment. As for any alignment problem, it is especially difficult to determine accurate alignments for distantly related proteins. For 30% sequence identity between a target and a template protein, the two proteins are likely to have a similar structure, however, this ratio under 20% may have very different structure [30].

③ Model construction. A variety of approaches are employed, such as rigid-body assembly and segment matching.

④ Model evaluation. Several principal types of errors such as errors in side-chain packing, distortions within correctly aligned regions, utility of incorrect templates may occur in comparative modeling [29]. Thus, the accuracy of protein structure prediction largely relies on the percentage of sequence identity between a target protein and its template. When the two proteins share 50% amino acid identity or more, the quality of the model is usually excellent. Model accuracy declines when comparative models rely on 30% to 50% identity, and the error rate rises rapidly below 30% identity. Several methods, tools or databases such as SWISS MODEL, Phyre and Phyre2, MODELLER [31, 32] can perform quality assessment for homology models. Some frequently used tools and the corresponding brief description and URL are listed in Table 5.1.

Table 5.1 Tools and servers for protein structure prediction based on Homology modeling

Name	Description	URL
MODELLER	From Andrej Sali's group	http://www.salilab.org/modeller/
SWISS MODEL	ExPASy	http://swissmodel.expasy.org/

Continued

Name	Description	URL
Phyre and Phyre2	Remote template detection, alignment, 3D modeling, multi-templates, abinitio	http://www.sbg.bio.ic.ac.uk/phyre2/html/page.cgi?id=index
PROCHECK	Quality assessment	http://www.biochem.ucl.ac.uk/_roman/procheck/procheck.html
GeneSilico	Consensus template search/fragment assembly	https://genesilico.pl/meta2/
WHATIF	Quality assessment	http://swift.cmbi.kun.nl/whatif/
Prime	Physics-based energy function	https://www.schrodinger.com/prime
RaptorX	Remote homology detection, protein 3D modeling, binding site prediction	http://raptorx.uchicago.edu/
CPHModel	Fragment assembly	http://www.cbs.dtu.dk/services/CPHmodels/
ESyPred3D	Template detection, alignment, 3D modeling	http://www.unamur.be/sciences/biologie/urbm/bioinfo/esypred/

5.3.2 Fold Recognition (Threading)

Fold recognition, also called threading, is useful when a target sequence may have common folds with proteins of known structure, but do not have their homologous protein structure deposited in databases [33]. There are four main steps for protein structure prediction using fold recognition method: the first step is template construction from a database such as PDB, FSSP, SCOP, or CATH. Then design a scoring function to measure the fitness between target sequence and template. Thirdly, align the target sequence with each of the structure templates base on designed scoring function. Lastly, perform threating prediction to select threading alignment with optimal statistical value [34-36]. As listed in Table 5.2, several web servers can provide automatic threading.

Table 5.2 Tools and servers for protein structure prediction based on fold recognition

Name	Description	URL
RaptorX	Remote template detection, single-template and multi-template threading, totally different from and much better than the old program RAPTOR designed by the same group	http://raptorx.uchicago.edu/

Continued

Name	Description	URL
NovaFold	Combination of threading and ab initio folding	http://www.dnastar.com/t-products-NovaFold.aspx
I-TASSER	Combination of ab initio folding and threading methods	https://zhanglab.ccmb.med.umich.edu/I-TASSER/
MUSTER	Profile-profile alignment	https://zhanglab.ccmb.med.umich.edu/MUSTER/
Phyre and Phyre2	Remote template detection, alignment, 3D modeling, multi-templates, ab initio.	http://www.sbg.bio.ic.ac.uk/phyre2/html/page.cgi?id=index
SUPERFAMILY	Hidden Markov modeling	http://supfam.org/SUPERFAMILY/
SPARKS-X	3D structure modeling by fold recognition according to sequence profiles and structural profiles	http://sparks-lab.org/yueyang/server/SPARKS-X/

5.3.3 Ab Initio Prediction (Template-Free Modeling)

Without detectable homologs, protein structure may be assessed by ab initio (or de novo) structure prediction. Ab initio prediction is the most difficult approach to structure prediction which refers to an algorithmic process by which protein tertiary structure is predicted from its amino acid primary sequence without the use of explicit templates [37, 38]. It is based on two assumptions: firstly, all the information about the structure of a protein is contained in its amino acid sequence; secondly, a globular protein folds into the structure with the lowest free energy. This method requires vast computational resources, and can only been carried out for relatively small proteins [39]. Ab initio protein structure modeling is distinguished from Template-based modeling (TBM) by the fact that no solved homolog to the protein of interest is used [40]. In Table 5.3, we listed some tools that may helpful for ab initio prediction.

Table 5.3 Tools and servers for protein ab initio structure prediction

Name	Description	URL
EVfold	Evolutionary couplings calculated from correlated mutations in a protein family, used to predict 3D structure from sequences alone and to predict functional residues from coupling strengths. Predicts both globular and transmembrane proteins	http://evfold.org/evfold-web/evfold.do

Continued

Name	Description	URL
FALCON	A position-specific hidden Markov model to predict protein structure by iteratively refining the distributions of dihedral angles	http://protein.ict.ac.cn/FALCON/
QUARK	Monte Carlo fragment assembly	https://zhanglab.ccmb.med.umich.edu/QUARK/
Selvita Protein Modeling Platform	Package of tools for protein modeling	https://selvita.com/
Rosetta@home	Distributed-computing implementation of Rosetta algorithm	http://boinc.bakerlab.org/rosetta/
NovaFold	Combination of threading and ab initio folding	http://www.dnastar.com/t-products-NovaFold.aspx
Bhageerath	A computational protocol for modeling and predicting protein structures at the atomic level	http://www.scfbio-iitd.res.in/bhageerath/index.jsp
PEP-FOLD	De novo approach, based on a HMM structural alphabet	http://bioserv.rpbs.univ-paris-diderot.fr/services/PEP-FOLD/

5.4 Protein Structure Determining and Analysis

In structural biology, X-ray crystallography and nuclear magnetic resonance spectroscopy (NMR) are two major methods used in protein structures determining and analysis. As a frequently used method, 80% of known structures were determined using X-ray crystallography. The workflow and basic steps of protein structure determining using X-ray crystallography are illustrated in Figure 5.6 and Figure 5.7. A protein must be obtained in high concentration and purification that meet the requirement of crystallization. The crystal scatters X-rays onto a detector, and the structure of the crystal is inferred from the diffraction pattern. The wavelength of X-rays (about 0.5 to 1.5 A°) is useful to measure the distance between atoms, making this technique suitable to trace the amino acid side chains of a protein [41, 42].

Nuclear magnetic resonance spectroscopy (NMR) is another important approach to crystallography. Protein NMR refers to application of NMR in obtaining structure and dynamic information of proteins [43, 44]. The principle of NMR in protein structure determining is described as follow: a magnetic field is applied to proteins in solution, then characteristic chemical shifts are observed. Protein structure can be deduced from these shifts. Protein determining using NMR does not require a protein to be crystallized. However, one limitation of NMR is that the detectable rang of protein molecular

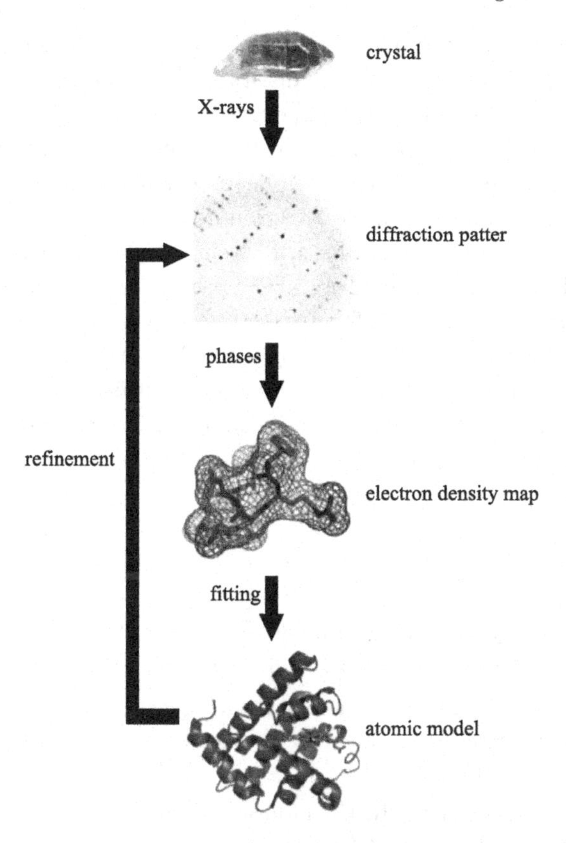

Fig. 5.6 Workflow of protein structure determining using X-ray crystallography

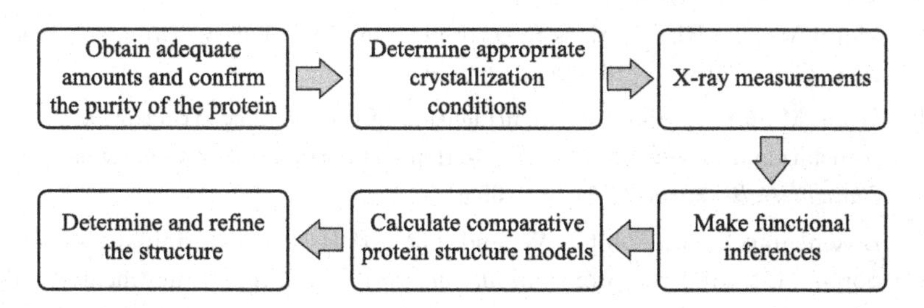

Fig. 5.7 Basic steps of protein structure determining using X-ray crystallography

weight is relatively narrow, the largest structures that have been determined by NMR are about 350 amino acids (40 kDa), considerably smaller than the size of proteins routinely studied by crystallography [43, 45].

In addition, Edman degradation and Mass Spectrometry (MS) can be used to analyze protein primary structure. Both have the capability to provide sequence information of specific amino acid. In the method of Edman degradation, the amino-terminal residue is labeled and cleaved from the peptide, then can be identified without disrupting the peptide bonds between other amino acid residues [46]. Edman degradation is a direct and cheap way to determine protein sequence, however, it has several limitations as follow: it will not work if the N-terminus has been chemically modified; it cannot handle complex samples; the final detection depends on the combined using of chromatography or electrophoresis. Protein identification base on MS is automatic and effective. As an advantage approach, it can deal with complex protein mixture that extracted form specific cell type or tissue. Meanwhile, it can provide high-throughput data by using MS in conjunction with corresponding database searching tools. Protein identification base on MS is always expensive and it can only identify known proteins that already been included in specific databases.

References

[1] Brocchieri L and Karlin S. Protein length in eukaryotic and prokaryotic proteomes. *Nucleic Acids Res.*, 2005, 33(10): 3390-3400.

[2] Miyashita M, et al. Attomole level protein sequencing by Edman degradation coupled with accelerator mass spectrometry. *Proc. Natl. Acad. Sci.* USA, 2001, 98(8): 4403-4408.

[3] Bridgen J and Morris H R. Use of mass spectrometry and quantitative Edman degradation for the determination of repeating amino-acid sequences. *Eur. J. Biochem*, 1974, 44(2): 333-334.

[4] Mann M. The Rise of Mass Spectrometry and the Fall of Edman Degradation. *Clin. Chem.*, 2016, 62(1): 293-294.

[5] Wada M and Shirahata A. Identification of the primary structure and post-translational modification of rat S-adenosylmethionine decarboxylase. *Biol. Pharm. Bull.*, 2010, 33(5): 891-894.

[6] Bokenkamp D, Jungblut P W, and Thole H H. The C-terminal half of the porcine estradiol receptor contains no post-translational modification: determination of the primary structure. *Mol. Cell Endocrinol*, 1994, 104(2): 163-172.

[7] Johansson M U, et al. Defining and searching for structural motifs using DeepView/Swiss-PdbViewer. *BMC Bioinformatics*, 2012(13): 173.

[8] Perticaroli S, et al. Secondary structure and rigidity in model proteins. *Soft Matter*, 2013, 9(40): 9548-9556.

[9] Kabsch W and Sander C. Dictionary of protein secondary structure: pattern recognition of hydrogen-bonded and geometrical features. *Biopolymers*, 1983, 22(12): 2577-2637.

[10] Pauling L, Corey R B, and Branson H R. The structure of proteins; two hydrogen-bonded helical configurations of the polypeptide chain. *Proc. Natl. Acad. Sci.* USA, 1951, 37(4): 205-211.

[11] Richards F M and Kundrot C E. Identification of structural motifs from protein coordinate data: secondary structure and first-level supersecondary structure. *Proteins*, 1988, 3(2): 71-84.

[12] Meiler J and Baker D. Rapid protein fold determination using unassigned NMR data. *Proc. Natl. Acad. Sci.* USA, 2003, 100(26): 15404-15409.

[13] Pelton J T and McLean L R. Spectroscopic methods for analysis of protein secondary structure. *Anal. Biochem.*, 2000, 277(2): 167-176.

[14] Pirovano W and Heringa J. Protein secondary structure prediction. *Methods Mol. Biol.*, 2010(609): 327-348.

[15] Frishman D and Argos P. Knowledge-based protein secondary structure assignment. *Proteins*, 1995, 23(4): 566-579.

[16] Lim V I and Efimov A V. The folding of protein chains. Prediction of tobacco mosaic virus protein tertiary structure. *FEBS Lett.*, 1976, 69(1): 41-44.

[17] Whisstock J C and Bottomley S P. Molecular gymnastics: serpin structure, folding and misfolding. *Curr. Opin. Struct. Biol.*, 2006, 16(6): 761-768.

[18] Willbold D, Hoffmann S, and Rosch P. Secondary structure and tertiary fold of the human immunodeficiency virus protein U (Vpu) cytoplasmic domain in solution. *Eur. J. Biochem.*, 1997, 245(3): 581-588.

[19] Chou K C and Cai Y D. Predicting protein quaternary structure by pseudo amino acid composition. *Proteins*, 2003, 53(2): 282-289.

[20] Oppel V V. [Primary structure of the globin component in hemoglobin. (A review)]. *Ukr. Biokhim. Zh.*, 1960, (32): 742-769.

[21] Marti H R. Results of the first two days of the symposium: review. International symposium: Synthesis, structure and function of hemoglobin. Molecular biology and clinical aspects. Bad Nauheim, 1971.

[22] Takeda-Shitaka M, et al. Protein structure prediction in structure based drug design. *Curr. Med. Chem.*, 2004, 11(5): 551-558.

[23] Lengauer T and Zimmer R. Protein structure prediction methods for drug design. *Brief Bioinform*, 2000, 1(3): 275-288.

[24] Capener C E, et al., Homology modeling and molecular dynamics simulation studies of an inward rectifier potassium channel. *Biophys J.*, 2000, 78(6): 2929-2942.

[25] Zhang Y and Skolnick J. The protein structure prediction problem could be solved using the current PDB library. *Proc. Natl. Acad. Sci.* USA, 2005, 102(4): 1029-1034.

[26] Gopal S, et al. Homology-based annotation yields 1,042 new candidate genes in the Drosophila melanogaster genome. *Nat. Genet*, 2001, 27(3): 337-340.

[27] Jones D T. Protein structure prediction in genomics. *Brief Bioinform*, 2001, 2(2): 111-125.

[28] Baker D and Sali A. Protein structure prediction and structural genomics. *Science*, 2001, 294(5540): 93-96.

[29] Marti-Renom M A, et al. Comparative protein structure modeling of genes and genomes. *Annu. Rev. Biophys. Biomol. Struct.*, 2000(29): 291-325.

[30] Chothia C and Lesk A M. The relation between the divergence of sequence and structure in proteins. *EMBO J.*, 1986, 5(4): 823-826.

[31] Fiser A, Do R K, and Sali A. Modeling of loops in protein structures. *Protein Sci.*, 2000, 9(9): 1753-1773.

[32] Sali A and Blundell T L. Comparative protein modelling by satisfaction of spatial restraints. *J. Mol. Biol.*, 1993, 234(3): 779-815.

[33] Bowie J U, Luthy R, and Eisenberg D. A method to identify protein sequences that fold into a known three-dimensional structure. *Science*, 1991, 253(5016): 164-170.

[34] Jones D T, Taylor W R, and Thornton J M. A new approach to protein fold recognition. *Nature*, 1992, 358(6381): 86-89.

[35] Yang Y, et al. Improving protein fold recognition and template-based modeling by employing probabilistic-based matching between predicted one-dimensional structural properties of query and corresponding native properties of templates. *Bioinformatics*, 2011, 27(15): 2076-2082.

[36] Ma J, et al. A conditional neural fields model for protein threading. *Bioinformatics*, 2012, 28(12): 59-66.

[37] Simons K T, Strauss C, and Baker D. Prospects for ab initio protein structural genomics. *J. Mol. Biol.*, 2001, 306(5): 1191-1199.

[38] Osguthorpe D J. Ab initio protein folding. *Curr. Opin. Struct. Biol.*, 2000, 10(2): 146-152.

[39] Rohl C A, et al. Protein structure prediction using Rosetta. *Methods Enzymol*, 2004, 383: 66-93.

[40] Bonneau R, et al. De novo prediction of three-dimensional structures for major protein families. *J. Mol. Biol.*, 2002, 322(1): 65-78.

[41] Ilari A and Savino C. Protein structure determination by X-ray crystallography. *Methods Mol. Biol.*, 2008, 452: 63-87.

[42] Papageorgiou A C and Mattsson J. Protein structure validation and analysis with X-ray crystallography. *Methods Mol. Biol.*, 2014, 1129: 397-421.

[43] Wuthrich K. The way to NMR structures of proteins. *Nat. Struct. Biol.*, 2001, 8(11): 923-925.

[44] Wuthrich K. Protein structure determination in solution by NMR spectroscopy. *J. Biol. Chem.*, 1990, 265(36): 22059-22062.

[45] Clore G M and Gronenborn A M. Structures of larger proteins in solution: three- and four-dimensional heteronuclear NMR spectroscopy. *Science*, 1991, 252(5011): 1390-1399.

[46] Edman P. A method for the determination of amino acid sequence in peptides. *Arch. Biochem*, 1949, 22(3): 475.

PART II
BIOINFORMATICS FOR OMICS DATA

Chapter 6 Human Genetic Variation and Human Disease

Qing Xu[1]

6.1 Human Genetic Variation

Human genetic variation is the genetic differences both within and among populations. There may be multiple variants of any given gene in the human population, leading to polymorphism. Many genes are not polymorphic, meaning that only a single allele is present in the population. No two humans are genetically identical. On average, in terms of DNA sequence, each human is 99.5% similar to any other human [1]. Even monozygotic twins have infrequent genetic differences due to mutations occurring during development and gene copy number variation [2]. The study of human genetic variation has both evolutionary significance and medical applications. It can help scientists understand ancient human population migrations as well as how different human groups are biologically related to one another. For medicine, study of human genetic variation may be important because some disease-causing alleles occur more often in people from specific geographic regions. New findings show that each human has on average 60 new mutations compared to their parents [3].

6.1.1 Types of Genetic Variation

1. Single nucleotide polymorphisms

A single nucleotide polymorphism (SNP) is difference in a single nucleotide between members of one species that occurs in at least 1% of the population. The 2,504 individuals characterized by the 1000 Genomes Project had 84.7 million SNPs among them [4]. SNPs are the most common type of sequence

1. Xu Qing, College of Life Sciences and Bioengineering, School of Science, Beijing Jiao Tong University, Beijing, China, 100044.

variation, estimated in 1998 to account for 90% of all sequence variants [5]. Other sequence variations are single base exchanges, deletions and insertions [6]. SNPs occur on average about every 100 to 300 bases and so are the major source of heterogeneity [7].

2. Structural variation

Structural variation is the variation in structure of an organism's chromosome. Structural variations, such as copy-number variation and deletions, inversions, insertions and duplications, account for much more human genetic variation than single nucleotide diversity. According to the 1,000 Genomes Project, a typical human has 2,100 to 2,500 structural variations, which include approximately 1,000 large deletions, 160 copy-number variants, 915 Alu insertions, 128 L1 insertions, 51 SVA insertions, 4 NUMTs, and 10 inversions [4].

Copy number variation

A copy-number variation (CNV) is a difference in the genome due to deleting or duplicating large regions of DNA on some chromosome. It is estimated that 0.4% of the genomes of unrelated humans differ with respect to copy number. When copy number variation is included, human-to-human genetic variation is estimated to be at least 0.5% (99.5% similarity) [8]. Copy number variations are inherited but can also arise during development [9].

Epigenetics

Epigenetic variation is the variation in the chemical tags that attach to DNA and affect how genes get read. At some alleles, the epigenetic state of the DNA, and associated phenotype, can be inherited across generations of individuals [10].

Haplogroups

A haplogroup is a group of similar haplotypes that share a common ancestor with a SNP mutation. Haplogroups pertain to deep ancestral origins dating back thousands of years [11]. The most commonly studied human haplogroups are Y-chromosome (Y-DNA) haplogroups and mitochondrial DNA (mtDNA) haplogroups. Y-DNA is passed from father to son, while mtDNA is passed from mother to both daughter and son. The Y-DNA and mtDNA may change by chance mutation at each generation.

Variable number tandem repeats

A variable number tandem repeat (VNTR) is the variation of length of a tandem repeat. Tandem repeats exist on many chromosomes, and their length varies between individuals. Each variant acts as an inherited allele, so they are used for personal or parental identification. Their analysis is useful in genetics and biology research, forensics, and DNA fingerprinting. Short tandem repeats (about 5 base pairs) are called microsatellites, while longer ones are called minisatellites.

6.1.2 Genetic Variation and Health

Differences in allele frequencies contribute to group differences in the incidence of some diseases, and they may contribute to differences in the incidence of some common diseases [12]. For the monogenic diseases, the frequency of causative alleles usually correlates best with ancestry, whether familial, ethnic, or geographical. With common diseases involving numerous genetic variants and environmental factors, investigators point to evidence suggesting the involvement of differentially distributed alleles with small to moderate effects. Frequently cited examples include hypertension [13], diabetes [14], obesity [15], and prostate cancer [16]. However, the role of genetic factors in generating these differences remains uncertain [17].

1. SNP

A single-nucleotide polymorphism (SNP) is a variation in a single nucleotide that occurs at a specific position in the genome, which each variation is present to some appreciable degree within a population (e.g. $> 1\%$). For example, at a specific base position in the human genome, the base C may appear in most individuals, but in a minority of individuals, the position is occupied by base A. There is a SNP at this specific base position, and the two possible nucleotide variations – C or A – are said to be alleles for these base positions. Variations in the DNA sequences of humans can affect how humans develop diseases and respond to pathogens, chemicals, drugs, vaccines, and other agents [18]. SNPs are also critical for personalized medicine [19].

SNP may fall within coding sequences of genes, non-coding regions of genes, or regions between genes. SNPs in the coding region have two types, synonymous and nonsynonymous SNPs. Synonymous SNPs do not affect the protein sequence while nonsynonymous SNPs (two types: missense and nonsense) change the amino acid sequence of protein.

SNPs located in non-coding regions may still affect gene splicing, transcription factor binding, messenger RNA degradation, or the sequence of non-coding RNA. SNP affecting gene expression may be upstream or downstream from the gene.

2. Applications of SNPs

Association studies can determine whether a genetic variant is associated with a disease or trait.

A tag SNP is a representative single-nucleotide polymorphism (SNP's) in a region of the genome with high linkage disequilibrium (the non-random association of alleles at two or more loci). Tag SNPs are useful in whole-genome SNP association studies.

Haplotype mapping: sets of alleles or DNA sequences can be clustered so that a single SNP can identify many linked SNPs.

Linkage disequilibrium (LD), a term used in population genetics, indicates non-random association of alleles at two or more loci, not necessarily on the same chromosome. It refers to the phenomenon that SNP allele or DNA sequence which are close together in the genome tend to be inherited together. LD is affected by two parameters: The distance between the SNPs, the larger the distance the lower the LD. Recombination rate, the lower the recombination rate the higher the LD.

3. SNP and Disease

A single SNP may cause a Mendelian disease, though for complex diseases, SNPs do not usually function individually, rather, they work in coordination with other SNPs to manifest a disease condition.

All types of SNPs can have an observable phenotype or can result in disease:

① SNPs in non-coding regions can manifest in a higher risk of cancer [20] and may affect mRNA structure and disease susceptibility [21]. Non-coding SNPs can also alter the level of expression of a gene.

② SNPs in coding regions:

• Synonymous do not result in a change of amino acid in the protein, but still can affect its function in other ways. An example would be a seemingly silent mutation in the multidrug resistance gene 1 (MDR1), which codes for a cellular membrane pump that expels drugs from the cell, can slow down translation and allow the peptide chain to fold into an unusual conformation, causing the mutant pump to be less functional (in MDR1 protein) [22].

• Nonsynonymous substitutions: ① Missense: single change in the base results in change in amino acid of protein and its malfunction which leads to disease. e.g. c. 1580 G>T SNP in LMNA gene causing the guanine to be replaced with the thymine, yielding CTT codon in the DNA sequence, results at the protein level in the replacement of the arginine by the leucine in the position 527, at the phenotype level this manifests in overlapping mandibuloacral dysplasia and progeria syndrome [23]. ② Nonsense: point mutation in a sequence of DNA results in a premature stop codon, or a nonsense codon in the transcribed mRNA, and in a truncated, incomplete, and usually nonfunctional protein product. e.g. c. cystic fibrosis is caused by the G542X mutation in the cystic fibrosis transmembrane conductance regulator gene [24].

4. Databases

dbSNP is a SNP database from the National Center for Biotechnology Information (NCBI). As of 8 June 2015, dbSNP listed 149,735,377 SNPs in humans (http://www.ncbi.nlm.nih.gov/SNP/index.html).

Kaviar is a compendium of SNPs from multiple data sources including dbSNP (http://db.systemsbiology.net/kaviar/).

The OMIM database describes the association between polymorphisms and diseases (http://www.omim.org/).

dbSAP is the single amino-acid polymorphism database for protein variation detection.

The Human Gene Mutation Database provides gene mutations causing or associated with human inherited diseases and functional SNPs (http://www.hgmd.cf.ac.uk/ac/index.php).

The International HapMap Project, where researchers are identifying Tag SNP to be able to determine the collection of haplotypes present in each subject (http://www.hapmap.org).

GWAS Central allows users to visually interrogate the actual summary-level association data in one or more genome-wide association studies (http://www.gwascentral.org/).

5. dbSNP menu

A different search able interface for NCBI's dbSNP is found at the URL: http://www.ncbi.nlm.nih.gov/snp/. This page allows you to search more efficiently for SNPs associated with a specific gene or disease in specific organisms or in a particular region of a chromosome. You can do this by using the Limits tab on this page.

Solution: go to the NCBI SNP homepage → enter gene name in the search box and click 'Search' → alternately, you can use the 'Advanced' link under the search box to specify exact fields to search → after the search, use the filters on the left to narrow your results → under 'Organism List' select 'Homo sapiens' → under 'Function Class List' select the types of functional outcomes you are interested in → go to Validation Status and select all options except 'no-info' → You can click on 'Search Details' and review your Query Translation (optional).

Filters for dbSNP searches: Organism, Variation Class, Clinical Significance, Annotation, Global Minor Allele Frequency, Validation Status, Chromosomes, Map Weight (how many times in genome), Chromosome Range, Variation Alleles, % Heterozygosity, Success Rate, Method Class, Individual SNP maps, Minor Allele Frequency (by HapMap population class), Search Fields (in which you can choose from a pull-down menu of all SNP record fields to search on).

6.2 Human Disease

Disease may be defined as maladaptive changes that afflict individuals within a population. Disease is also defined as an abnormal condition in which physiological function is impaired. Mutations affect all parts of the human genome. There are many mechanisms by which mutations can cause disease. These include disruptions of gene function by point mutations that change the identity of amino acid residues; by deletions or insertions of DNA, ranging in size from one nucleotide to an entire chromosome that is over 100 million base pairs (Mb); or inversions of the orientation of a DNA fragment. In many cases, different kinds of mutations affecting the same gene cause distinct phenotypes. A disease causing mutation in a gene results in the failure to produce the gene product with normal function. This has profound consequences on the ability of the cells in which the gene product is normally expressed to function.

6.2.1 A Bioinformatics Perspective on Human Disease

Our approach to human disease is reductionist, in which we seek to describe genes and gene products that cause disease. However, an appreciation of the molecular basis of disease may be in tegrated with a holistic approach to uncover the logic of disease in the entire human population. As we explore bioinformatics approaches to human disease, we are constantly facing the complexity of all biological systems. Even when we uncover the gene that when mutated causes a disease, our challenge is to attempt to connect the genotype to the phenotype. We can only accomplish this by synthesizing information about the biological context in which each gene functions and in which each gene products contributes to cellular function [25].

The field of bioinformatics offers approaches to human disease that may help us to understand basic questions about the influence of genes and the environment on all aspects of the disease process. Some examples of ways in which this field can have an impact on our knowledge of disease will be highlighted throughout the chapter, and include the following.

To the extent that the genetic basis of disease is a function of variation in DNA sequences, DNA databases offer us the basic material necessary to compare DNA sequences. These databases include major, general repositories of DNA sequence such as GenBank/EMBL/DDBJ, general resources such as Online Mendelian Inheritance in Man (OMIM), and locus-specific databases that provide data on sequence variations at individual loci.

Geneticists who search for disease-causing genes through linkage studies, association studies, or other tests (described below) depend on physical and genetic maps in their efforts to identify mutant genes.

When a protein-coding gene is mutated, there is a consequence on the three dimensional structure of the protein product. Bioinformatics tools described allow us to predict the structure of protein variants, and from such analyses we may infer changes in function.

Once a mutant gene is identified, we want to understand the consequence of that mutation on cellular function. There are a variety of approaches to understand protein function. Gene expression studies have been employed to study the transcriptional response to disease states.

6.2.2 Genome-Wide Association Studies

Genome-wide association studies (GWAS) represent a recently developed research technique with many implications on both a global and an individual scale. GWAS seek to identify the single nucleotide polymorphisms that are common to the human genome and to determine how these polymorphisms are distributed across different populations. On a broad scale, these studies help scientists uncover associations between individual SNPs and disorders that are passed from one generation to the next in Mendelian fashion. On a small scale, GWAS can be used to determine an individual's risk of developing a particular disorder. Although the impact of GWAS on medical genetics is undeniable, the true usefulness of these studies largely depends upon researchers' understanding of the interacting factors behind common genetic disorders.

Variations in the Human Genome (SNPs): with the completion of the Human Genome Project in 2003, researchers began to pinpoint areas of the genome that varied between individuals. Shortly thereafter, they discovered that the most common type of DNA sequence variation found in the genome is the single nucleotide polymorphism. There are an estimated 10 million SNPs that commonly occur in the human genome. A worldwide effort known as the HapMap Project seeks to identify and localize these and other genetic variants, and to learn how the variants are distributed within and among populations from different parts of the world. To date, the project has identified over 3.1 million SNPs across the human genome which are common to individuals of African, Asian, and European ancestry.

The HapMap information, which is available to the public, has facilitated a new type of research effort: the genome-wide association study (GWAS). In such a study, the distribution of SNPs is determined in hundreds or even thousands of people with and without a particular disease. By tallying which SNPs co-occur with disease symptoms, researchers can make a statistical estimate regarding the level of increased risk associated with each SNP. For instance, in a landmark 2007 study conducted in the United Kingdom, researchers identified people affected by seven common disorders, and they then genotyped 2,000 people in each disease category (for a total of 14,000 individuals studied). Next, these individuals were compared to 3,000 genotyped controls who did not have the disorders in question. As a result of these comparisons, the researchers were able to identify new genetic markers that point to an increased risk for multifactorial disorders such as heart disease. It was announced in July 2008 that this study was expanded to include an additional 36,000 individuals and focused on examining the genetic contributions to a total of 14 common disorders, as well as to individuals' responses to certain drugs.

6.2.3 Using GWAS to Estimate Disease Risk

The practicality of identifying a correlation between a genetic change and the incidence of a complex disease is limited to statistical estimation of increased risk for developing the disorder, rather than a hard-and-fast prediction. This is due to the significant number of genetic and environmental variables that interact to cause the onset of a complex disease. Therefore, any genetic variant, such as a SNP, makes only a small contribution to an individual's overall risk.

In most cases, findings from a GWAS cannot be directly applied to the prevention or treatment of disease. Rather, before doctors are able to recommend medicinal, behavioral, and environmental interventions based on a SNP profile, the full pathway of disease development and the involvement of all variables must be understood. For example, one situation that makes the link between SNPs and disease difficult to understand is these SNPs are not located within exons of genes. In such instances, studies are required to investigate the possibility that the SNP lies in a promoter or enhancer region and somehow affects regulation of the causal gene.

Occasionally, the results of a GWAS seem relatively straightforward; this is often the case when a gene is identified that contains a variant that confers susceptibility to a multifactorial disorder [26]. However, the effect of that variant or other alleles of the gene is still just one of many factors influencing disease risk; therefore, the predictive power of the various alleles is not absolute. A well-known example of this is the link between certain alleles of the apolipoprotein E (*ApoE*) gene found on chromosome 19 and the development of Alzheimer's disease. *ApoE* codes for a protein that helps carry cholesterol in the bloodstream, and it has three common alleles: e2, e3, and e4. Research has shown that having one or two copies of the *ApoE* e4 allele significantly increases a person's risk for developing Alzheimer's disease, but it does not guarantee development of this disorder. It also remains unclear how certain forms of *ApoE* influence cerebral plaque formation, the hallmark of Alzheimer's disease [27].

References

[1] Levy S, Sutton G, Ng P C, Feuk L, Halpern A L, Walenz B P, Axelrod N, Huang J, Kirkness E F, Denisov G, et al. The diploid genome sequence of an individual human. *PLoS. Biol.* 2007, 5(10): e254.

[2] Bruder C E, Piotrowski A, Gijsbers A A, Andersson R, Erickson S, Diaz de Stahl T, Menzel U, Sandgren J, von Tell D, Poplawski A, et al. Phenotypically concordant and discordant monozygotic twins display different DNA copy-number-variation profiles. *Am. J. Hum. Genet.*, 2008, 82(3): 763-771.

[3] Conrad D F, Keebler J E M, DePristo M A, Lindsay S J, Zhang Y J, Casals F, Idaghdour Y, Hartl C L, Torroja C, Garimella K V, et al. Variation in genome-wide mutation rates within and between human families. *Nat. Genet.*, 2011, 43(7): 712-U137.

[4] Altshuler D M, Durbin R M, Abecasis G R, Bentley D R, Chakravarti A, Clark A G, Donnelly P, Eichler E E, Flicek P, Gabriel S B, et al. A global reference for human genetic variation. *Nature*, 2015, 526(7571): 68-+.

[5] Collins F S, Brooks L D, Chakravarti A. A DNA polymorphism discovery resource for research on human genetic variation. *Genome. Res.*, 1998, 8(12): 1229-1231.

[6] Thomas P E, Klinger R, Furlong L I, Hofmann-Apitius M, Friedrich C M. Challenges in the association of human single nucleotide polymorphism mentions with unique database identifiers. *BMC Bioinformatics*, 2011, 12 Suppl 4: S4.

[7] Ke X, Taylor M S, Cardon L R. Singleton SNPs in the human genome and implications for genome-wide association studies. *Eur. J. Hum. Genet.*, 2008, 16(4): 506-515.

[8] Levy S, Sutton G, Ng P C, Feuk L, Halpern A L, Walenz B P, Axelrod N, Huang J, Kirkness E F, Denisov G, et al. The diploid genome sequence of an individual human. *Plos. Biology.*, 2007, 5(10): 2113-2144.

[9] Redon R, Ishikawa S, Fitch K R, Feuk L, Perry G H, Andrews T D, Fiegler H, Shapero M H, Carson A R, Chen W W, et al. Global variation in copy number in the human genome. *Nature*, 2006, 444(7118): 444-454.

[10] Rakyan V, Whitelaw E. Transgenerational epigenetic inheritance. *Curr. Biol.*, 2003, 13(1): R6.

[11] Arora D, Singh A, Sharma V, Bhaduria H S, Patel R B. HgsDb: Haplogroups Database to understand migration and molecular risk assessment. *Bioinformation*, 2015, 11(6): 272-275.

[12] Risch N, Burchard E, Ziv E, Tang H. Categorization of humans in biomedical research: genes, race and disease. *Genome. Biol.*, 2002, 3(7): comment 2007.

[13] Douglas J G, Thibonnier M, Wright J T, Jr.: Essential hypertension: racial/ ethnic differences in pathophysiology. *J. Assoc. Acad. Minor. Phys.*, 1996, 7(1): 16-21.

[14] Goran M I, Coronges K, Bergman R N, Cruz M L, Gower B A. Influence of family history of type 2 diabetes on insulin sensitivity in prepubertal children. *J. Clin. Endocrinol Metab.*, 2003, 88(1): 192-195.

[15] Campos G, Ryder E, Diez-Ewald M, Rivero F, Fernandez V, Raleigh X, Arocha-Pinango C L. Prevalence of obesity and hyperinsulinemia: its association with serum lipid and lipoprotein concentrations in healthy individuals from Maracaibo, Venezuela. *Invest. Clin.*, 2003, 44(1): 5-19.

[16] Platz E A, Rimm E B, Willett W C, Kantoff P W, Giovannucci E. Racial variation in prostate cancer incidence and in hormonal system markers among male health professionals. *J. Natl. Cancer Inst.*, 2000, 92(24): 2009-2017.

[17] Mountain J L, Risch N. Assessing genetic contributions to phenotypic differences among 'racial' and 'ethnic' groups. *Nat. Genet.*, 2004, 36(11 Suppl): S48-53.

[18] Kidd K K, Pakstis A J, Speed W C, Kidd J R. Understanding human DNA sequence variation. *J. Hered.*, 2004, 95(5): 406-420.

[19] Wolf A B, Caselli R J, Reiman E M, Valla J. APOE and neuroenergetics: an emerging paradigm in Alzheimer's disease. *Neurobiol. Aging.*, 2013, 34(4): 1007-1017.

[20] Li G, Pan T, Guo D, Li L C. Regulatory Variants and Disease: The E-Cadherin -160C/A SNP as an Example. *Mol. Biol. Int.*, 2014, 2014: 967565.

[21] Lu Y F, Mauger D M, Goldstein D B, Urban T J, Weeks K M, Bradrick S S. IFNL3 mRNA structure is remodeled by a functional non-coding polymorphism associated with hepatitis C virus clearance. *Sci. Rep.*, 2015, 5: 16037.

[22] Kimchi-Sarfaty C, Oh J M, Kim I W, Sauna Z E, Calcagno A M, Ambudkar S V, Gottesman M M. A "silent" polymorphism in the MDR1 gene changes substrate specificity. *Science*, 2007, 315(5811): 525-528.

[23] Al-Haggar M, Madej-Pilarczyk A, Kozlowski L, Bujnicki J M, Yahia S, Abdel-Hadi D, Shams A, Ahmad N, Hamed S, Puzianowska-Kuznicka M. A novel homozygous p.Arg527Leu LMNA mutation in two unrelated Egyptian families causes overlapping mandibuloacral dysplasia and progeria syndrome. *Eur. J. Hum. Genet.*, 2012, 20(11): 1134-1140.

[24] Cordovado S K, Hendrix M, Greene C N, Mochal S, Earley M C, Farrell P M, Kharrazi M, Hannon W H, Mueller P W. CFTR mutation analysis and haplotype associations in CF patients. *Mol. Genet. Metab.*, 2012, 105(2): 249-254.

[25] Childs B, Valle D. Genetics, biology and disease. *Annu. Rev. Genomics Hum. Genet*, 2000(1): 1-19.

[26] Steinthorsdottir V, Thorleifsson G, Reynisdottir I, Benediktsson R, Jonsdottir T, Walters G B, Styrkarsdottir U, Gretarsdottir S, Emilsson V, Ghosh S, et al. A variant in CDKAL1 influences insulin response and risk of type 2 diabetes. *Nat. Genet*, 2007, 39(6): 770-775.

[27] Huang W, Qiu C, von Strauss E, Winblad B, Fratiglioni L. APOE genotype, family history of dementia, and Alzheimer disease risk: a 6-year follow-up study. *Arch. Neurol*, 2004, 61(12): 1930-1934.

Chapter 7 Gene Expression Profiling with Microarray: Online Resources and Data Management

Qing Xu[1]

Gene expression profiling is a technique used in molecular biology to measure the expression of thousands of genes simultaneously and create a whole picture of cellular function. In cancer, gene expression profiling has been used to more accurately classify tumors and help in predicting the patient's clinical outcome.

Gene expression profile can distinguish between cells that are actively dividing, or show how the cells react to a particular treatment. While almost all cells in an organism contain the entire genome of the organism, only a small part of genes is expressed as mRNA at any given time. So the genome sequence tells us what the cell could possibly do, while the expression profile tells us what it is actually doing at a point. Many factors determine whether a gene expresses or not, such as local environment and chemical signals from other cells. Therefore, an expression profile may let one to know a cell's type, state, environment, and so on. DNA microarrays, evolved from Southern blotting which may detect a specific DNA sequence, is one of important techniques to evaluate relative expression of genes.

Expression profile experiments often measure the relative amount of mRNA expressed in two or more experimental conditions because altered mRNA levels suggest a changed need for the protein coded by the mRNA. For example, higher levels of mRNA coding for alcohol dehydrogenase suggest that the cells or tissues under study are responding to increased levels of ethanol in their environment. Similarly, if breast cancer cells express higher levels of mRNA of a particular receptor than normal cells do, it might be that this receptor plays a role in breast cancer. A drug that interferes with this receptor may prevent or treat breast cancer.

1. Xu Qing, College of Life Sciences and Bioengineering, School of Science, Beijing Jiao Tong University, Beijing, China, 100044.

7.1 Microarray Data Analysis Software

Three main types of software are available for microarray data analysis: commercial software packages associated with microarray manufacturers; other commercial software packages include Bio Discovery, Gene Sifter, and Spot fire, statistics packages include STATA and SAS; open source software in the bioinformatics community.

• BASE (BioArray Software Environment) is the Microarray database server with normalization and some analysis facilities (http://base.thep.lu.se/).

• Bengtsson, Henrik. Develops aroma, an elegant R package for microarray normalization, diagnostics and data analysis using a custom object-orientated system (http://www.braju.com/R/).

• Bioconductor is an R-based development project for open-source genomic software, includes a number of important packages, and coordinated by Robert Gentleman, Fred Hutchinson Cancer Center (http://www.bioconductor.org/).

• BioSieve is the produce ExpressionSieve, a microarray data analysis, data mining and data visualization software package written in Java.

• Broad Institute of MIT and Harvard produce the GenePattern package for genomics and microarray analysis (http://software.broadinstitute.org/cancer/software/genepattern).

• Eisen Lab distribute ScanAlyze, Cluster and TreeView free for non-profit researchers.

• Biodiscovery GeneSight (http://www.biodiscovery.com/nexus-expression/).

• Genedata produce Expressionist software for large scale gene expression analysis (https://www.genedata.com/).

• GeneSifter.Net is a web-based data analysis tool for Affymetrix microarrays created by VizX Labs, Seattle (http://www.cambridgesoft.com/services/).

• Gerstein Lab (Yale University). Express Yourself is an online platform for pre-preprocessing of microarray data (http://www.gersteinlab.org/).

• InforMax (Array Pro, Vector Xpression).

• Iobian Informatics. GeneTraffic is microarray database and data analysis software based on open software.

• Molecular Devices. Acuity for data warehousing, analysis and visualization (https://www.moleculardevices.com/).

• Molmine (J-Express) is a Java software (http://jexpress.bioinfo.no/site/).

• National Center for Genome Resources (GeneX) (http://www.ncgr.org/).

• PAM: Prediction Analysis for Microarrays. Robert Tibshirani, Stanford University (http://statweb.stanford.edu/~tibs/PAM/).

• Pevsner Lab, (Kennedy Kreiger Institute). SNOMAD (Standardization and normalization of MicroArray Data, http://pevsnerlab.kennedykrieger.org/).

• Partek. Produce Partek Pro software for statistical analysis and visualisation including gene expression (http://www.partek.com/).

• PermutMatrix. MS Windows software for clustering and seriation analysis of gene expression data. Gilles Caraux, LIRMM, France (http://www.atgc-montpellier.fr/permutmatrix/).

• Qlucore Omics Explorer (http://www.qlucore.com/).

• Churchhill Lab (Jackson Laboratories). maanova for analysis of variance with microarray data.

• Rosetta Inpharmatics. No links to software on their website.

• Silicon Genetics. Genespring is a very popular program for general data analysis of microarray data (http://genespring-support.com/).

• SilicoCyte. Integrated commercial analysis solution.

• Spotfire. DecisionSite software for functional analysis (https://spotfire.tibco.com/).

• Speed Group (University of California, Berkeley). R package sma (Statistical Microarray Analysis), Windows application RMAExpress and contributions to the Bioconductor project (https://www.stat.berkeley.edu/users/terry/Group/home.html).

• TIGR. SpotFinder image analysis, MIDAS data management, MeV differential expression analysis (http://www.jcvi.org/cms/research/software/).

• Probability of Expression (POE). An approach to the analysis of gene expression microarrays using three-component mixtures. Giovanni Parmigiani, Johns Hopkins.

• SuperArray (http://superarray.com/).

• Townsend Lab, University of Connecticut. Produces BAGEL (Bayesian Analysis of Gene Expression Levels) for the statistical analysis of spotted microarray data, Pathway Processor to test for overrepresentation of differentially expressed genes within known pathways of S. cerevisiae or B. subtilis, and SeqPop for computing population genetics statistics on sequence data.

• University of Pittsburgh & UPMC Bioinformatics Web-Tools Collection includes expression analysis tools.

• Walter and Eliza Hall Institute of Medical Research. Produce R package limma (linear models and differential expression for microarray data) and associated user-interfaces limmaGUI and affylmGUI (http://bioinf.wehi.edu.au/).

• Wong Lab (Harvard University). DNA-Chip Analyzer (https://www.hsph.harvard.edu/biostatistics/complab/).

• Yang, Jean (University of Sydney). R packages marray, DEDS and stepNorm (http://www.maths.usyd.edu.au/u/jeany/).

7.2 Microarray Databases

A microarray database is a repository containing microarray gene expression data. The key uses of a microarray database are to store the measurement data, manage a searchable index, and make the data available to other applications for analysis and interpretation (either directly, or via user downloads).

• ArrayTrack is a multi-purpose bioinformatics tool primarily used for microarray data management, analysis, and interpretation (https://www.fda.gov/ScienceResearch/BioinformaticsTools/Arraytrack/).

• NCI mAdb hosts NCI data with integrated analysis and statistics tools.

• ImmGen database is the open access acrossing all immune system cells; expression data, differential expression, coregulated clusters, regulation (http://www.immgen.org/index_content.html).

• Genevestigator: Gene expression search engine based on manually curated, well annotated public and proprietary microarray and RNA-seq datasets (https://genevestigator.com/gv/).

• Gene Expression Omnibus – NCBI: any curated MIAME compliant molecular abundance study (https://www.ncbi.nlm.nih.gov/geo/).

• ArrayExpress at EBI: Any curated MIAME or MINSEQE compliant transcriptomics data (https://www.ebi.ac.uk/arrayexpress/).

• Stanford Microarray database: private and published microarray and molecule abundance database (now defunct, http://smd.princeton.edu/).

• The Cancer Genome Atlas (TCGA) is the collection of expression data for different cancers (https://tcga-data.nci.nih.gov/).

• GeneNetwork system is the open access acrossing standard arrays, exons arrays, and RNA-seq data for genetic analysis (eQTL studies) with analysis suite (http://www.genenetwork.org/webqtl/main.py).

• UNC modENCODE Microarray database: Nimblegen customer 2.1 million array (https://genome.unc.edu:8443/nimblegen).

• UPSC-BASE: data generated by microarray analysis within Umeå Plant Science Centre (UPSC, http://www.upscbase.db.umu.se/).

• UPenn RAD database: MIAME compliant public and private studies, associated with ArrayExpress (http://www.cbil.upenn.edu/RAD).

• UNC Microarray database: provides the service for microarray data storage, retrieval, analysis, and visualization (https://genome.unc.edu/).

• MUSC database: The database is a repository for DNA microarray data generated by MUSC investigators as well as researchers in the global research community (http://proteogenomics.musc.edu/).

• caArray at NCI includes cancer data, which can be prepared for analysis on caBIG (https://wiki.nci.nih.gov/display/caArray2/).

Many journals now require that investigators submit raw microarray data upon publishing articles. Two main public repositories of microarray data:

• Gene Expression Omnibus at the National Center for Biotechnology Information (GEO at NCBI, http://www.ncbi.nlm.nih.gov/geo/).

• Array Express at the European Bioinformatics Institute (EBI, http://www.ebi.ac.uk/arrayexpress/).

7.3 Microarray Data Analysis

Microarray data analysis involves several distinct steps, as outlined below.

7.3.1 Creating Raw Data

Most microarray manufacturers, such as Affymetrix and Agilent provide commercial data analysis software with microarray equipment such as plate readers.

7.3.2 Background Correction

This step aims to remove non-biological contributions to the measured signal. Nonspecific signal such as unspecific binding of transcripts, background signal from incoplete washing of the microarray, background patterns across arrays, etc can be subtracted to achieve better results. One commonly applied techniques for background adjusment is included in affymetrix software. A variety of tools for background correction and further analysis are available from TIGR, Agilent (GeneSpring), and Ocimum Bio Solutions (Genowiz).

7.3.3 Aggregation and Normalization

The term 'normalization' in microarray data refers to the process of correcting two or more data sets prior to comparing their gene expression values. Normalization is required for both one- and two-channel microarray experiments. Normalization is essential to allow the comparison of gene expression across multiple microarray experiments in one-channel microarray experiments. In two-channel microarray experiments, the Cy3 and Cy5 dyes are incorporated into cDNA with different efficiencies. Without normalization, it would not be possible to accurately assess the relative expression of samples that are labeled with those dyes.

Most investigators apply a global normalization to raw array element intensities so that the average ratio for gene expression is one. The main assumption of microarray data normalization is that the average gene does not change in its expression level in the biological samples being tested. The procedure for global normalization can be applied to both two-channel data and one-channel data sets. Two-channel data are treated as two individual one-channel data sets such that each element signal intensity is divided by a correction factor specific to the channel from which it was derived. For the two or more data sets being normalized, the intensity for all the gene expression measurements in one channel are multiplied by a constant factor so that the total red and green intensity measurements are equal.

Other approaches to global normalization are possible. Some investigators normalize all expression values to a set of housekeeping genes that are represented on the array. This approach is also used by Affymetrix software. Housekeeping genes might include b-actin, *GAPDH* and others. Then each gene expression value in a single array experiment is divided by the mean expression value of these housekeeping genes. A major assumption of this approach is that such genes do not change in their expression values between two conditions.

Quantile normalization is an approach that produces the same overall distribution for all the arrays within an experiment [1]. It is a nonparametric method which ranks the gene expression variable from high to low. In quantile normalization, for each array each signal intensity value is assigned to a quantile. Then we consider a pooled distribution of each probe across all chip. Normalization is performed for each chip by converting an original probe set value to that quantile's value.

Quantile normalization is incorporated into Robust Multiarray Analysis (RMA) which is a normalization approach that does not take advantage of these mismatch spots, but still must summarize the perfect matches through median polish [2]. The median polish algorithm, although robust, behaves differently depending on the number of samples analyzed. Quantile normalization, also part of RMA, is one sensible approach to normalize a batch of arrays in order to make further comparisons meaningful.

7.3.4 Identification of Significant Differential Expression

Many strategies exist to identify which genes were regulated by at least twofold between treatment groups. More sophisticated approaches are often related to t-tests or other mechanisms that take both effect size and variability into account. Curiously, the p-values associated with particular genes do not reproduce well between replicate experiments, and lists generated by straight fold change perform much better [3, 4].

Applying p-values to microarrays is complicated by the large number of multiple comparisons involved. A p-value of 0.05 is typically thought to indicate significance, since it estimates a 5% probability of observing the data by chance. But with 10,000 genes on a microarray, 500 genes would be identified as significant at $p < 0.05$ even if there were no difference between the experimental groups. One obvious solution is to consider significant only those genes meeting a much more stringent p value criterion. One could perform a Bonferroni correction on the p-values, or use a false discovery rate calculation to adjust p-values in proportion to the number of parallel tests involved. Commonly cited methods include the Significance Analysis of Microarrays (SAM) and a wide variety of methods are available from Bioconductor and a variety of analysis packages from bioinformatics companies.

Selecting a different test usually identifies a different list of significant genes [5] since each test operates under a specific set of assumptions, and places a different emphasis on certain features in the data. Many tests begin with the assumption of a normal distribution in the data, because that seems like a sensible starting point and often produces results that appear more significant. Some tests consider the joint distribution of all gene observations to estimate general variability in measurements [6], while others look at each gene in isolation.

As the number of replicate measurements in a microarray experiment increases, various statistical approaches yield increasingly similar results, but lack of concordance between different statistical methods makes array results appear less trust worthy. The MAQC Project makes recommendations to guide researchers in selecting more standard methods (e.g. using p-value and fold-change together for selecting the differentially expressed genes) so that experiments performed in different laboratories will agree better.

Different from the analysis on differentially expressed individual genes, another type of analysis focuses on differential expression or perturbation of pre-defined gene sets and is called gene set analysis [4, 7]. Gene set analysis demonstrated several major advantages over individual gene differential expression analysis. Gene sets are groups of genes that are functionally related according to current knowledge. Therefore, gene set analysis is considered a knowledge based analysis approach. Commonly used gene sets include those derived from KEGG pathways, Gene Ontology terms, gene groups that share some other functional annotations, such as common transcriptional regulators etc. Representative gene set analysis methods include GSEA which estimates significance of gene sets based on permutation of sample labels [7], and GAGE which tests the significance of gene sets based on permutation of gene labels or a parametric distribution.

7.3.5 Significance Analysis of Microarrays

Significance analysis of microarrays (SAM) is a modified t-test that finds significantly regulated genes in microarray experiments [8]. SAM assigns a score to each gene in a microarray experiment based on its change in gene expression. Statistical significance is assessed using a permutation test in which observed scores are compared to the results of repeated measurements fromashuffleddataset. SAM offers several useful features. It accepts microarray data from experiments using a variety of experimental designs. Prior to operating SAM, the user must normalize and scale expression data. The SAM input data can be in a raw or log-transformed format. SAM uses a modified t-statistic to test the null hypothesis [8]. A key feature of SAM is its ability to provide information on the FDR. The user can adjust a parameter called delta to adjust the false positive rate. The SAM algorithm calculates a 'q value,' which is the lowest false discovery rate at which a gene is described as significantly regulated.

7.3.6 Cluster Analysis of Microarray Data

Clustering is a commonly used tool to find patterns of gene expression in microarray experiments [9]. Genes, samples, or both may be clustered in trees. Clustering is the representation of distance measurements between objects. The main goal of clustering is to use similarity or distance measurements between objects to represent them. Data points within a cluster are more similar, and those in separate clusters are less similar. It is common to use a distance matrix for clustering based on Euclidean distances. There are several kinds of clustering techniques.

1. Hierarchical clustering

The most common form for microarray analysis is hierarchical clustering, in which a sequence of nested partitions is identified resulting in a dendrogram. Hierarchical clustering can be performed using agglomerative or divisive approaches which generally produce similar results, although large differences can occur.

2. k-Means clustering

Some times we know into how many clusters our data should fit. For example, we may have treatment conditions we are evaluating, or a set number of time points. An alternative type of unsupervised clustering algorithm is a partitioning method that constructs k clusters [10]. The steps are as follows: choose samples and/or genes to be analyzed; choose a distance metric such as Euclidean; choose k; data are classified into k groups as specified by the user.

Each group must contain at least gene expression value, and each object must belong to exactly one group. Two different clusters cannot have any objects in common, and the k groups together constitute the full data set; perform clustering; assess cluster fit.

3. Self-organizing maps

The self-organizing map (SOM) algorithm resembles k-means clustering in that it partitions data into a two-dimensional matrix. For SOMs and other structured clustering techniques, you can estimate the numberof clusters you expect based on the number of experimental conditions in order to decide on the initial number of clusters to use. SOMs impose a partial structure on the clusters [11]. Clusters are calculated in an iterative process, as in k-means clustering, with additional information from the profiles in adjacent clusters.

7.3.7 Principal Components Analysis

Principal components analysis (PCA) is an exploratory technique used to find patterns in gene expression data from microarray experiments. It is both easy to use and powerful in its ability to represent complex data sets succinctly. PCA is used to reduce the dimensionality of data sets in order to create a two- or three-dimensional plot that reflects the relatedness of the objects. PCA has been used to analyze expression data in yeast and mammalian systems [12-15]. The central idea behind PCA is to transform a number of variables into a smaller number of uncorrelated variables called principal components. The variables that are operatedon by PCA maybe the expression of many genes, or the results of gene expression across various samples, or even both gene expression values and samples. In a typical microarray experiment, PCA detects and removes redundancies in the data (such as genes whose expression values do not change and thus are not informative about differences in how the samples behave).

7.3.8 Supervised Data Analysis for Classification of Genes

The distances and similarities among gene expression values can be described using two types of analysis: supervised or unsupervised. The unsupervised approaches we have described so far are especially useful for finding patterns in large data sets. In supervised analyses, the approach is different because the experimenter assumes some prior knowledge of the genes and/or samples in the experiment. For example, transcriptional profiling has been performed on cell lines or biopsy samples that are either normal or cancerous [16-18]. The goal of supervised microarray data analysis algorithms is to define a rule that can be used to assign genes (or conditions) into groups. In each case, we begin with gene expression values from known groups and train an algorithm to learn a rule. Positive and negative examples are used to train the algorithm. The algorithm is then applied to unknown samples, and its accuracy as a predictor or classifier is assessed. It is critical that the data used for building a classifier are entirely separate from the data used to assess its predictive accuracy.

7.3.9 Gene Pathway and Network Analysis

Specialized software tools for statistical analysis to determine the extent of over- or under-expression of a gene in a microarray experiment relative to a reference state have also been developed to aid in identifying genes or gene sets associated with particular phenotypes. Commercial systems for gene network analysis such as Ingenuity and Pathway studio create visual representations of differentially expressed genes based on current scientific literature.

1. Gene set enrichment analysis (GSEA)

GSEA offers the user information on the genes or gene sets of interest, including links to entries in databases such as NCBI's GenBank and curated databases such as Biocarta and Gene Ontology. GSEA represents one increasingly popular approach to identifying regulated sets of genes [19]. That procedure includes over 1,000 defined gene sets in which members of each set share common features (such as biological pathways). With all these annotation procedures it is important to keep in mind that the product of mRNAs is protein. Identification of a set of mRNAs encoding proteins in a particular cellular pathway does not mean that the proteins themselves are present in altered levels, nor does it mean that the function of that pathway has been perturbed. Such conclusions can only be drawn from experiments on proteins and pathways performed at the cellular level.

2. Protein complex enrichment analysis tool (COMPLEAT)

COMPLEAT provides similar enrichment analysis at the level of protein complexes. The tool can identify the dynamic protein complex regulation under different condition or time points [20].

3. PAINT and SCOPE

PAINT and SCOPE perform a statistical analysis on gene promoter regions, identifying over and under representation of previously identified transcription factor response elements.

4. Rank sum statistics for gene set collections (RssGsc)

RssGsc uses rank sum probability distribution functions to find gene sets that explain experimental data.

5. Genevestigator

Genevestigator is a public tool to perform contextual meta-analysis across contexts such as anatomical parts, stages of development, and response to diseases, chemicals, stresses, and neoplasms.

References

[1] Bolstad B M, Irizarry R A, Astrand M, Speed T P. A comparison of normalization methods for high density oligonucleotide array data based on variance and bias. *Bioinformatics*, 2003, 19(2): 185-193.

[2] Irizarry R A, Hobbs B, Collin F, Beazer-Barclay Y D, Antonellis K J, Scherf U, Speed T P. Exploration, normalization, and summaries of high density oligonucleotide array probe level data. *Biostatistics*, 2003, 4(2): 249-264.

[3] Shi L M, Reid L H, Jones W D, Shippy R, Warrington J A, Baker S C, Collins P J, de Longueville F, Kawasaki E S, Lee K Y, et al. The MicroArray Quality Control (MAQC) project shows inter- and intraplatform reproducibility of gene expression measurements. *Nat. Biotechnol.*, 2006, 24(9): 1151-1161.

[4] Guo L, Lobenhofer E K, Wang C, Shippy R, Harris S C, Zhang L, Mei N, Chen T, Herman D, Goodsaid F M, et al. Rat toxicogenomic study reveals analytical consistency across microarray platforms. *Nat. Biotechnol.*, 2006, 24(9): 1162-1169.

[5] Yauk C L, Berndt M L. Review of the literature examining the correlation among DNA microarray technologies. *Environ. Mol. Mutagen*, 2007, 48(5): 380-394.

[6] Breitling R. Biological microarray interpretation: The rules of engagement. *Bba-Gene Struct. Expr.*, 2006, 1759(7): 319-327.

[7] Subramanian A, Tamayo P, Mootha V K, Mukherjee S, Ebert B L, Gillette M A, Paulovich A, Pomeroy S L, Golub T R, Lander E S, et al. Gene set enrichment analysis: A knowledge-based approach for interpreting genome-wide expression profiles. *Proc. Natl. Acad. Sci.* 2005, 102(43): 15545-15550.

[8] Tusher V G, Tibshirani R, Chu G. Significance analysis of microarrays applied to the ionizing radiation response. *Proc. Natl. Acad. Sci.* 2001, 98(9): 5116-5121.

[9] Gollub J, Sherlock G. Clustering microarray data. *Methods Enzymol*, 2006, 411: 194-213.

[10] Tavazoie S, Hughes J D, Campbell M J, Cho R J, Church G M. Systematic determination of genetic network architecture. *Nat. Genet.*, 1999, 22(3): 281-285.

[11] Tamayo P, Slonim D, Mesirov J, Zhu Q, Kitareewan S, Dmitrovsky E, Lander E S, Golub T R. Interpreting patterns of gene expression with self-organizing maps: methods and application to hematopoietic differentiation.

Proc. Natl. Acad. Sci. 1999, 96(6): 2907-2912.

[12] Landgrebe J, Welzl G, Metz T, van Gaalen M M, Ropers H, Wurst W, Holsboer F. Molecular characterisation of antidepressant effects in the mouse brain using gene expression profiling. *J. Psychiatr. Res.*, 2002, 36(3): 119-129.

[13] Misra J, Schmitt W, Hwang D, Hsiao L L, Gullans S, Stephanopoulos G, Stephanopoulos G. Interactive exploration of microarray gene expression patterns in a reduced dimensional space. *Genome. Res.*, 2002, 12(7): 1112-1120.

[14] Alter O, Brown P O, Botstein D. Singular value decomposition for genome-wide expression data processing and modeling. *Proc. Natl. Acad. Sci.* 2000, 97(18): 10101-10106.

[15] Bouton C M, Hossain M A, Frelin L P, Laterra J, Pevsner J. Microarray analysis of differential gene expression in lead-exposed astrocytes. *Toxicol. Appl. Pharmacol.*, 2001, 176(1): 34-53.

[16] Alizadeh A A, Staudt L M. Genomic-scale gene expression profiling of normal and malignant immune cells. *Curr. Opin. Immunol.*, 2000, 12(2): 219-225.

[17] Shipp M A, Ross K N, Tamayo P, Weng A P, Kutok J L, Aguiar R C, Gaasenbeek M, Angelo M, Reich M, Pinkus G S, et al. Diffuse large B-cell lymphoma outcome prediction by gene-expression profiling and supervised machine learning. *Nat. Med.*, 2002, 8(1): 68-74.

[18] Perou C M, Jeffrey S S, van de Rijn M, Rees C A, Eisen M B, Ross D T, Pergamenschikov A, Williams C F, Zhu S X, Lee J C, et al. Distinctive gene expression patterns in human mammary epithelial cells and breast cancers. *Proc. Natl. Acad. Sci.* 1999, 96(16): 9212-9217.

[19] Subramanian A, Tamayo P, Mootha V K, Mukherjee S, Ebert B L, Gillette M A, Paulovich A, Pomeroy S L, Golub T R, Lander E S, et al. Gene set enrichment analysis: a knowledge-based approach for interpreting genome-wide expression profiles. *Proc. Natl. Acad. Sci.* 2005, 102(43): 15545-15550.

[20] Vinayagam A, Hu Y, Kulkarni M, Roesel C, Sopko R, Mohr S E, Perrimon N. Protein complex-based analysis framework for high-throughput data sets. *Sci. Signal.*, 2013, 6(264): rs5.

Chapter 8 Bioinformatics for Qualitative and Quantitative Proteomics

Yong Zeng[1], Lan Zhang[2], Wei Zhu[3]

Proteomics is the large-scale study of entire proteins. As a complementary approach to genomics and transcriptomics, it can provide comprehensive and direct information in exploring biological process. Mass Spectrometry (MS)-based technologies are the main approaches in modern proteomics studies. Protein identification and quantification are two major tasks in proteomics study, which could advance our understanding of the complex and dynamic nature of proteins. Proteomics studies have been widely applied in biological and clinical fields in identifying biomarkers, monitoring disease status, and assessing treatment effect. However, researchers have to face a pressing problem that how to deal with the high-throughput data generated form MS-based proteomics platform. In this chapter, we will introduce some bioinformatics databases and tools which are widely used in protein identification and quantification, proteomics data analysis, storage and sharing.

8.1 Protein Identification and Quantification from MS Raw Data

Benefit from the application of mass spectrometry analysis to protein samples, proteomics is developing rapidly during the past decade. Large volumes of experimental data was generated from MS proteomic platform [1]. To systematically identify and quantify whole proteins from the raw data, multiple search engines in conjunction with several well-established protein sequence databases are applied. Herein, we will briefly introduce some com-

1. Zeng Yong, College of Life Sciences and Bioengineering, School of Science, Beijing Jiao Tong University, Beijing, China, 100044.

2. Zhang Lan, Center for Bioinformatics and Genomics, Department of Biostatistics and Bioinformatics, Tulane University, New Orleans, LA, USA, 70112.

3. Zhu Wei, College of Life Sciences, Hunan Normal University, Changsha, China, 410081.

mon databases and search engines in MS-based proteomics studies.

8.1.1 Common Protein Sequence Databases

1. UniProt

UniProt (http://www.uniprot.org/) is an authoritative, high-quality and freely accessible database of protein sequence and functional information which combines the Swiss-Prot, TrEMBL and PIR-PSD databases into a single resource [2]. It contains four core sub-databases including Protein knowledgebase (UniProtKB), Sequence clusters (UniRef), Sequence archive (UniParc) and Proteomes. It has excellent tools for dataset retrieval such as BLAST (Basic Local Alignment Search Tool), Align and Retrieve/ID mapping. Many entries with large amounts of biological and sequencing information are derived from research literature or genome sequencing projects. This database serves as a basic and curated sequence resource for protein prediction and for planning new experiments [3]. Small datasets can be directly downloaded from the UniProt web site. However, for downloading complete datasets, UniProt FTP site (ftp://ftp.uniprot.org/) is recommended.

2. Swiss-Prot

Swiss-Prot is a leading and comprehensive universal protein sequence database which was integrated into UniProt and can be accessed via the official website of UniProt (http://www.uniprot.org/). Currently, it contains more than 540,000 manually reviewed and annotated entries from numerous species. It is a non-redundant database, which means that all reports for a given protein are merged into a single entry. It is also highly integrated with other databases [2, 4].

3. TrEMBL

As an essential part of UniProt database, TrEMBL (http://www.uniprot.org/) currently contains more than 50 million automatically annotated and not reviewed entries. It is a computer-annotated protein sequence database complementing the Swiss-Prot Protein Knowledgebase. TrEMBL consists of computer annotated translation of the coding sequences (CDs) incorporated in public databases such as EMBL/GenBank/DDBJ Nucleotide Sequence Databases and also protein sequences extracted from the literature or submitted to UniProtKB/Swiss-Prot [5]. TrEMBL strictly follows the Swiss-Prot format and conventions [2].

4. RefSeq

NCBI's Reference Sequence database (RefSeq) (http://www.ncbi.nlm.nih.gov/refseq/) contains large amounts of sequencing information for DNA, RNA and protein from plasmids, organelles, viruses, archaea, bacteria, and eukaryotes. The aim of the project is to provide a non-redundant collection of references for nucleotide and protein sequences. Each reference sequence is constructed wholly from sequence data submitted to the International Nucleotide Sequence Database Collaboration. As a fundamental database with genetic and functional information, RefSeq provides reference standards for multiple purposes, such as genome annotation or reporting locations of sequence variation in medical records. The RefSeq database can be easily retrieved in several different ways including Nucleotide, Protein, and Map Viewer. Sequence information included in RefSeq database can be searched via BLAST and downloaded from the RefSeq FTP site.

8.1.2 Frequently Used Software for Protein Identification and Quantification

1. Mascot

Mascot is a database search engine, which can identify proteins by matching mass spectrometry raw data to known peptide sequence databases [6, 7]. Mascot is widely used in proteomics research. Customers can easily get access to a series of Mascot software and temporarily use them on the Matrix Science website (http://www.matrixscience.com/). However, for large scale data analysis and routine work, researchers must purchase the software and obtain a license to run in-house. Mascot is based on the Molecular Weight Search (MOWSE) algorithm [8] as well as probability-based scoring [9].

2. SEQUEST

SEQUEST is a tandem MS data analysis program for protein identification. SEQUEST identifies proteins by matching experimental tandem mass spectra to known protein/peptide sequences database. A cross-correlation function is calculated between the measured fragment mass spectrum and the protein sequences in the database, and it is used to score the proteins in the database [10]. SEQUEST supports the use of information from several fragment mass spectra in the database search and shows good sensitivity and flexibility in handling data generated by different types of mass spectrometers [11, 12]. Modern proteome research based on tandem mass spectrometer will produce a large scale of tandem mass spectra data. Identifying such a data collection requires automation and stability, and SEQUEST is the first software to fill that need.

3. PLGS

ProteinLynx Global SERVER (PLGS) is developed by Waters Corporation. As a fully integrated Mass-Informatics platform for quantitative and qualitative proteomics research, it plays a central role in data analysis in the

Waters proteomics system. PLGS always comes together with instruments from Waters, and it has the selectivity and specificity required for MSE data analysis. Based on open system architecture, PLGS has automatic workflow for high throughput data processing. It can minimize the false positive rate by setting strict statistical filters. The friendly operation interface of PLGS allows users to set individual parameters including database importing, protein modification selecting, and threshold setting. The project and database management tools included in PLGS provide functions for results visualization and reporting.

4. PEAKS

PEAKS is commonly used for peptide identification through de novo peptide sequencing, thus extracting amino acid sequence information without the use of databases. Based on a progressive model and algorithm, PEAKS computes the best peptide sequences whose fragment ions can best match the peaks in the MS/MS spectrum. PEAKS provides a complete sequence for each peptide, confidence scores on individual amino acid assignments, and simple reporting for high-throughput analysis. As a well-known de novo sequencing software, PEAKS has the ability to compare results of multiple search engines [13].

8.2 Proteomics Data Analysis

MS-based proteomics is widely applied for various purpose in biological or clinical area. It becomes an invaluable tool for system-wide protein identification and quantification. Proteomics studies and publications increased rapidly in recent years. With the rising of large-scale proteomics data, how to perform in-depth data analysis becomes an urgent problem to each researcher. Thus, multiple bioinformatics and biostatistics techniques, tools and databases were developed to interpret proteomics data.

8.2.1 Functional Annotation of Single Protein

Proteins identified from MS-base proteomics platform need to be annotated in different ways for diverse research topics. The first step of functional analysis is to map the protein name to a unique identifier. Usually, we use the classical databases such as Uniprot [14] and Ensembl [15] for protein names normalizing. Then, we check the protein lists and remove the redundant names, as repetitive entries may lead to biased result in some analysis tools. Searching extensive amounts of literature to interpret a given protein list is still common, and several comprehensive databases like Uniprot [14] can be easily accessed to explore the associated function for each protein entry. Moreover, several available repositories databases like PRIDE [16], PepSeeker [17], PeptideAtlas [18], GPM [19] can be used to search for previously carried-out proteomics experiments to compare the proteomics results to other relevant experiments. Functional annotation of single protein can provide direct information and visual associations between proteins and phenotypes. However, it has its own limitations. Due to the intricate hypotheses for different experiment design and proteomics datasets, merely comparing experimental findings to manually collected information could introduce bias. In addition, it unable to reveal the essential hidden relationships between the members of a protein list.

8.2.2 Comparative Proteomics Analysis

To identify significant differentially expressed proteins between two or more experimental states, usually between case and control groups, comparative proteomics has become a powerful tool to explore protein expression level response to diverse biological systems, such as different cell developmental stages or disease conditions. However, to enhance the reliability and utility of comparative proteomics analysis, we must perform data quality control including data normalization, missing value remediation before subsequent statistical analysis. Since proteomics data is usually subject to a number of challenging analytical issues like experimental noise and low coverage rate, cares should be taken when dealing with such data [20].

1. Data normalization

Data normalization is a commonly used method to remove systematic biases by adjusting the variability due to sample preparation or equipment conditions. Several normalization methods have been used in proteomics studies. The simplest one is the global normalization method, in which the raw protein expression level is normalized by a well-conserved protein [21]. In addition, owing to the similarity of data features between gene expression and protein data, numerous normalization approaches such as locally weighted scatterplot smoothing (LOWESS) regression and quantile normalization, have been widely borrowed from gene expression studies [21].

2. Missing values

Missing values are common in MS expression data, and considerable proportions of missing observations have been observed in proteomics studies [21]. There are several causes for missing events, such as low concentrations (hard to be detected) of the peptide, low resolution of the equipment or absence of the peptide in the sample. It is hard to determine the exact missing mechanism, which further complicates the problem. There are two major categories: missing at random (MAR) and missing not at random (MNAR). For the MAR situation, a simple imputation process is commonly used that replace missing values by a constant or a randomly selected value (e.g., impute with mean, median, etc.). For more complex situations, several other methods are applied, like KNN (K-nearest neighbors) [22], LSA (Least-squares adaptive) [23], which estimate missing values based on the expression profiles of other proteins with similar intensity within the dataset. Because of the complicated missing mechanisms in proteomics, no individual method is guaranteed to be the best solution.

3. Differential expression analysis

Conventional proteomics studies often include comparison of protein expression profiles under two or more different conditions (e.g., normal versus disease). Statistical approaches such as ANOVA (analysis of variance), student's t test or regression are commonly applied in identifying differentially expressed proteins (DEPs) with a pre-defined statistical threshold [24]. Typically, a test statistic, which reflects how much a feature discriminates between classes, will be generated by a specific test, and the p values for DEPs adjusted for multiple testing be calculated based on the null distribution (theoretical or permutation-based) [25]. A variety of methods originally designed to compare microarray datasets have been used in proteomics data as well. For instance, the SAM (Significance Analysis of Microarray) method and the Power Law Global Error Model (PLGEM) are now widely applied in comparative proteomics analysis [26, 27].

8.2.3 Enrichment Analysis

Various functional databases contain experimentally proven or otherwise inferred connections between the genes or proteins and their specific functions. Generally, these functions belong to a certain controlled vocabulary, which means they have a clearly described meaning defined and recognized by domain specialists. The Gene Ontology (GO) [28] is a project for consistent descriptions of gene and gene product attributes across species. Each GO contains an identifier and a term belongs to one of the three GO categories: biological process, molecular function or cellular component. It is worth mentioning that GO is organized in a hierarchical way, which can be represented as a tree structure. Normally, an initial step for functional interpretation of the acquired protein list is to link the protein identifier with its associated GO. For instance, one can use the AmiGO [29], which is a user-friendly platform, to search the GO associated with a particular gene/protein.

1. Annotation term enrichment analysis

Generally, the subsequent step after GO annotation is GO enrichment analysis, which determines whether a specific GO term is enriched in a particular group of biological processes, functions or cellular compartments. It involves comparing the frequency of individual functional annotations within a reference list, and the enrichment score can be statistically tested (e.g., hyper-geometric, binomial or Chi-square tests, etc.). The result will be a *p* value, which can be used as the cutoff to measure significance of over- or under-representation of a GO term. Since the number of functional terms for enrichment test is usually large, an adjusted multiple *p*-value (e.g. Benjamini-Hochberg correction [30]) is often provided. A wide range of software is available for GO enrichment analysis, and an extensive list can be found at (http://geneontology.org/). DAVID is a popular meta-tool [31] that aids with GO enrichment analysis. It applies a modified Fisher exact *p*-value to determine whether a GO term is enriched in a given proteomics dataset with reference dataset as background [32]. It enables the researcher to gather information about over- or under-representation, and to understand the biological meaning behind large lists of proteins. In proteomics studies, the GO-term enrichment analysis has been applied in numerous contexts. The advantage of annotation term enrichment compared to the singular analysis of individual protein lists is that it can summarize the functional properties on a global scale instead of individual protein entries.

2. Set enrichment analysis

The classical overrepresentation analyses strictly rely on a pre-defined quantitative threshold selecting proteins to be included, and ignore the expression level alterations when calculating the functional enrichment score. Thus, set-based enrichment analyses were developed to evaluate the significance of predefined protein lists. As a powerful method to determine functional significance of gene groups, gene set enrichment analysis (GSEA) was originally developed in the gene expression data [33]. Without modification, this method has been applied in the proteomics field [34], and a modified approach called PSEA (Protein set enrichment analysis) was developed to study differential protein expression based on Spl (Spectral index) from label-free quantitative proteomics in breast cancer research [35].

The basis of set-based enrichment analysis is to determine whether the pre-defined gene subsets (e.g., based on a common functional annotation) are distributed in a ranked list (e.g., generated by quantitative feature of expression data) randomly. The pre-defined annotation terms are commonly gained from libraries such as GO [28] or MSigDB [36]. Then, the ranked protein list is generated based on the quantitative differences between case and control group and the enrichment score (ES) is calculated based on specified algorithms. Usually, empirical distribution of ES is determined by permutations, and it is used to test the statistical significance of observed ES.

The set-based enrichment analysis has many advantages compared with traditional overrepresentation analysis. There is no requirement of an arbitrary threshold to distinguish significantly differential proteins, and it provides more statistical power. In addition, set enrichment analysis could detect coordinated changes of gene products, which may reveal some noteworthy proteins and protein groups important for disease status.

3. Pathway analysis

Biological pathways usually describe a series of chemical reactions in the cell that lead to a specific biological outcome. Proteins involved in these chemical interactions and those that have a regulatory effect play important roles in pathway analysis. Rather than merely taking the gene-centric view of GO-based analyses, pathway analyses provide us with more insight into biological mechanisms [37].

Several comprehensive databases such as KEGG [38], Reactome [39] and Ingenuity Pathways Knowledge Base [40] contain a large amount of pathway information. These instrumental tools are typically derived from intracellular reactions like metabolic signaling pathways. In addition to the above comprehensive resources, numerous highly specific databases have been developed as well. For instance, the PANTHER [41], GenMAPP [42] and PID [43] mainly focus on signal transduction processes. Lately, several databases were created which include pathways active in disease. Netpath [44], for example, can help researchers to extract cancer relevant pathways from a complex dataset.

A comprehensive overview of hundreds biological pathway-related resources and molecular interaction-related resources can be found on the Pathguide website (http://pathguide.org) [45]. This powerful resource could help researchers to select the optimal database for their studied biological systems, and it can be recommended as a starting point for proteomics pathway analysis.

8.2.4 Protein-Protein Interaction Analysis

Protein-protein interactions (PPIs) represent non-random physical contact between two or more proteins, and thus form smaller or larger complexes in a space- and time-dependent manner. Since physical contact between proteins could trigger conformational changes or PTMs that modulate the activity of those proteins, PPIs can help us gain further insight into biological processes. PPIs are not static or permanent, but experience continuous reassembly and turnover. PPIs are usually regulated based on many factors, such as specific cell-type, developmental stage of the cell, cell-cycle phase, external stimulus or signal and the presence of other proteins. Therefore, the network of PPIs can provide powerful information to reveal and explain important functional modules within proteins. For instance, if proteins are neighbors in the PPI network and found to be co-regulated in the differential expression list, it may suggest that they work together to play important roles in the affected biological processes. PPIs are essential to development and homeostasis of biological mechanisms, and many human diseases can be traced to aberrant PPIs. Thus, the inhibition of these aberrant associations is of great clinical significance. Because of the diverse nature of PPIs, the successful design of therapeutics requires detailed knowledge of each system at a molecular level. Several recent studies have identified and characterized specific interactions from various disease systems, and many of the key PPIs are known to participate in disease-associated signaling pathways [46].

1. Commonly used databases for PPIs analyses

Large databases documenting protein interactions are publicly available for several organisms, HPRD [47], IntAct [48], MINT [49], MIPS [50], DIP [51] and BioGRID [52] are some of the most commonly used resources. Among these, HPRD (http://www.hprd.org/) can depict and incorporate interaction networks, domain structure, post-translational modifications, and associated disease for human proteins. IntAct (http://www.ebi.ac.uk/intact/) is an analysis tool for molecular interaction data, which gathers useful information from previously published results or user-submitted data. MINT (http://mint.bio.uniroma2.it/mint/Welcome.do) is an interactive database based on experimentally verified PPIs, usually extracting data from technical literature. MIPS (http://mips.helmholtz-muenchen.de/proj/ppi/) contains manually curated high-quality PPI data collected from the scientific literature by expert curators. DIP (http://dip.doe-mbi.ucla.edu/dip/Main.cgi) interprets signaling or regulatory pathways, and can also detect protein interactions at the cellular level. BioGRID (http://thebiogrid.org/) is a curated biological database of protein-protein interactions, genetic interactions, chemical interactions, and post-translational modifications. In addition, several meta-bases such as APID [53], I2D [54] and STRING [55] are widely used in PPIs analyses. For instance, STRING (http://string-db.org/), a powerful meta-database including data from many curated databases, serves as a very popular tool for interaction network analysis of proteomic data, incorporating both interactions and pathway information to form an easy-to-use web interface. Moreover, tools like Cytoscape [56], Osprey [57] and VisANT [58] are popular open-source programs for visualizing protein-protein interaction networks (PPINs).

Since the covered experimental data sets and criteria for PPIs vary widely, cares should be taken in selecting the most suitable database for proteomic analysis. In addition, one must be very careful to select appropriate parameters and types of interaction data, especially when incorporating the predicted interactions. Checking GO term coherence is a good way to assess the reliability of an edge in a PPIN, that is, whether the two proteins on both ends of the edge could be annotated to an informative GO term in common [59]. Another way to assess the reliability of an edge in a PPIN is based on the hypothesis that proteins are more likely to share common neighbors in the PPIN if they interact.

2. PPIs analysis methods

Most PPIs analysis methods include identifying protein complexes by mining modular or dense sub-networks from PPI networks. In disease-related research, four sequential steps are usually conducted to generate a biological hypothesis on target cells (e.g. cancer cells) through PPIN construction and analysis [60]. The starting point is to define the seed proteins, which should be the molecules of major interest and the skeleton of the PPIN. Typical choices are differentially expressed proteins observed in a given experiment, or proteins known to be involved in the disease process. The second step is to determine interactions between proteins. The interacting partners of proteins are commonly identified from curated databases as described above, and the reliability of these interactions needs to be assessed. The third step is the PPIN construction and visualization process. The PPIN is constructed based on high-confidence evidence, and several algorithms allow the creation of a visual representation of the built network. The final step is to use bioinformatics tools to extract meaningful biological information from the network. Although the strategy seems straightforward, these methods are restricted by limitations in existing PPI datasets, particularly the lack of sufficient interactions between proteins and the presence of a large number of false-positive interactions. Therefore, increasing interaction coverage by integrating PPI datasets from multiple studies and reducing noise by assessing the reliabilities of interactions are crucial for accurate PPIN analysis.

8.3 Proteomics Data Storage, Exchange and Sharing

In order to facilitate data comparison, exchange and verification in or between public repositories, the common standards for data representation in proteomics must be established. In the fields of mass spectrometry and protein-protein interaction study, much progress has been made in developing common standards for data storage, sharing and exchange [61]. As the most well-known international consortium of proteomics research associations, the Human Proteome Organization (HUPO) was launched in 2001. The organization promotes the development and awareness of proteomics research and facilitates scientific collaborations between members and initiative. Its goal is to gain a comprehensive understanding of the human proteome [62].

8.3.1 Public Proteomics Repositories

1. PRIDE

The PRIDE (http://www.ebi.ac.uk/pride/) is a centralized, standards compliant, public data repository, which includes information on protein and peptide identifications, post-translational modifications, and supporting spectral evidence. Protein and peptide identifications in this database have been described in previous scientific literature. Data generated from different species, tissues and subcellular locations (perhaps under specific disease conditions) can be uploaded, downloaded or viewed via a single, centralized web interface [16]. PRIDE supports the submission of data from different platforms. However, data prepared to be uploaded to PRIDE database should obey strict proteomics data standards [1]. Many tools allow researchers to achieve standards compliance for data generated by many different platforms. By the end of 2014, PRIDE had accumulated data for 41,835 proteins, 269,806 unique peptides, and about 101 million spectra [63]. Currently, datasets from a total of 51,922 assays and 3,233 projects were centralized in PRIDE. It is one of the most popular proteomic data repositories and has played an important role in Human Proteome Project (HPP) [64].

2. Global Proteome Machine Database

The Global Proteome Machine Database (GPMDB) is a resource for collecting diverse tandem proteomics data and open source software, and it also includes peptide and protein identifications that are important for further MS computational research [19]. The database allows worldwide research scientists to use its proteomics data and tools for the purpose of proteome research. Raw data submitted by researchers or downloaded from other databases are reprocessed. Extensible Markup Language (XML) files including protein or peptide identification information are uploaded and stored in GPMDB. By the end of 2014, GPMDB database spanned a total of 136,373 proteins, 1,786,698 peptides, and 1,020 million spectra [63, 64]. The GPMDB plays an important role in proteome research and can be accessed at (http://gpmdb.thegpm.org/index.html).

3. PeptideAtlas

The PeptideAtlas (http://www.peptideatlas.org/) is one of the largest and most well-curated protein expression data resources. It serves as a compendium of peptides observed with tandem mass spectrometry methods from multiple species. It also contains a growing set of software tools and underlying platforms for proteomics data analysis and visualization. It stores various formats of output files and metadata from MS-based experiments [65]. PeptideAtlas supports raw data submissions from users, which will be reprocessed through a uniform analysis and validation pipeline. The results are loaded into a database, and the information derived from the raw data is returned to the community for identification and statistical analysis purposes. Users can search the PeptideAtlas web interface by protein accession, peptide sequence, gene name, keyword, or phrase [63]. Search results are displayed on a page with associated summary statistics. PeptideAtlas can help plan targeted proteomics experiments, improve genome annotation, and support data mining projects [64, 65].

4. MOPED

Model Organism Protein Expression Database (MOPED) is a proteomics repository that integrates protein expression information from human specimens and several other model organisms [66]. It provides protein-level expression data, meta-analysis capabilities, and quantitative data from standard analyses based on mass spectrometry. It also provides new estimates of protein abundance and concentration, and statistical summaries from experiments. The web interface contains six main panels: protein absolute expression, protein relative expression, gene relative expression, pathways, experiments, and visualization for different data handling purposes. Additionally, a suite of tools for data searching and visualization are available. With rapid development in recent years, MOPED has grown into a repository containing more than 17,000 proteins, 250,000 unique peptides, and approximately 15 million spectra [63, 64, 66]. As a significant public proteomics database, MOPED provides abundant information on complex biological processes and thus benefits fundamental biological or medical investigation. The MOPED database can be accessed at (http://moped.proteinspire.org).

5. Tranche

Tranche (http://www.proteomexchange.org/databases/tranche) is a data repository for storing, sharing information for proteomics researchers. As a widely-used database, Tranche hosts several kinds of data. Indeed, it plays a crucial role in proteomics field. It allows researchers to use and disseminate both data and software. A client tool is required to upload and download datasets. Tranche provides interfaces for PRIDE, Human Proteinpedia, and PeptideAtlas to store and disseminate large MS-based data files [67].

8.3.2 Public Proteomics Standards

1. MS raw data unification

Different models of instruments were applied for MS data acquisition. An important consideration is how to get the data provided by multiple MS software into mzML format. The mzML is a XML-based common file format for proteomics data, and it provides a standard container for MS and MS/MS data. There are many free and commercial software packages that support mzML format. However, in order to export the acquired MS data as mzML format, some applications must be used in conjunction with additional translators and transformation utilities. ProteoWizard and OpenMS are two popular programs for MS raw data handling.

2. Qualitative and quantitative proteomics

The purpose of qualitative proteomics studies is to identify peptides or proteins, which can normally be identified by analyzing raw MS data. As mentioned in the previous section, there are many tools used for peptide and protein identification. Mascot, SEQUEST or PEAKS are frequently-used programs for peptide and protein identification. Furthermore, based on rapidly developing experimental technology and software packages, quantitative proteomics study can be achieved based on variety of quantitative techniques including multiple labeling or label-free approaches. For instance, MaxQuant and PLGS are well equipped for handling quantitative proteomics data based on label or label-free methods, respectively.

3. Integration of public standards and local or third-party tools

Many researchers have their own bioinformatics capabilities and homemade tools for specific data analysis in proteomics study. We will briefly introduce how public standards can be integrated into the tools produced by proteome bioinformatics individuals or groups. Many popular commercial or free software packages have their own native file formats for data storage or reporting. Since 2002, when the HUPO Proteomics Standards Initiative (HUPO-PSI) was founded, the uniform public standards have become more carefully established for data reporting and exchange. More and more existing software tools for proteomics are following HUPO-PSI standards. In order to improve the efficiency of proteomics data use and exchange, in-house software development in many proteomics labs should adhere to the HUPO-PSI standards as well [68].

4. Reporting, uploading and exchanging data

Minimum Information about a Proteomics Experiment (MIAPE) guidelines were published by HUPO-PSI in 2007. According to MIAPE, formalized information should be reported when publishing a dataset. Two websites provide assistance to guide users to create a MIAPE compliant report: MIAPEGelDB [69] and Proteo-Red MIAPE web repository [70]. Proteo-Red MIAPE web toolkit [71] is a website capable of linking the latest versions of the HUPO-PSI XML schemas to the Proteo-Red MIAPE web repository in an automated, accessible, and comprehensive way. It covers multiple data formats such as mzML, mzIdentML, and PRIDE XML [68]. When uploading data to either a public repository or local database, the most important thing is that the experiment has been reported using data standards so it can be shared and verified.

References

[1] Riffle M and Eng J K. Proteomics data repositories. *Proteomics*, 2009, 9(20): 4653-4663.

[2] Apweiler R, Bairoch A, and Wu C H. Protein sequence databases. *Current Opinion in Chemical Biology*, 2004, 8(1): 76-80.

[3] Hinz U and UniProt C. From protein sequences to 3D-structures and beyond: the example of the UniProt knowledgebase. *Cell Mol. Life. Sci.*, 2010, 67(7): 1049-1064.

[4] Gasteiger E, Jung E, and Bairoch A. SWISS-PROT: connecting biomolecular knowledge via a protein database. *Curr. Issues Mol. Biol.*, 2001, 3(3): 47-55.

[5] Schneider M, Tognolli M, and Bairoch A. The Swiss-Prot protein knowledgebase and ExPASy: providing the plant community with high quality proteomic data and tools. *Plant Physiol. Biochem.*, 2004, 42(12): 1013-1021.

[6] Perkins D N, et al. Probability-based protein identification by searching sequence databases using mass spectrometry data. *Electrophoresis*, 1999, 20(18): 3551-3567.

[7] Koenig T, et al. Robust prediction of the MASCOT score for an improved quality assessment in mass spectrometric proteomics. *J. Proteome Res.*, 2008, 7(9): 3708-3717.

[8] Pappin D J C, Hojrup P, and Bleasby A J. Rapid Identification of Proteins by Peptide-Mass Finger Printing (Vol. 3, Pg. 327, 1993). *Current Biology*, 1993, 3(7): 487.

[9] Fenyo D. Identifying the proteome: software tools. *Current Opinion in Biotechnology*, 2000, 11(4): 391-395.

[10] Eng J K, McCormack A L, and Yates J R. An approach to correlate tandem mass spectral data of peptides with amino acid sequences in a protein database. *J. Am. Soc. Mass Spectrom*, 1994, 5(11): 976-989.

[11] Griffin P R, et al. Direct database searching with MALDI-PSD spectra of peptides. *Rapid Commun. Mass Spectrom*, 1995, 9(15): 1546-1551.

[12] Sadygov R G, Cociorva D, and Yates 3rd J R. Large-scale database searching using tandem mass spectra: looking up the answer in the back of the book. *Nat. Methods*, 2004, 1(3): 195-202.

[13] Ma B, et al. PEAKS: powerful software for peptide de novo sequencing by tandem mass spectrometry. *Rapid Commun. Mass Spectrom*, 2003, 17(20): 2337-2342.

[14] Bairoch A, et al. The Universal Protein Resource (UniProt) 2009. *Nucleic Acids Research*, 2009(37): D169-D174.

[15] Hubbard T J P, et al. Ensembl. *Nucleic Acids Research*, 2009(37): D690-D697.

[16] Jones P, et al. PRIDE: a public repository of protein and peptide identifications for the proteomics community. *Nucleic Acids Res.*, 2006(34) (Database issue): D659-D663.

[17] McLaughlin T, et al. PepSeeker: a database of proteome peptide identifications for investigating fragmentation patterns. *Nucleic Acids Research*, 2006(34): D649-D654.

[18] Deutsch E W, Lam H, and Aebersold R. PeptideAtlas: a resource for target selection for emerging targeted proteomics workflows. *Embo. Reports*, 2008, 9(5): 429-434.

[19] Craig R, Cortens J P, and Beavis R C. Open source system for analyzing, validating, and storing protein identification data. *Journal of Proteome Research*, 2004, 3(6): 1234-1242.

[20] Listgarten J and Emili A. Statistical and computational methods for comparative proteomic profiling using liquid chromatography-tandem mass spectrometry. *Molecular & Cellular Proteomics*, 2005, 4(4): 419-434.

[21] Karpievitch Y V, Dabney A R, and Smith R D. Normalization and missing value imputation for label-free LC-MS analysis. *BMC Bioinformatics*, 2012, 13(Suppl 16): S5.

[22] Troyanskaya O, et al. Missing value estimation methods for DNA microarrays. *Bioinformatics*, 2001, 17(6): 520-525.

[23] Bo T H, Dysvik J, and Jonassen I L. Simpute: accurate estimation of missing values in microarray data with least squares methods. *Nucleic Acids Research*, 2004, 32(3).

[24] Wang X, et al. A hybrid approach to protein differential expression in mass spectrometry-based proteomics. *Bioinformatics*, 2012, 28(12): 1586-1591.

[25] Pendarvis K, et al. An automated proteomic data analysis workflow for mass spectrometry. *BMC Bioinformatics*, 2009, 10.

[26] Bin Goh W W and Wong L. Computational proteomics: designing a comprehensive analytical strategy. *Drug Discovery Today*, 2014, 19(3): 266-274.

[27] Roxas B A P and Li Q. Significance analysis of microarray for relative quantitation of LC/MS data in proteomics. *BMC Bioinformatics*, 2008, 9.

[28] Ashburner M, et al. Gene Ontology: tool for the unification of biology. *Nature Genetics*, 2000, 25(1): 25-29.

[29] Carbon S, et al. AmiGO: online access to ontology and annotation data. *Bioinformatics*, 2009, 25(2): 288-289.

[30] Benjamini Y and Hochberg Y. Controlling the false discovery rate - a practical and powerful approach to multiple testing. *Journal of the Royal Statistical Society Series B-Methodological*, 1995, 57(1): 289-300.

[31] Dennis G, et al. DAVID: Database for annotation, visualization, and integrated discovery. *Genome Biology*, 2003, 4(9).

[32] Malik R, et al. From proteome lists to biological impact–tools and strategies for the analysis of large MS data sets. *Proteomics*, 2010, 10(6): 1270-1283.

[33] Subramanian A, et al. Gene set enrichment analysis: A knowledge-based approach for interpreting genome-wide expression profiles. *Proceedings of the National Academy of Sciences of the United States of America*, 2005, 102(43): 15545-15550.

[34] Clutterbuck A L, et al. High throughput proteomic analysis of the secretome in an explant model of articular cartilage inflammation. *Journal of Proteomics*, 2011, 74(5): 704-715.

[35] Cha S, et al. In Situ Proteomic Analysis of Human Breast Cancer Epithelial Cells Using Laser Capture Microdissection: Annotation by Protein Set Enrichment Analysis and Gene Ontology. *Molecular & Cellular Proteomics*, 2010, 9(11): 2529-2544.

[36] Liberzon A, et al. Molecular signatures database (MSigDB) 3.0. *Bioinformatics*, 2011, 27(12): 1739-1740.

[37] Schmidt A, Forne I, and Imhof A. Bioinformatic analysis of proteomics data. *BMC Syst. Biol.*, 2014, 8 Suppl 2: S3.

[38] Kanehisa M, et al. Data, information, knowledge and principle: back to metabolism in KEGG. *Nucleic Acids Research*, 2014, 42(D1): D199-D205.

[39] Croft D, et al. Reactome: a database of reactions, pathways and biological processes. *Nucleic Acids Research*, 2011(39): D691-D697.

[40] Ficenec D, et al. Computational knowledge integration in biopharmaceutical research. *Brief Bioinform*, 2003, 4(3): 260-278.

[41] Thomas P D, et al. PANTHER: A library of protein families and subfamilies indexed by function. *Genome Research*, 2003, 13(9): 2129-2141.

[42] Salomonis N, et al. GenMAPP 2: new features and resources for pathway analysis. *BMC Bioinformatics*, 2007, 8.

[43] Schaefer C F, et al. PID: the Pathway Interaction Database. *Nucleic Acids Research*, 2009(37): D674-D679.

[44] Kandasamy K, et al. NetPath: a public resource of curated signal transduction pathways. *Genome Biology*, 2010, 11(1).

[45] Bader G D, Cary M P, and Sander C. Pathguide: a Pathway Resource List. *Nucleic Acids Research*, 2006(34): D504-D506.

[46] Ryan D P and Matthews J M. Protein-protein interactions in human disease. *Current Opinion in Structural Biology*, 2005, 15(4): 441-446.

[47] Prasad T S K, et al. Human Protein Reference Database-2009 update. *Nucleic Acids Research*, 2009(37): D767-D772.

[48] Kerrien S, et al. IntAct - open source resource for molecular interaction data. *Nucleic Acids Research*, 2007(35): D561-D565.

[49] Persico M, et al. HomoMINT: an inferred human network based on orthology mapping of protein interactions discovered in model organisms. *BMC Bioinformatics*, 2005, 6.

[50] Mewes H W, et al. MIPS: analysis and annotation of proteins from whole genomes. *Nucleic Acids Research*, 2004(32): D41-D44.

[51] Xenarios I, et al. DIP, the Database of Interacting Proteins: a research tool for studying cellular networks of protein interactions. *Nucleic Acids Research*, 2002, 30(1): 303-305.

[52] Breitkreutz B-J, et al. The BioGRID interaction database: 2008 update. *Nucleic Acids Research*, 2008(36): D637-D640.

[53] Prieto C and Rivas J D L. APID: Agile Protein Interaction Data Analyzer. *Nucleic Acids Research*, 2006(34): W298-W302.

[54] Brown K R and Jurisica I. Unequal evolutionary conservation of human protein interactions in interologous networks. *Genome. Biol.*, 2007, 8(5): R95.

[55] Jensen L J, et al. STRING 8-a global view on proteins and their functional interactions in 630 organisms. *Nucleic Acids Research*, 2009(37): D412-D416.

[56] Shannon P, et al. Cytoscape: A software environment for integrated models of biomolecular interaction networks. *Genome Research*, 2003, 13(11): 2498-2504.

[57] Breitkreutz B J, Stark C, and Tyers M. Osprey: a network visualization system. *Genome Biology*, 2003, 4(3).

[58] Hu Z J, et al. VisANT: an online visualization and analysis tool for biological interaction data. *BMC Bioinformatics*, 2004(5).

[59] Chua H N and Wong L. Increasing the reliability of protein interactomes. *Drug Discovery Today*, 2008, 13(15-16): 652-658.

[60] Srihari S, et al. Methods for Protein Complex Prediction and Their Contributions Towards Understanding the Organisation, Function and Dynamics of Complexes. *FEBS Lett*, 2015, 589(19 Pt A): 2590-2602.

[61] Orchard S, Hermjakob H, and Apweiler R. The Proteomics Standards Initiative. *Proteomics*, 2003, 3(7): 1374-1376.

[62] Huber L A. Is Proteomics Heading in the Wrong Direction. *Nature Reviews Molecular Cell Biology*, 2003, 4(1): 74-80.

[63] Perez-Riverol Y, et al. Making proteomics data accessible and reusable: current state of proteomics databases and repositories. *Proteomics*, 2015, 15(5-6): 930-949.

[64] Chen T, et al. Web resources for mass spectrometry-based proteomics. *Genomics Proteomics Bioinformatics*, 2015, 13(1): 36-39.

[65] Deutsch E W. The PeptideAtlas Project. *Methods Mol. Biol.*, 2010(604): 285-296.

[66] Kolker E, et al. MOPED: Model Organism Protein Expression Database. *Nucleic Acids Res.*, 2012, 40(Database issue): D1093-D1099.

[67] Smith B E, et al. Tranche distributed repository and ProteomeCommons.org. *Methods Mol. Biol.*, 2011(696): 123-145.

[68] Medina-Aunon J A, et al. A Guide for Integration of Proteomic Data Standards into Laboratory Workflows. *Proteomics*, 2013, 13(3-4): 480-492.

[69] Robin X, et al. MIAPEGelDB, a Web-Based Submission Tool and Public Repository for MIAPE Gel Electrophoresis Documents. *J. Proteomics*, 2008, 71(2): 249-251.

[70] Martinez-Bartolome S, et al. Semi-Automatic Tool to Describe, Store and Compare Proteomics Experiments Based on MIAPE Compliant Reports. *Proteomics*, 2010, 10(6): 1256-1260.

[71] Medina-Aunon J A, et al. The ProteoRed MIAPE Web Toolkit: a User-Friendly Framework to Connect and Share Proteomics Standards. *Mol. Cell Proteomics*, 2011, 10(10): M111 008334.

Chapter 9 Bioinformatics for Metabolomics

Yuxiang Wen[1], Ziling Wang[2]

9.1 Metabonomics and Metabolomics

9.1.1 Brief Introduction of Metabonomics and Metabolomics

Metabonomics or Metabolomics, is the comprehensive study of the metabolome, the repertoire of biochemicals (or small molecules) present in cells, tissues, and body fluids [1].

The small molecules involved in biochemical processes provide a great deal of information on the status and functioning of a living system under study both from effects caused by changes in gene expression, and also by differences in life style and diet in humans and other mammals. The process of monitoring and evaluating such changes is termed metabonomics. A parallel approach has also been under way, mostly in model organisms and in plant systems, and that has led to the term metabolomics. Metabonomics really grew out of work using NMR spectroscopy of biofluids going back to the mid-1980s, and which was subsequently combined with the use of pattern recognition and multivariate statistics investigation of the complex data sets. The term was not coined until much later and was formally defined in 1999 by Jeremy Nicholson and colleagues as 'the quantitative measurement of the dynamic multiparametric metabolic response of living systems to pathophysiological stimuli or genetic modification'. A little later, in 2001, the term metabolomics was introduced by Oliver Fiehn and defined somewhat differently as 'a comprehensive and quantitative analysis of all metabolites' in a system.

Although there remain some differences in concept, there is now a great

1. Wen Yuxiang, College of Life Sciences and Bioengineering, School of Science, Beijing Jiao Tong University, Beijing, China, 100044.

2. Wang Ziling, College of Life Sciences and Bioengineering, School of Science, Beijing Jiao Tong University, Beijing, China, 100044.

deal of overlap in the philosophies and methodologies, and the two terms are often used interchangeably by scientists and organisations [2]. Therefore we will use 'Metabonomics' in the subsequent content.

At the center of metabolomics, is the concept that a person's metabolic state provides a close representation of that individual's overall health status. This metabolic state reflects what has been encoded by the genome, and modified by diet, environmental factors, and the gut microbiome. The metabolic profile provides a quantifiable readout of biochemical state from normal physiology to diverse pathophysiologies in a manner that is often not obvious from gene expression analyses.

9.1.2 Metabonomics in Relation to Other 'Omics' Approaches

The Human Genome Project and the subsequent publication of a reference human genome sequence represent milestones in biology research in general and in cancer in particular. Since the 1990s, there has been a revolution in the techniques and approaches used in molecular biology. This followed from the decoding of the human and other genomes, and as a consequence, the principal emphasis in biomedical studies has now largely switched to simultaneous determination of gene expression changes between subjects mainly carried out using micro-array technology [4], and this type of study has been given the name of transcriptomics. An equivalent impetus to map out all protein expression changes in a cell or tissue has also subsequently emerged and has been termed proteomics. In analogy to genomics, transcriptomics and proteomics, the term metabolomics can be defined as the study of the complete ensemble of all small molecules (molecular weight (MW) < 1500 Da) formed by numerous biosynthetic and catabolic pathways within a biological system or originating from host-specific microbes and the intake of food nutrients and pharmaceuticals, which are present in a cell, tissue, or biofluids such as urine, blood, or saliva, in the context of a physiological or pathological condition [5].

Genes code for proteins. Epigenetics regulates gene expression. Transcriptomics regulates RNA synthesis in turn regulating protein synthesis. Proteomics deals with the protein synthesis and function. The most dynamic functions of proteins are served as enzymes taking part in various metabolic pathways, using up substrates to form end products. Each reaction is therefore governed by epigenomics, genomics, transcriptomics and proteomics. Thus, metabolomics as defined by Oliver Fiehn is 'A comprehensive analysis in which all the metabolites of a biological system are identified and quantified' [6]. Quantitatively analyzing each metabolite in samples of biological fluids is what metabolomics deals with, in a way being the integration of all 'omics'.

9.2 Basic Approaches to Study Metabonomics

9.2.1 Metabonomics Techniques

The revolution in the study of metabolites in the last 15 years and the development of the field of metabolomics have resulted from innovative advances in scientific instrumentation and advances in computational resources available. The continued development of chromatography coupled to mass spectrometry (MS) and nuclear magnetic resonance (NMR) spectroscopy have advanced our capabilities from monitoring only a small number of metabolites in a traditional hypothesis-testing study, to being able to simultaneously quantify hundreds to thousands of metabolites in a biological sample with an analysis time of less than 20 min (Cajka and Fiehn 2014; Fiehn 2016).

The main analytical techniques that are employed for metabonomic studies are based on nuclear magnetic resonance (NMR) spectroscopy and mass spectrometry (MS). The latter technique requires a pre-separation of the metabolic components using either gas chromatography (GC) after chemical derivatisation, or liquid chromatography (LC), with the newer method of ultra-high-pressure LC (UPLC) being used increasingly. The use of capillary electrophoresis (CE) coupled to MS has also shown some promises. Other more specialised techniques such as Fourier transform infra-red (FTIR) spectroscopy and arrayed electrochemical detection have been used in some cases [7]. The main limitation of the use of FTIR is the low level of detailed molecular identification that can be achieved, and indeed in the case quoted above, MS was also employed for metabolite identification. Similarly, although an array of coulometric detectors following high pressure LC (HPLC) separation does not identify compounds directly, the combination of retention time and redox properties can serve as a basis for database searching of libraries of standard compounds. The separation output can also be directed to a mass spectrometer for additional identification experiments [8].

All metabonomics studies result in complex multivariate data sets that require visualisation software and chemometric and bioinformatic methods for interpretation. The aim of these procedures is to produce biochemically based fingerprints that are of diagnostic or other classification value. A second stage, crucial in such studies, is to identify the substances causing the diagnosis or classification, and these become the combination of biomarkers that define the biological or clinical context.

9.2.2 Nuclear Magnetic Resonance Spectroscopy

Nuclear magnetic resonance (NMR) is a physical phenomenon in which nuclei in a magnetic field absorb and re-emit electromagnetic radiation. This energy is at a specific resonance frequency which depends on the strength of the magnetic field and the magnetic properties of the isotope of the atoms; in practical applications, the frequency is similar to VHF and UHF television broadcasts (60~1000 MHz). NMR allows the observation of specific quantum mechanical magnetic properties of the atomic nucleus. Many scientific techniques exploit NMR phenomena to study molecular physics, crystals, and non-crystalline materials through nuclear magnetic resonance spectroscopy. NMR is also routinely used in advanced medical imaging techniques, such as in magnetic resonance imaging (MRI). NMR detects molecules based on their magnetic properties. Multiple nuclei can be used for NMR analysis, including ^{1}H, ^{31}P, and ^{19}F with a natural abundance of 100% and isotopes such as ^{13}C and ^{15}N that require stable isotope labeling, because of their low natural abundance (1.1% and 0.4% respectively). ^{1}H NMR is the most predominant in the field and it produces high-resolution spectra with good sensitivity because of its inherent high sensitivity and its high abundance in biological systems [9].

NMR spectroscopy is a non-destructive technique, widely used in chemistry, that provides detailed information on molecular structure, both for pure compounds and in complex mixtures as well as information on absolute or relative concentrations [10]. The NMR spectroscopic methods can also be used to probe metabolite molecular dynamics and mobility as well as substance concentrations through the interpretation of NMR spin relaxation times and by the determination of molecular diffusion coefficients [11]. Automatic sample preparation is possible for NMR spectroscopy involving buffering and addition of D_2O as a magnetic field lock signal for the spectrometer and standard NMR spectra typically take only a few minutes to acquire using robotic flow-injection methods. For large-scale studies, bar-coded vials containing the biofluid can be used and the contents of these can be transferred and prepared for analysis using robotic liquid-handling technology into 96-well plates under LIMS system control. Currently, using such approaches, well over 100 samples per day can be measured on one spectrometer, each taking a total data acquisition time of only around 5 min. Alternatively, for more precious samples or for those of limited volume, conventional 5 mm or capillary NMR tubes are usually used, either individually or using a commercial sample tube changer and automatic data acquisition. A typical ^{1}H NMR spectrum of urine contains thousands of sharp lines from predominantly low molecular weight metabolites. The large interfering NMR signal arising from water in all biofluids is easily eliminated by use of appropriate standard NMR solvent suppression methods, either by secondary RF irradiation at the water peak chemical shift or by use of a specialised NMR pulse sequence that does not excite the water resonance. The position of each spectral band (known as its chemical shift and measured in frequency terms, in parts per million, from that of an added standard reference substance) gives information on molecular group identity and its molecular environment. The reference compound used in aqueous media is usually the sodium salt of 3-trimethylsilylpropionic acid (TSP) with the methylene groups deuterated to avoid giving rise to peaks in the ^{1}H NMR spectrum. The multiplicity of the splitting pattern on each NMR band and the magnitudes of the splittings (caused by nuclear spin–spin interactions mediated through the electrons of the chemical bonds, and known as J-coupling) provide knowledge about nearby protons, their throughbond connectivities, the relative orientation of nearby C–H bonds and hence also molecular conformations. The band areas relate directly to the number of protons giving rise to the peak and hence to the relative concentrations of the substances in the sample. Absolute concentrations can be obtained if the sample contains an added internal standard of known concentration, or if a standard addition of the analyte of interest is added to the sample, or if the concentration of a substance is known by independent means (e.g. glucose in plasma can be quantified by a conventional biochemical assay). Blood plasma and serum contain both low and high molecular weight components, and these give a wide range of signal line widths. Broad bands from protein and lipoprotein signals contribute strongly to the ^{1}H NMR spectra, with sharp peaks from small molecules superimposed on them [12]. Standard NMR pulse sequences, where the observed peak intensities are edited on the basis of molecular diffusion coefficients or on NMR relaxation times, can be used to select only the contributions from macromolecules, or alternatively to select only the signals from the small molecule metabolites, respectively. It is also possible to use these approaches to investigate molecular mobility and flexibility, and to study inter-molecular interactions such as the reversible binding between small molecules and proteins. Identification of biomarkers can involve the application of a range of techniques including two-dimensional NMR experiments. The ^{1}H NMR spectra of urine and other biofluids, even though they are very complex, allow many resonances to be assigned directly based on their chemical shifts, signal multiplicities, and by adding authentic material, further information can be obtained by using spectral editing and two-dimensional techniques. Two-dimensional NMR spectroscopy can be useful for increasing signal dispersion and for elucidating the connectivities between signals, thereby enhancing the information content and helping to identify biochemical substances. These include the ^{1}H–^{1}H 2-D J-resolved experiment, which attenuates the peaks from macromolecules and yields information on the multiplicity and coupling patterns of resonances, a good aid to molecule identification. The appropriate projection of such a spectrum on to the chemical shift axis yields a fingerprint of peaks from only the most highly mobile small molecules, with the added benefit that all of the spincoupling peak multiplicities have been removed. Other 2-D experiments such as correlation spectroscopy (COSY) and total correlation spectroscopy (TOCSY) provide ^{1}H–^{1}H spin–spin coupling connectivities, giving information on which hydrogens in a molecule are close in chemical bond terms. Use of other types of nuclei, such as naturally abundant ^{13}C or ^{15}N, or where present ^{31}P, can be important to help assign NMR peaks and here such heteronuclear correlation NMR experiments are achievable. These now benefit from the use of so-called inverse detection, where the lower sensitivity or less abundant nucleus NMR spectrum (such as ^{13}C) is detected indirectly using the more sensitive/abundant nucleus (^{1}H) by making use of spin–spin interactions such as the one-bond ^{13}C–^{1}H spin–spin coupling between the nuclei to effect the connection. These yield both ^{1}H and ^{13}C NMR chemical shifts of CH, CH_2 and CH_3 groups, useful again for identification purposes. There is also a sequence that allows correlation of protons to quaternary carbons based on long-range ^{13}C–^{1}H spin–spin coupling between the nuclei. A very useful recent advance in NMR technology has been the development of cryogenic probes where the detector coil and pre-amplifier (but not the samples) are cooled to around 20 K. This has provided an improvement in spectral signal–noise ratios of up to a factor of 5 by reducing the thermal noise in the electronics of the spectrometer. Conversely, because the NMR signal-to-noise ratio is proportional to the square root of the number of co-added scans, shorter data acquisition times by 14 Metabonomics and Metabolomics Overview up to a factor of 25 become possible for the same amount of sample. The NMR spectroscopy of biofluids detecting the much less sensitive ^{13}C nuclei which only have a natural abundance (1.1%) also becomes possible because of this reduction in spectrometer noise [13]. This technology also makes the use of tissue-specific microdialysis samples more feasible [14]. Within the last few years, the development of a technique called high resolution ^{1}H magic angle spinning (MAS) NMR spectroscopy has made feasible the acquisition of high resolution NMR data on small pieces of intact tissues with no pretreatment [15-17]. Rapid spinning of the sample (typically at 4∼6 kHz) at an angle of 54.7° relative to the applied magnetic field serves to reduce the loss of information caused by line broadening effects seen in non-liquid samples such as tissues. These broadenings are caused by sample heterogeneity, and residual anisotropic NMR parameters that are normally averaged out in free solution where molecules can tumble isotropically and rapidly. MAS NMR spectroscopy has straightforward, but manual, sample preparation. NMR spectroscopy on a tissue sample in an MAS experiment is the same as solution state NMR and all common pulse techniques can be employed in order to study metabolic changes and to perform molecular structure elucidation and molecular dynamics studies. Some typical ^{1}H NMR spectra are given in Figure 9.1 showing the varied profiles from mouse blood plasma.

In conclusion, NMR spectroscopy has been used extensively for Metabonomics studies, mainly for human health.

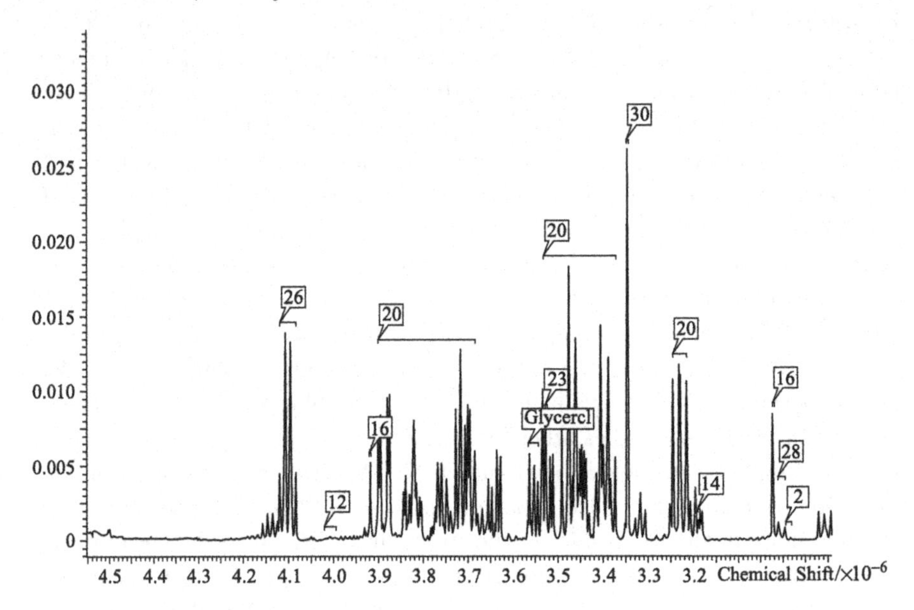

Fig. 9.1 NMR spectra of animal plasma samples (3.0×10^{-6}—5.0×10^{-6}). Different number represents different metabolites; metabolites are identified using Chenomx NMR suit (version 8.0, Chenomx, Edmonton, Canada) software for all ^{1}H NMR spectra and ppm range

9.2.3 Mass Spectrometry

Mass spectrometry (MS) has also been widely used in metabolic fingerprinting and metabolite identification as well as being a mainstay technique in the pharmaceutical industry for identification and quantitation of drug metabolites. Although most MS-based studies to date have been on plant extracts and model cell system extracts, its application to mammalian studies is increasing. In general, a prior separation of the complex mixture sample using chromatography is required. MS is inherently considerably more sensitive than NMR spectroscopy, but it is necessary generally to employ different separation techniques for different classes of substances. MS is also a major technique for molecular identification purposes, especially through the use of tandem MS methods for fragment ion studies or using Fourier transform MS for very accurate mass determination. Gas Chromatography-Mass Spectrometry (GC-MS) and Liquid Chromatography-Mass Spectrometr (LC-MS) are the most widely used in metabonomics [18-19]. Analyte quantitation by MS in complex mixtures of highly variable composition can be impaired by variable ionisation and ion suppression effects. For plant metabolic studies, most investigations have used chemical derivatisation to ensure volatility and analytical reproducibility, followed by GC-MS analysis. More recently, the use of GC-MS and GC-GC-MS has been exploited for mammalian metabonomics applications [20]. Some approaches using MS rely on more targeted studies, for example by detailed analysis of lipids [21]. For metabonomics applications on biofluids such as urine, an HPLC chromatogram generated with MS detection, usually using electrospray ionisation, and both positive and negative ion chromatograms can be measured. At each sampling point in the chromatgram, there is a full mass spectrum and so the data is three dimensional in nature, that is retention time, mass and intensity. Given this very high resolution, it is possible to cut out any mass peaks from interfering substances such as drug metabolites, without unduly affecting the integrity of the data set. Recently introduced, UPLC is a combination of a 1.7 μm reversed-phase packing material, and a chromatographic system, operating at around 12,000 psi. This has enabled better chromatographic peak resolution and increased speed and sensitivity to be obtained for complex mixture separation. UPLC provides around a 10-fold increase in speed and a threefold to fivefold increase in sensitivity compared to a conventional stationary phase. Because of the much improved chromatographic resolution of UPLC, the problem of ion suppression from co-eluting peaks is greatly reduced. UPLC-MS has already been used for metabolic profiling of urines from males and females of two groups of phenotypically normal mouse strains and a nude mouse strain [22]. Recently, CE coupled to mass spectrometry has also been explored as a suitable technology for metabonomics studies [23]. Metabolites are first separated by CE based on their charge and size and then selectively detected using MS monitoring. This method has been used to measure 352 metabolic standards and then employed for the analysis of 1,692 metabolites from *Bacillus subtilis* extracts, revealing changes in metabolite levels during the bacterial growth. For biomarker identification, it is also possible to separate out substances of interest on a larger scale from a complex biofluid sample using techniques such as solid-phase-extraction or HPLC. For metabolite identification, directly coupled chromatography-NMR spectroscopy methods can also be used. The most general of these 'hyphenated' approaches is HPLC-NMR-MS [24] in which the eluting HPLC peak is split, with parallel analysis by directly coupled NMR and MS techniques. This can be operated in on-flow, stopped-flow and loop-storage modes and thus can provide the full array of NMR and MS-based molecular identification tools. These include two-dimensional NMR spectroscopy as well as MS-MS for identification of fragment ions and Fourier transform-MS (FT-MS) or time-of-flight-MS (TOF-MS) for accurate mass measurement and hence derivation of molecular empirical formulae.

In summary, NMR and MS approaches are highly complementary, and use of both is often necessary for full molecular characterisation. MS can be more sensitive with lower detection limits provided the substance of interest can be ionised, but NMR spectroscopy is particularly useful for distinguishing isomers, for obtaining molecular conformation information and for studies of molecular dynamics and compartmentation, and given the now increasing use of cryoprobes, it is becoming ever more sensitive.

9.3 Data Analysis Methods

An NMR spectrum or a mass spectrum of a biofluid sample can be thought of as an object with a multi-dimensional set of metabolic coordinates, the values of which are the spectral intensities at each data point and the spectrum is therefore a point in a multi-dimensional metabolic hyperspace. The initial objective in metabonomics is to classify a spectrum based on identification of its inherent patterns of peaks and secondly to identify those spectral features responsible for the classification. The approach can also be used for reducing the dimensionality of complex data sets, for example by two-dimensional or three-dimensional mapping procedures, to enable easy visualisation of any clustering or similarity of the various samples.

9.3.1 Unsupervised Methods

Unsupervised machine learning is the machine learning task of inferring a function to describe hidden structure from 'unlabeled' data (a classification or categorization is not included in the observations). Since the examples given to the learner are unlabeled, there is no evaluation of the accuracy of the structure that is output by the relevant algorithm—which is one way of distinguishing unsupervised learning from supervised learning and reinforcement learning.

A central case of unsupervised learning is the problem of density estimation in statistics [25], though unsupervised learning encompasses many other problems (and solutions) involving summarizing and explaining key features of the data.

As we discussed before, metabonomics datas can be thought of as an object with a multi-dimensional set, thus many unsupervised machine learning techniques can be used for metabonomics, such as principal components analysis (PCA), hierarchical cluster analysis (HCA), Nonlinear mapping (NML) and so on.

One of the simplest unsupervised techniques that has been used extensively in metabonomics is principal components analysis (PCA). PCA is a statistical procedure that uses an orthogonal transformation to convert a set of observations of possibly correlated variables into a set of values of linearly uncorrelated variables called principal components (PC). The number of principal components is less than or equal to the smaller of the number of original variables or the number of observations. This transformation is defined in such a way that the first principal component has the largest possible variance, and each succeeding component in turn has the highest variance possible under the constraint that it is orthogonal to the preceding components. The resulting vectors are an uncorrelated orthogonal basis set. PCA is sensitive to the relative scaling of the original variables. Conversion of the data matrix to PC results in two matrices known as scores and loadings. Scores, the linear combinations of the original variables, are the coordinates for the samples in the established model and may be regarded as the new variables. In a scores plot, each point represents a single sample spectrum. The PC loadings define the way in which the old variables are linearly combined to form the new variables and indicate those variables carrying the greatest weight in transforming the position of the original samples from the data matrix into their new position in the scores matrix. In the loadings plot, each point represents a different spectral intensity. Thus the cause of any spectral clustering observed in a PC scores plot is interpreted by examination of the loadings that cause any cluster separation.

Here we will give an example of PCA to show the main processes of PCA. Figure 9.2 list the data (part) of 59 metabolites' absolute concentration (measured by NMR) of totally 176 human plasma samples which belong to two separate groups. Group H represents healthy group, and the group P represents non-small cell lung cancer patients. Before operate PCA, the data should be normalized, the common normalization methods include Min-max normalization, Decimal scaling and z-score normalization. Z-score is used in this case, Figure 9.3 list the data normalized by z-score.

Then the normalized data is imported to the Matlab R2015a (The MathWorks, Inc.) for further analyze. Then import the origin data to the Matlab as a numeric matrix, and named the matrix as X. The PCA can be carried out by the codes:

		2-Hydroxy	2-Hydroxy	2-Oxoglut	2-Oxoisoc	3-Hydroxy	3-Hydroxy	Acetate	Acetoacet	Acetone	Adenine	Alanine	Arginine	Aspartate	Betaine	Cadaverir	Carnitine
410884	H	0.0211	0.0076	0.0096	0.0033	0.0146	0.0139	0.0514	0.0076	0.0113	0.1518	0.3674	0.0909	0	0.0558	0.0481	0.0254
410885	H	0.0264	0.0097	0	0.0032	0.0311	0.0172	0.1077	0.0141	0.0214	0.103	0.2824	0.0521	0	0.065	0.025	0.0363
410886	H	0.0231	0.017	0.0099	0.0033	0.0187	0.0129	0.0512	0.0161	0.0172	0.0523	0.3698	0.0462	0	0.0552	0.0258	0.0249
410887	H	0.0252	0.0068	0.0107	0.0034	0.0273	0.0206	0.0788	0.0111	0.0188	0.1336	0.3061	0.0541	0	0.0459	0.0309	0.0276
410888	H	0.0194	0.0102	0.0049	0.0031	0.0126	0.0155	0.0761	0.0108	0.0119	0.0907	0.3607	0.05	0	0.0581	0.0202	0.0467
410889	H	0.0127	0.0111	0.0054	0.0028	0.0146	0.0116	0.0597	0.0058	0.0128	0.0989	0.3233	0	0	0.0442	0.0255	0.0281
410890	H	0.0255	0.0066	0.0105	0	0.0185	0.0099	0.0404	0.0113	0.0141	0.1047	0.3565	0	0	0.0462	0.0359	0.0182
410891	H	0.0228	0.0048	0.0127	0.0025	0.0161	0.0178	0.0743	0.0091	0.0149	0.1132	0.2845	0.043	0	0.0429	0.0195	0.0358
410892	H	0.0151	0.0071	0.0056	0.0022	0.0114	0.011	0.0696	0.0084	0.0124	0.1138	0.2617	0.0976	0	0.0593	0.0252	0.0344
410893	H	0.0156	0.0098	0.006	0.0032	0.0232	0.018	0.0444	0.0075	0.0137	0.0035	0.3166	0.0634	0	0.0437	0.0277	0.0202
410894	H	0.0166	0.0062	0.0055	0.0037	0.0243	0.0181	0.0698	0.0127	0.0189	0.0837	0.2707	0.0824	0	0.0447	0.028	0.0211
410895	H	0.0126	0.0079	0.0081	0.0025	0.0211	0.0129	0.0538	0.0104	0.0204	0.0885	0.2289	0.0616	0	0.0459	0.0224	0.0296
410896	H	0.0206	0.0095	0.0094	0.0052	0.0171	0.0191	0.0493	0.0082	0.0113	0.0822	0.4106	0.0663	0	0.0526	0.0298	0.0345
410980	P	0.0386	0.0092	0.025	0.0037	0.0171	0.0089	0.0115	0	0.0168	0	0.386	0.0587	0.0108	0.0342	0.0477	0.0396
410981	P	0.0283	0.0089	0.0098	0.0036	0.0146	0.0097	0.0157	0	0.015	0	0.4112	0.0561	0.0113	0.0365	0.0613	0.0328
410982	P	0.0547	0.0089	0.0115	0.0036	0.143	0.0091	0.0221	0.0193	0.0937	0.0079	0.1395	0.0514	0.0159	0.0835	0.0418	0.0287
410983	P	0.0652	0.0069	0.0113	0.0048	0.0272	0.0148	0.022	0.0041	0.0213	0	0.4551	0.0489	0.0057	0.0384	0.0244	0.0427
410984	P	0.0263	0.0034	0.0074	0.0029	0.0101	0.0065	0.018	0.0031	0.0099	0	0.2985	0.0429	0.014	0.0597	0.0137	0.0363
410985	P	0.0381	0.0096	0.0066	0.0027	0.0198	0.0081	0.0156	0	0.014	0.0432	0.523	0.0419	0.0141	0.0398	0.0365	0.0368
410986	P	0.0411	0.0072	0.0144	0.0028	0.026	0.0088	0.0199	0	0.0304	0	0.2526	0.0513	0.0057	0.0415	0.0452	0.0564
410987	P	0.0269	0.0049	0.0123	0.0024	0.0081	0.0114	0.0162	0	0.0131	0.0154	0.4711	0.0438	0.0076	0.0427	0.0427	0.0403
410988	P	0.0325	0.0058	0.0082	0.0034	0.013	0.0055	0.0541	0	0.0164	0.0873	0.3621	0.0308	0.0131	0.039	0.0247	0.0302
410989	P	0.0395	0.0049	0.008	0.0039	0.012	0.0094	0.0325	0	0.0115	0.0058	0.4011	0.0344	0.0063	0.0605	0.0445	0.04
410990	P	0.0208	0.0035	0.0053	0.0017	0.0122	0.0072	0.0513	0	0.0146	0	0.3056	0.0278	0.0075	0.0458	0.0325	0.0364
410991	P	0.0334	0.0062	0.0076	0.003	0.0309	0.0083	0.0718	0	0.0241	0	0.2147	0.0352	0	0.0827	0.0315	0.0187
410992	P	0.0369	0.0035	0.0061	0.005	0.0125	0.013	0.3354	0	0.0132	0	0.4459	0.0549	0.0063	0.0544	0.0275	0.034

Fig. 9.2 Part of 59 metabolites absolute concentration (measured by NMR) of totally 176 human plasma samples which belong two separate groups (group H and group P)

		2-Hydroxy	2-Hydroxy	2-Oxoglut	2-Oxoisoc	3-Hydroxy	3-Hydroxy	Acetate	Acetoacet	Acetone	Adenine	Alanine	Arginine	Aspartate	Betaine	Cadaverir	Carnitine(
410884	H	-0.7431	0.380517	-0.27965	-0.35116	-0.37229	0.548095	0.134149	1.096474	-0.57232	3.504346	-0.13955	1.910917	-0.70508	0.26822	0.35427	-0.72967
410885	H	-0.37812	1.081259	-0.7702	-0.37361	0.094742	1.297997	1.220675	2.584629	0.031847	2.214043	-0.68424	0.164518	-0.70508	0.87734	-0.59813	0.419874
410886	H	-0.60537	3.517174	-0.26432	-0.35116	-0.25624	0.320852	0.130289	3.042523	-0.21939	0.873502	-0.12417	-0.10104	-0.70508	0.228495	-0.56515	-0.7824
410887	H	-0.46076	0.113567	-0.22344	-0.32871	-0.01282	2.070623	0.662938	1.897788	-0.12368	3.023126	-0.53237	0.254538	-0.70508	-0.38725	-0.35488	-0.49765
410888	H	-0.86017	1.248103	-0.51981	-0.39606	-0.4289	0.911684	0.610831	1.829104	-0.53643	1.888823	-0.18249	0.069996	-0.70508	0.4205	-0.79603	1.516689
410889	H	-1.32155	1.548421	-0.49427	-0.46341	-0.37229	0.025436	0.29433	0.684369	-0.4826	2.105636	-0.42215	-2.18052	-0.70508	-0.4998	-0.57752	-0.44492
410890	H	-0.4401	0.04683	-0.23366	-1.092	-0.2619	-0.36088	-0.07814	1.943578	-0.40483	2.258992	-0.2094	-2.18052	-0.70508	-0.36738	-0.14873	-1.489
410891	H	-0.62603	-0.55381	-0.12124	-0.53076	-0.32984	1.434343	0.576093	1.439894	-0.35698	2.483737	-0.67078	-0.24508	-0.70508	-0.58587	-0.8249	0.367143
410892	H	-1.15628	0.213673	-0.48405	-0.59811	-0.46287	-0.11091	0.485388	1.279631	-0.50652	2.499602	-0.81689	2.212486	-0.70508	0.499951	-0.58989	0.219495
410893	H	-1.12185	1.114628	-0.46361	-0.37361	-0.12887	1.479791	-0.00094	1.073579	-0.42876	-0.4168	-0.46508	0.673134	-0.70508	-0.5329	-0.48681	-1.27808
410894	H	-1.05298	-0.08664	-0.48916	-0.26136	-0.09773	1.502516	0.489248	2.264103	-0.1177	1.703738	-0.75921	1.528329	-0.70508	-0.4667	-0.47444	-1.18316
410895	H	-1.32844	0.480623	-0.3563	-0.53076	-0.18831	0.320852	0.180466	1.737525	-0.02797	1.830653	-1.02707	0.592115	-0.70508	-0.38725	-0.70533	-0.28673
410896	H	-0.77753	1.014522	-0.28987	0.075385	-0.30153	1.729759	0.093621	1.233842	-0.57232	1.664077	0.137275	0.803664	-0.70508	0.056353	-0.40023	0.230041
410980	P	0.46201	0.914416	0.507272	-0.26136	-0.30153	-0.58812	-0.63588	-0.64352	-0.24332	-0.50934	-0.02036	0.461586	0.264336	-1.16189	0.337778	0.767902
410981	P	-0.24728	0.81431	-0.26943	-0.28381	-0.37229	-0.40633	-0.55482	-0.64352	-0.35099	-0.50934	0.14112	0.344559	0.309217	-1.00961	0.8985	0.050754
410982	P	1.57071	0.81431	-0.18256	-0.28381	3.262095	-0.54267	-0.43131	3.775154	4.356747	-0.30046	-1.59995	0.133011	0.722118	2.1022	0.094523	-0.69803
410983	P	2.293776	0.146936	-0.19278	-0.01441	-0.01565	0.752613	-0.43324	0.295159	0.025865	-0.50934	0.422434	0.020485	-0.19345	-0.88381	-0.62287	1.094837
410984	P	-0.38501	-1.02097	-0.39207	-0.44096	-0.49967	-1.1335	-0.51043	0.066213	-0.65607	-0.50934	-0.58107	-0.24958	0.551571	0.526434	-1.06403	0.419874
410985	P	0.427579	1.047891	-0.43295	-0.48586	-0.22511	-0.76991	-0.55675	-0.64352	-0.41081	0.632893	0.857542	-0.29459	0.560548	-0.79112	-0.12399	0.472606
410986	P	0.634169	0.247042	-0.03438	-0.46341	-0.04961	-0.61084	-0.47377	-0.64352	0.570216	-0.50934	-0.8752	0.128509	-0.19345	-0.67856	0.234704	2.53968
410987	P	-0.34369	-0.52044	-0.14168	-0.55321	-0.55628	-0.02001	-0.54517	-0.64352	-0.46465	-0.10216	0.524963	-0.20907	-0.0229	-0.59911	0.13163	0.841726
410988	P	0.041944	-0.22012	-0.35119	-0.32871	-0.41758	-1.36075	0.186256	-0.64352	-0.26725	1.798925	-0.17352	-0.7942	0.470787	-0.84409	-0.6105	-0.22345
410989	P	0.523988	-0.52044	-0.36141	-0.21646	-0.44589	-0.4745	-0.2306	-0.64352	-0.56036	-0.35599	0.076398	-0.63216	-0.13959	0.579401	0.205843	0.810087
410990	P	-0.76376	-0.9876	-0.49938	-0.71036	-0.44023	-0.97443	0.132219	-0.64352	-0.37492	-0.50934	-0.53557	-0.92923	-0.03188	-0.39387	-0.28891	0.430421
410991	P	0.103921	-0.08664	-0.38185	-0.41851	0.089081	-0.72447	0.527846	-0.64352	0.193358	-0.50934	-1.11807	-0.59616	-0.70508	2.049233	-0.33014	-1.43627
410992	P	0.344943	-0.9876	-0.4585	0.030486	-0.43173	0.343576	5.615026	-0.64352	-0.45867	-0.50934	0.36348	0.290547	-0.13959	0.175528	-0.49506	0.17731

Fig. 9.3 Part of 59 metabolites data after z-score normalization of totally 176 human plasma samples which belong two separate groups (group H and group P)

```
for i=1:59
A(:,i)=zscore(X(:,i));
[COEFF,SCORE, latent]=princomp(A);
cumsum(latent)./sum(latent)
plot(SCORE(:,1),SCORE(:,2),'.')
xlabel('1st Principal Component')
ylabel('2nd Principal Component')
plot3(SCORE(:,1),SCORE(:,2),SCORE(:,3),'.')
xlabel('1st Principal Component')
ylabel('2nd Principal Component')
zlabel('3rd Principal Component')
plot(SCORE(1:55,1),SCORE(1:55,2),'r.')
hold on
```

```
plot(SCORE(56:85,1),SCORE(56:85,2),'g.')
hold on
plot(SCORE(86:176,1),SCORE(86:176,2),'r.')
xlabel('1st Principal Component')
ylabel('2nd Principal Component')
legend('P ', 'H')
```

Performing this code can get the result of PCA and plot of PC scores.

A clear classification between group H and Group P is shown in Figure 9.4, the other results of PCA include COEFF, SCORE and latent. COEFF = princomp(X) performs principal components analysis (PCA) on the n-by-p data matrix X, and returns the principal component coefficients, also known as loadings. Rows of X correspond to observations, columns to variables. COEFF is a p-by-p matrix, each column containing coefficients for one principal component. The columns are in order of decreasing component variance. [COEFF, SCORE] = princomp(X) returns SCORE, the principal component scores; that is, the representation of X in the principal component space. Rows of SCORE correspond to observations, columns to components. [COEFF, SCORE, latent]= princomp(X) returns latent, a vector containing the eigenvalues of the covariance matrix of X.

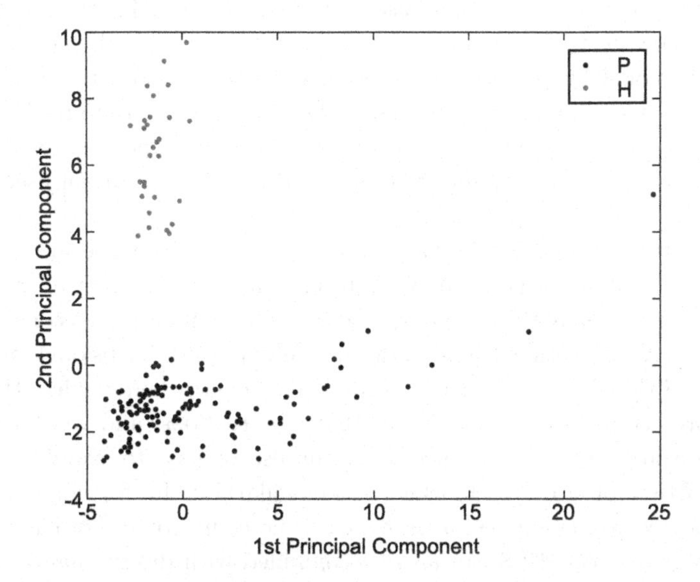

Fig. 9.4 Group H vs group P multivariate statistical analysis. PCA score plot (PC1 vs PC2) obtained by performing PCA on the NMR data

9.3.2 Supervised Methods

Supervised learning is the machine learning task of inferring a function from labeled training data [26]. The training data consist of a set of training examples. In supervised learning, each example is a pair consisting of an input object (typically a vector) and a desired output value (also called the supervisory signal). A supervised learning algorithm analyzes the training data and produces an inferred function, which can be used for mapping new examples. An optimal scenario will allow for the algorithm to correctly determine the class labels for unseen instances. This requires the learning algorithm to generalize from the training data to unseen situations in a reasonable way.

As we discussed before, metabonomics datas can be thought of as an object with a multi-dimensional set, thus many supervised machine learning techniques can be used for metabonomics, such as partial least squares (PLS) and Partial least squares Discriminant Analysis (PLS-DA), k-Nearest Neighbour (kNN), Artificial neural networks (ANN) and so on.

One widely used supervised method (i.e. using a training set of data with known outcomes) is partial least squares (PLS) [27]. Partial least squares regression (PLS regression) is a statistical method that bears some relation to principal components regression; instead of finding hyperplanes of maximum variance between the response and independent variables, it finds a linear regression model by projecting the predicted variables and the observable variables to a new space. Because both the X and Y data are projected to new spaces, the PLS family of methods are known as bilinear factor models. Partial least squares Discriminant Analysis (PLS-DA) is a variant used when the Y is categorical.

PLS is used to find the fundamental relations between two matrices (X and Y), i.e. a latent variable approach to modeling the covariance structures in these two spaces. A PLS model will try to find the multidimensional direction in the X space that explains the maximum multidimensional variance direction in the Y space. PLS regression is particularly suited when the matrix of predictors has more variables than observations, and when there is multicollinearity among X values. PLS can also be used to examine the influence of time on a data set, which is particularly useful for biofluid NMR data collected from samples taken over a time course of the progression of a pathological effect. PLS can also be combined with discriminant analysis (DA) to establish the optimal position to place a discriminant surface which best separates classes. It is possible to use such supervised models to provide classification probabilities and quantitative response factors for a wide range of sample types, but given the strong possibility of chance correlations when the number of descriptors is large, it is important to build and test such chemometric models using independent training data and validation data sets, cross-validation and permutations are usually used to test the validation of PLS models.

Here we will give an example of PLS-DA to show the main processes of PLS-DA. Figure 9.5 list the data (part) of 59 metabolites' absolute concentration (measured by NMR) of totally 146 non-small cell lung cancer patients' plasma samples which belong to two separate groups. Group A represents untreated group, the group A have not received any therapy yet, and the group B represents treated group, the group B had some chemotherapy to treat the disease for a period of time. Before operating PLS-DA, the data should be normalized, z-score is used in this case, Figure 9.6 list the data normalized by z-score.

		2-Hydroxy	2-Hydroxy	2-Oxoglut	2-Oxoisoc	3-Hydroxy	3-Hydroxy	Acetate	Acetoacet	Acetone	Adenine	Alanine	Arginine	As
410930	A	0.0208	0.0056	0.0092	0.0014	0.0144	0.0096	0.0256	0	0.0204	0	0.2632	0.0664	
410931	A	0.0343	0.0052	0.0102	0.004	0.0265	0.0121	0.0136	0	0.0233	0	0.3652	0.0546	
410932	A	0.026	0.0064	0.0136	0.003	0.0298	0.0094	0.0225	0	0.024	0	0.2437	0.0646	
410933	A	0.0412	0.0094	0.0139	0.0068	0.0244	0.0138	0.0228	0	0.0239	0	0.4287	0.031	
410934	A	0.0247	0.0047	0.0109	0.0042	0.0122	0.0109	0.0245	0.0023	0.0095	0	0.499	0.0479	
410935	A	0.0288	0.0109	0.0099	0.0038	0.031	0.0101	0.0146	0.0064	0.0281	0	0.2487	0.0411	
410936	A	0.0356	0.0052	0.0083	0.0025	0.0194	0.0176	0.0179	0.0024	0.0191	0	0.3133	0.048	
410937	A	0.0477	0.0078	0.02	0.0035	0.0228	0.0184	0.0253	0.0023	0.0196	0	0.4978	0.0388	
410938	A	0.0503	0.0046	0.006	0.0029	0.0169	0.0165	0.023	0.0027	0.0153	0	0.3547	0.0506	
410939	A	0.0343	0.0042	0.0074	0.0026	0.0098	0.0127	0.0228	0	0.0162	0	0.2583	0.0465	
410940	A	0.0496	0.0078	0.0147	0.0052	0.0371	0.0124	0.0221	0	0.0266	0	0.452	0.0499	
410941	A	0.0397	0.0188	0.1763	0.0445	0.0218	0.021	0.3232	0.0056	0.0396	0	1.4097	0.1438	
410942	A	0.0533	0.0068	0.0054	0.0035	0.1244	0.0138	0.0175	0.0021	0.0784	0	0.2605	0.0637	
410943	A	0.0339	0.0091	0.0073	0.0044	0.0144	0.0084	0.0239	0.0014	0.0119	0	0.4748	0.0265	
411016	B	0.0471	0.0077	0.0459	0.0085	0.0279	0.0135	0.0337	0	0.0299	0	0.4806	0.027	
411033	B	0.059	0.0067	0.0133	0.005	0.0297	0.0173	0.018	0	0.0272	0	0.2688	0.0459	
411034	B	0.0273	0.0018	0.0085	0.0049	0.0205	0.0091	0.0292	0	0.0192	0	0.2953	0.0536	
411035	B	0.0567	0.0049	0.012	0.0051	0.0255	0.0132	0.0241	0	0.0187	0	0.3316	0.0361	
411036	B	0.0425	0.0036	0.0156	0.0031	0.0875	0.0078	0.016	0.0076	0.0314	0	0.2566	0.0353	
411037	B	0.0304	0.0052	0.0236	0.0088	0.0145	0.0059	0.0211	0	0.0152	0	0.7598	0.0346	
411038	B	0.0133	0.0037	0.0066	0.0025	0.0483	0.0064	0.0331	0	0.0258	0	0.2231	0.0254	
411039	B	0.0113	0.0021	0.0083	0.0021	0.0135	0.0045	0.0835	0	0.0106	0	0.338	0.0351	
411040	B	0.0291	0.0022	0.0058	0.0021	0.0177	0.0043	0.0166	0	0.0096	0.0729	0.2753	0.0283	
411041	B	0.0358	0.0036	0.0115	0.0042	0.0186	0.0125	0.0228	0.0035	0.0124	0	0.4043	0.04	
411042	B	0.0277	0.0037	0.0087	0.0042	0.0137	0.0104	0.075	0	0.0097	0.0025	0.4423	0.0354	
411043	B	0.0321	0.0075	0.0465	0.0116	0.0225	0.0107	0.0961	0	0.0179	0	0.7587	0.0311	

Fig. 9.5 Part of 59 metabolites absolute concentration (measured by NMR) of totally 146 human plasma samples which belong two separate groups (group A and group B)

Then the normalized data is imported to the software SIMCA-P (Umetricto.Inc.) for further analyze. The task consists of the following steps:

- First, creat a new blank regular project and import the normalized data, then we will get an unfitted PCA-X model (Figure 9.7).
- Second, click the right mouse button on the unfitted PCA-X model and select 'Edit model' (Figure 9.8).
- Third, select the fitted parameters of variables and observations (Figure 9.9).
- At last, select the PLS-DA model and finish (Figure 9.10).

After these steps, SIMCA-P will autofit the PLS-DA model and output the summary of fit plot which shows us cross validation correlation coefficients Q2=0.473 and R2=0.687 (Figure 9.11). R2 is the percent of variation of the training set – Y with PLS – explained by the model. R2 is a measure of

		2-Hydroxy	2-Hydroxy	2-Oxoglut	2-Oxoisoc	3-Hydroxy	3-Hydroxy	Acetate	Acetoacet	Acetone	Adenine	Alanine	Arginine	Aspartate	Betaine	C:
410930	A	-1.00187	-0.10697	-0.3471	-0.79709	-0.40196	-0.33594	-0.28045	-0.45745	-0.11093	-0.3735	-0.83193	1.211245	-0.19573	-0.05826	-
410931	A	-0.04053	-0.25361	-0.29985	-0.25513	-0.08686	0.218645	-0.4954	-0.45745	0.051197	-0.3735	-0.22048	0.57388	-0.00597	-0.95879	0
410932	A	-0.63158	0.186313	-0.13919	-0.46358	-0.00093	-0.38031	-0.33598	-0.45745	0.090333	-0.3735	-0.94882	1.11402	0.244177	-0.86011	-
410933	A	0.450821	1.28611	-0.12502	0.328518	-0.14155	0.595767	-0.33061	-0.45745	0.084742	-0.3735	0.160174	-0.70085	-0.1526	0.540043	-
410934	A	-0.72415	-0.43691	-0.26677	-0.21344	-0.45925	-0.04756	-0.30016	0.238571	-0.72033	-0.3735	0.581592	0.211986	0.425316	-0.78609	-
410935	A	-0.43219	1.836009	-0.31402	-0.29682	0.030322	-0.22503	-0.47749	1.479307	0.319553	-0.3735	-0.91885	-0.15531	-0.10948	-0.2433	-
410936	A	0.052042	-0.25361	-0.38962	-0.5678	-0.27175	1.438744	-0.41838	0.268833	-0.18361	-0.3735	-0.5316	0.217388	-0.16123	-0.90328	-
410937	A	0.91369	0.699551	0.163208	-0.35936	-0.18321	1.616213	-0.28583	0.238571	-0.15566	-0.3735	0.574398	-0.27954	-0.05772	0.219304	0
410938	A	1.098837	-0.47357	-0.4983	-0.48442	-0.33685	1.194724	-0.32702	0.359619	-0.39606	-0.3735	-0.28342	0.357824	-0.81678	-0.47152	-
410939	A	-0.04053	-0.62021	-0.43215	-0.54696	-0.52174	0.351747	-0.33061	-0.45745	-0.34575	-0.3735	-0.8613	0.136367	-0.27336	-0.31732	-
410940	A	1.04899	0.699551	-0.08722	-0.005	0.189171	0.285196	-0.34315	-0.45745	0.235692	-0.3735	0.299847	0.320014	1.469023	-1.53859	-
410941	A	0.344005	4.732141	7.548463	8.18696	-0.20925	2.192987	5.050291	1.237212	0.962489	-0.3735	6.04084	5.391927	-0.81678	0.558547	1
410942	A	1.312469	0.332952	-0.52665	-0.35936	2.462537	0.595767	-0.42554	0.178047	3.131697	-0.3735	-0.84811	1.065407	1.581156	2.519989	-
411033	B	1.718369	0.296292	-0.15337	-0.04669	-0.00353	1.372193	-0.41659	-0.45745	0.269236	-0.3735	-0.79836	0.103958	-0.81678	-0.54554	-
411034	B	-0.53901	-1.50004	-0.38017	-0.06753	-0.24311	-0.44686	-0.21597	-0.45745	-0.17802	-0.3735	-0.6395	0.519866	-0.81678	-0.23096	-
411035	B	1.554585	-0.36359	-0.2148	-0.02584	-0.1129	0.462665	-0.30732	-0.45745	-0.20598	-0.3735	-0.4219	-0.42538	-0.81678	-0.18162	-
411036	B	0.543395	-0.84016	-0.04469	-0.44274	1.50163	-0.73525	-0.45241	1.842449	0.504048	-0.3735	-0.87149	-0.46859	-0.81678	-0.27414	-
411037	B	-0.31825	-0.25361	0.33331	0.745411	-0.39935	-1.15674	-0.36106	-0.45745	-0.40165	-0.3735	2.144974	-0.5064	3.116526	0.108279	
411038	B	-1.53595	-0.8035	-0.46995	-0.5678	0.480828	-1.04582	-0.14611	-0.45745	0.190966	-0.3735	-1.07231	-1.00333	-0.28199	-0.379	-
411039	B	-1.67837	-1.39006	-0.38962	-0.65118	-0.42539	-1.46731	0.756678	-0.45745	-0.65883	-0.3735	-0.38353	-0.47939	-0.81678	-0.2248	-
411040	B	-0.41083	-1.3534	-0.50775	-0.65118	-0.31602	-1.51168	-0.44166	-0.45745	-0.71473	2.991154	-0.75939	-0.84669	-0.19573	-0.2433	-
411041	B	0.066284	-0.84016	-0.23842	-0.21344	-0.29259	0.30738	-0.33061	0.601713	-0.55819	-0.3735	0.013907	-0.21472	-0.81678	-0.39133	-
411042	B	-0.51052	-0.8035	-0.37072	-0.21344	-0.42019	-0.15848	0.604422	-0.45745	-0.70914	-0.25812	0.2417	-0.46319	-0.81678	-0.98347	-
411043	B	-0.19719	0.589572	1.415347	1.329062	-0.19103	-0.09193	0.982375	-0.45745	-0.2507	-0.3735	2.13838	-0.69545	2.107322	0.990311	-
411044	B	-0.06902	0.003013	-0.06832	-0.02584	-0.26134	0.640134	0.28379	-0.45745	-0.30102	-0.26735	-0.56157	-0.97092	-0.81678	0.521539	-
411045	B	-0.56749	0.186313	1.561824	-0.005	-0.35769	-0.11411	-0.53123	-0.45745	-0.53583	-0.3735	1.075544	-0.86829	0.408065	-1.00197	-

Fig. 9.6 Part of 59 metabolites data after z-score normalization of totally 146 human plasma samples which belong two separate groups (group A and group B)

fit, i.e. how well the model fits the data. A large R2 (close to 1) is a necessary condition for a good model, but it is not sufficient. We can have poor models (models that cannot predict) even with a large R2. We will get a poor R2 when you have poor reproducibility (much noise) in the training data set, or when for other reasons X does not explain Y. Q2 is the percent of variation of the training set – Y with PLS – predicted by the model according to cross validation. Q2 indicates how well the model predicts new data. A large Q2 (Q2 > 0.5) indicates good predictivity. We will get a poor Q2 when the data have much noise, or when the relationship $X- > Y$ is poor, or when the model is dominated by a few scattered outliers. Because of the not very sat-

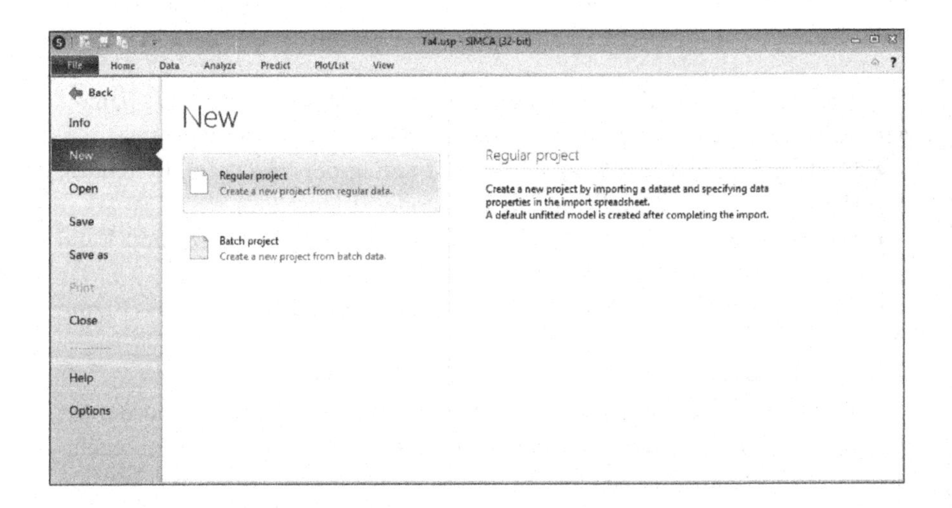

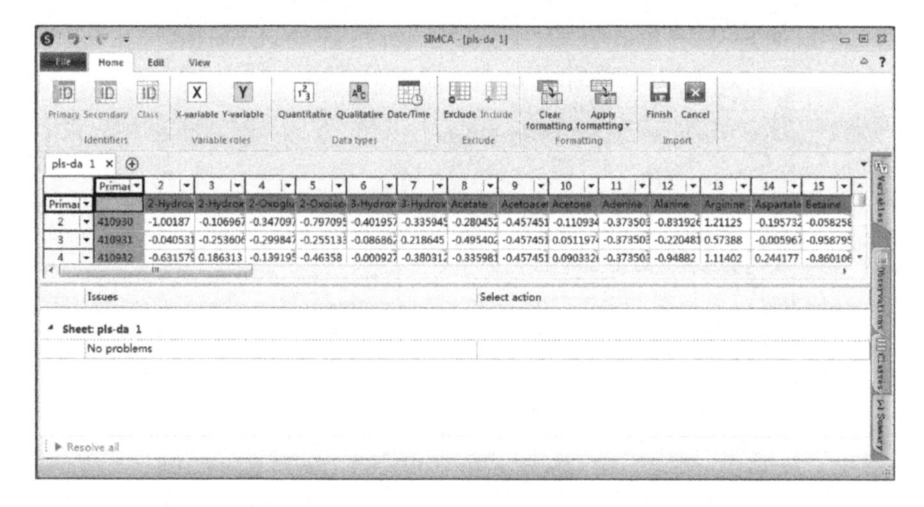

Fig. 9.7 Creat a new blank regular project and import the normalized data to SIMCA-P

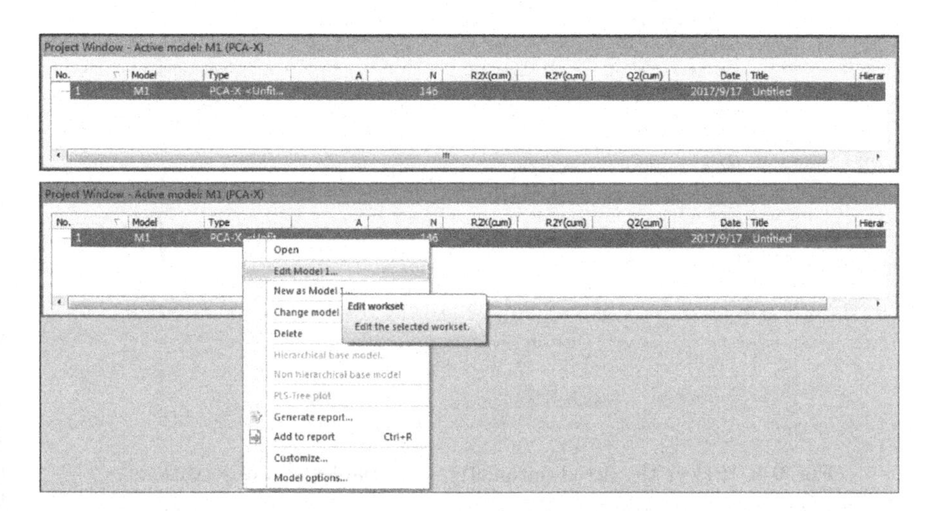

Fig. 9.8 Click the right mouse button on the unfitted PCA-X model and select Edit model

isfied cross validation correlation coefficients, further test of model validation is necessary, permutation is one common used method. By using the function permutations in the Analyze selection, we can obtain the permutations plot of PLS-DA model. The Permutations Plot (Figure 9.12) helps to assess the risk that the current PLS or PLS-DA model is spurious, i.e., the model just fits the training set well but does not predict Y well for new observations. The idea of this validation is to compare the goodness of fit (R2 and Q2) of

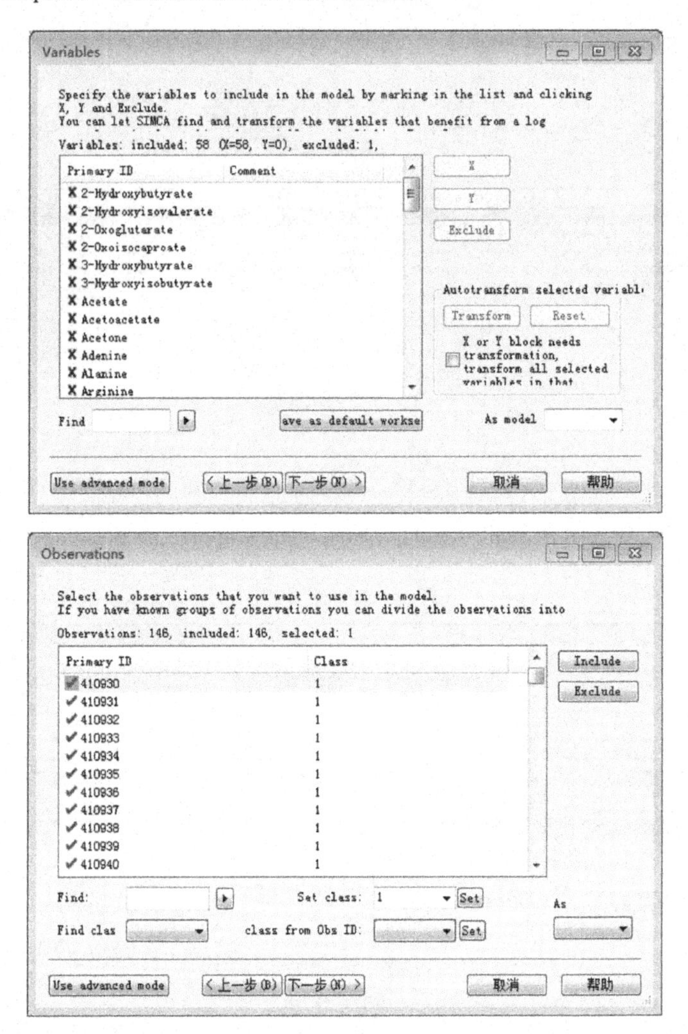

Fig. 9.9 Select the fitted parameters of variables and observations

the original model with the goodness of fit of several models based on data where the order of the Y-observations has been randomly permuted, while the X-matrix has been kept intact. The plot shows, for a selected Y-variable, on the vertical axis the values of R2 and Q2 for the original model (far to the right) and of the Y-permuted models further to the left. The horizontal axis shows the correlation between the permuted Y-vectors and the original Y-vector for the selected Y. The original Y has the correlation 1.0 with itself, defining the high point on the horizontal axis. The plot above strongly indicates that the original model is valid. The criteria for validity are:

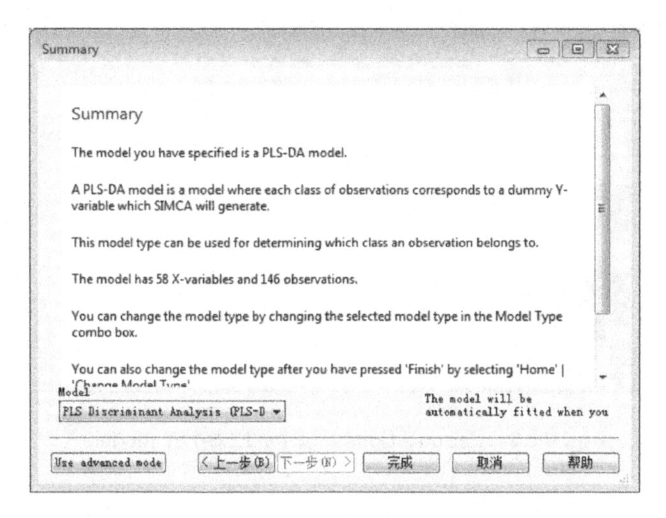

Fig. 9.10 Select the PLS-DA model and finish

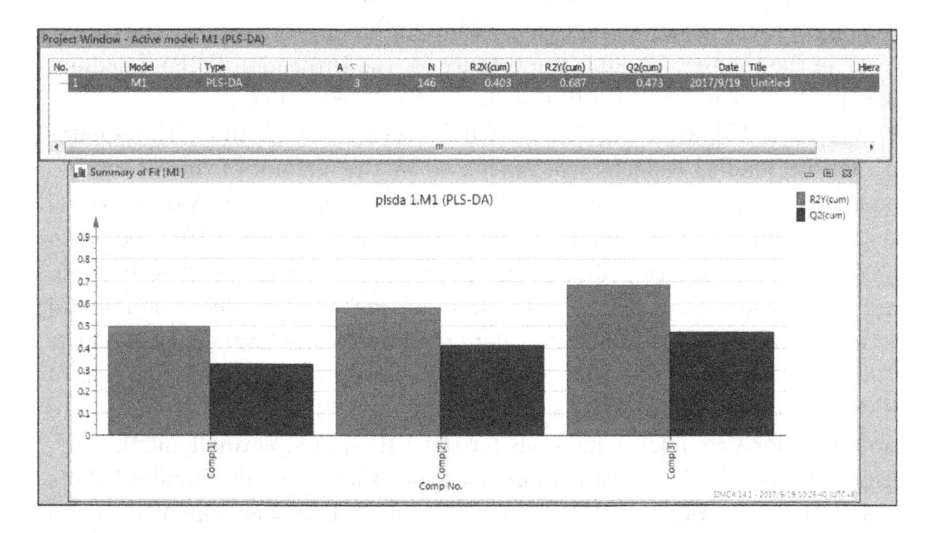

Fig. 9.11 Cross validation correlation coefficients Q2 and R2 of PLS-DA model

① All blue Q2-values to the left are lower than the original points to the right, or;

② The blue regression line of the Q2-points intersects the vertical axis (on the left) at, or below zero. Note that the R2-values always show some degree of optimism. However, when all green R2-values to the left are lower than the original point to the right, this is also an indication for the validity of the original model.

From the Figure 9.12 Permutations plot for PLS-DA model, the validity

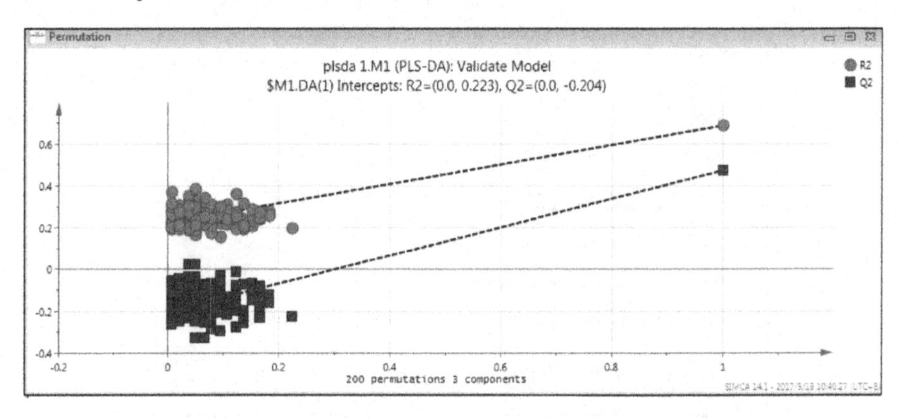

Fig. 9.12 Permutations plot for PLS-DA model

of this PLS-DA is powerfully proved, further analysis can be carried out.

Further analyses include Variable Importance for the Projection (VIP), Loadings Scatter Plot, Scores Scatter Plot and so on, all the functions can be found in the toolbar. Variable Importance for the Projection (VIP), Loadings Scatter Plot, Scores Scatter Plot will be introduced in detail.

VIP: The VIP values are calculated for each x_k by summing the squares of the PLS loading weights, wa_k, weighted by the amount of sum of squares explained in each model component. The sum of squares of all VIP's is equal to the number of terms in the model. Hence, the average VIP is equal to 1. VIP-values larger than 1 indicate 'important' X-variables (metabolites), and values lower than 0.5 indicate 'unimportant' X-variables. The interval between 1 and 0.5 is a gray area, where the importance level depends on the size of the data set. The VIP plot can be plotted through VIP function of SIMCA-P (Figure 9.13). The VIP plot (Figure 9.14) is sorted from high to low, and shows confidence intervals for the VIP values, normally at the 95% level. An unsorted VIP plot is available under Plot/List tab, Standard Plots group. VIP list (Figure 9.15) can be also created to show the VIP values explicitly.

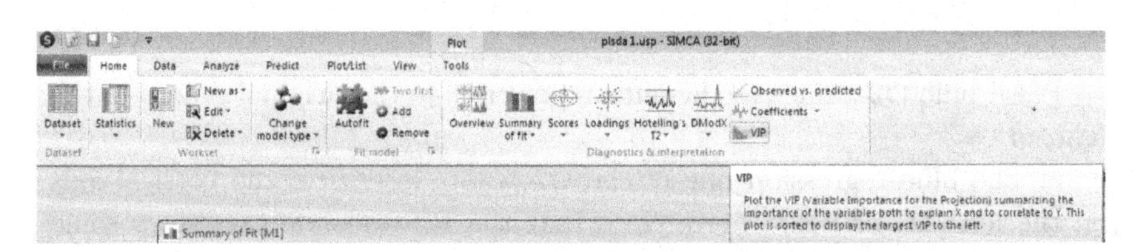

Fig. 9.13 The toolbar of SIMCA-P integrates many functions: VIP, scores, loadings, etc.

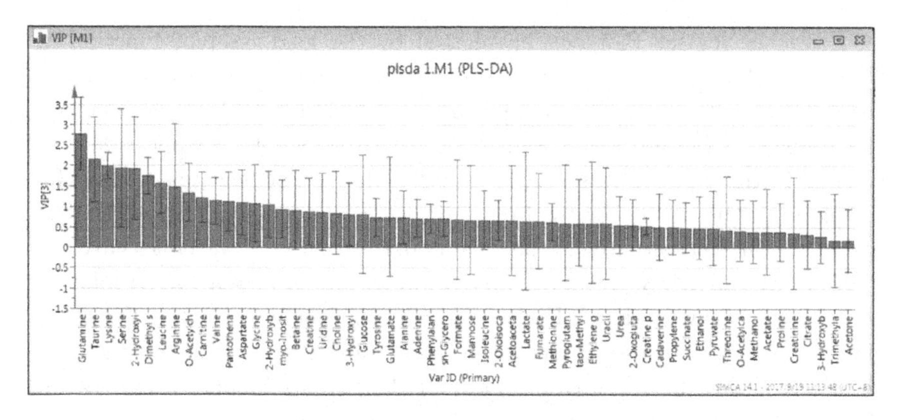

Fig. 9.14 VIP plot for PLS-DA model

General List [M1]

	1	2
1	Var ID (Primary)	M1.VIP[3]
2	Glutamine	2.78268
3	Taurine	2.1549
4	Lysine	1.997
5	Serine	1.94124
6	2-Hydroxyisovalerate	1.93654
7	Dimethyl sulfone	1.74911
8	Leucine	1.58278
9	Arginine	1.46761
10	O-Acetylcholine	1.34428
11	Carnitine	1.22862
12	Valine	1.1444
13	Pantothenate	1.12326
14	Aspartate	1.10378
15	Glycine	1.08711
16	2-Hydroxybutyrate	1.05302

Fig. 9.15 VIP list for PLS-DA model

Scores Scatter Plot can be plotted by the scores plot function of SIMCA-P (Figure 9.13). The scores t_1, t_2, etc., are new variables summarizing the X-variables. The scores are orthogonal, i.e., completely independent of each other. There are as many score vectors as there are components in the model. The score t_1 (first component) explains the largest variation of the X space, followed by t_2 etc. Hence the scatter plot of t_1 vs t_2 is a window in the X space, displaying how the X observations are situated with respect to each other. This plot shows the possible presence of outliers, groups, similarities and other patterns in the data. The score plot is a map of the observations. A clear classification between group A and Group B is shown in Figure 9.16, which means the PLS-DA is advisable, the two groups can be separated by PLS-DA.

Loadings Scatter Plot can be plotted by the loadings plot function of SIMCA-P (Figure 9.13). The PLS-DA loading weights scatter plot displays the relation between the X-variables (metabolites) and the Y-variables (classifications). To facilitate interpretation this plot is by default color coded according to the model terms. The above w^*c plot is a superimposition of the w^* plot and the c plot, for the first and second components of the PLS-DA model. X-variables situated in the vicinity of the dummy Y-variables have the highest discriminatory power between the classes. From Figure 9.17 we found glutamine, taurine, serine, 2-Hydroxyisovalerate, leucine, arginine, Dimethyl sulfone, lysine, Pantothenate have the highest discriminatory power between the classes, these metabolites can probably be potential biomarkers between the classes.

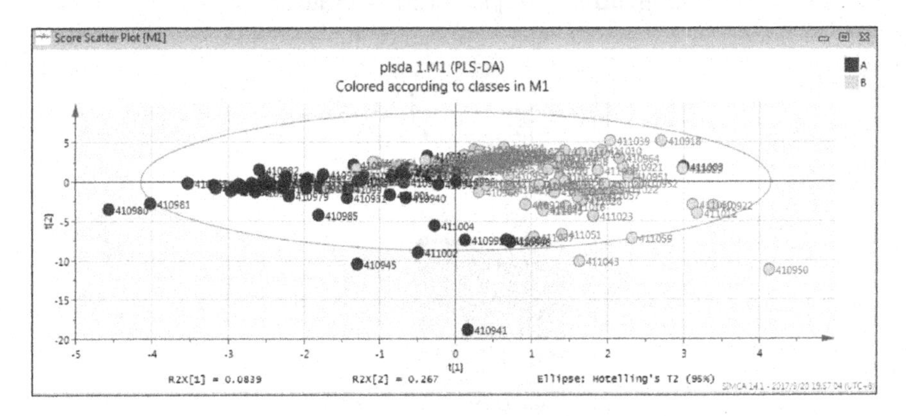

Fig. 9.16 Group A vs group B multivariate statistical analysis. PLS-DA score plot obtained by performing PLS-DA on the NMR data

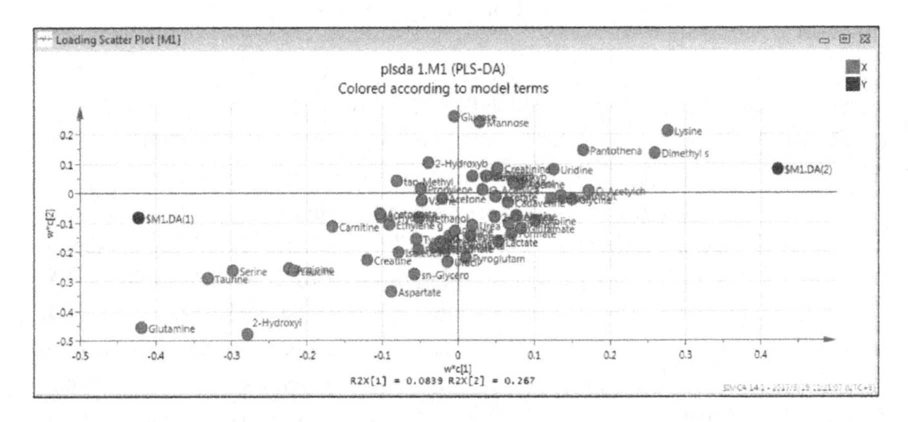

Fig. 9.17 Group A vs group B multivariate statistical analysis. PLS-DA loading plot obtained by performing PLS-DA on the NMR data

9.4 Metabonomics Databases

Metabonomics is still in the progress of innovation, development, and accumulation. Different laboratories used different analytical instruments to analyze biological samples and generated a lot of metabonomics data. The data exchange and assimilation require a proper database platform, which can store, manage, publish, search, and annotate a variety of data. The establishment of database also helps to join metabonomics and other branch platforms of systems biology.

Wiley Registry of Mass Spectral Data and NIST/EPA/NIH Mass Spectral Database are the two largest commercial and comprehensive GC-MS database with popular usages. The latest integrated version of Wiley 9th Edition/NIST 2008 contains a total of 796,000 mass spectra and 667,000 compounds, of which 746,000 spectra have chemical structures and 2,900,000 compound names. NIST MS/MS library contains 5,308 precursor ions. The NIST GC retention index database contains 43,000 retention indices[28].

The Golm Metabolome Database (GMD) is a metabonomics database developed by scholars in German Max-Planck and can be obtained free of charge, which including GC-MS of derivative metabolites and GC-TOF-MS spectra library. The current GMD contains about 2,000 evaluated mass spectra, including 1,089 nonredundant MS tags (MSTs) and 360 identified MSTs provided by both quadrupole and TOF-MS platform. In addition, GMD database also contains MS retention index (MSRI) libraries, which greatly improve the identification of structurally similar compounds. GMD also provides experimental methods and chromatography and mass spectrometry conditions.

MassBank is a mass spectra database jointly established by many Japanese universities and research institutions. It mainly collects data from high-resolution mass spectrometers including ESI-QqTOF-MS/MS, ESI-QqQ-MS/MS, ESI-IT-(MS)n, GC-EI-TOF-MS, LC-ESI-TOF-MS, FAB-CID-EBEB-MS/MS, FAB-MS, FD-MS, CI-MS, and LC-ESI-FT-MS. The reference spectra contain mass spectrometry information of multistage mass spectrometry. So far, it contains more than 24,993 mass spectra including positive and negative ion modes for more than 12,000 primary and secondary metabolites. MassBank allows free Web search, mass spectra comparing and three-dimensional visualization comparing by inputting mass spectra in text format.

The METLIN metabolite database is established by the Biological Mass Spectrometry Center of The SCRIPPS Research Institute (TSRI), which contains LCMS, MS/MS, and FTMS data of 23,000 endogenous and exogenous metabolites of human, small molecule drugs and drug metabolites, small peptides (8,000 di-and tripeptides), etc., which can be retrieved by mass and molecular formula and structure. METLIN provides free online search.

The Fiehn GC-MS database is established by the Olive Fiehn laboratory and currently contains about 1,050 entries of about 713 common metabolites, including entries of partial derivative metabolites. Every entry includes searchable EI mass spectra and retention indices.

MoTo DB is a metabolite database of tomato fruit obtained from Q-TOF-MS, which contains retention times, accurate masses, UV absorption spectrum, MS/MS fragment ions, and reference literatures of metabolites. The MeT-RO plant and microbial metabolomics database are established by the Rothamsted Institute, which includes GC-MS, LC-MS, and NMR spectra.

The Kyoto Encyclopedia of Genes and Genomes (KEGG) is a database of systematic analyzing gene functions and genomic information. The KEGG with integrated metabolic pathways' query is very excellent, including the metabolism of carbohydrates, nucleosides, and amino acids and the biodegradation of organics. It not only provides all possible metabolic pathways, but also comprehensively annotates the enzymes for each step of catalytic reaction, including amino acid sequence and links to PDB library. KEGG is a powerful tool for in vivo metabolism analysis and metabolic network research. KEGG now consists of six separate databases, including gene database (GENES database), pathway database (PATHWAY database), ligand chemical reaction database (NGAND database), sequence similarity database (SSDB), gene expression database (EXPRESSION), and protein molecular relationship database (BRITE). The PATHWAY database contains about 90 reference metabolic pathway patterns, and each reference pathway is a network composed of enzyme or EC number.

The MetaCyc is a sub-database of the BioCyc database, which is a database of metabolic pathways and enzymes. It expounded metabolic pathways of more than 1,600 organisms and contains metabolic pathways, reactions, enzymes, and substrates obtained from a large number of documents and online resources. Over 1,200 metabolic pathways, 5,500 enzymes, more than 5,100 genes, and 7,700 metabolites are included.

9.5 Summary

In the chapter, the definition of metabonomics or metabolomics, the relations between metabonomics and other 'omics', basic approaches to study metabonomics which includes metabonomics techniques and data analysis methods, and metabonomics datebases are briefly introduced.

NMR- and MS-based metabonomics and metabolomics are now recognised as widely used techniques for evaluating the biochemical consequences of drug action, and they have been adopted by a number of pharmaceutical companies into their drug development protocols. For drug safety studies, it is possible to identify the target organ of toxicity, derive the biochemical mechanism of the toxicity and determine the combination of biochemical biomarkers for the onset, progression and regression of the lesion. Additionally, the technique has been shown to be able to provide a metabolic fingerprint of an organism as an adjunct to functional genomics and hence has applications in design of drug clinical trials and for evaluation of genetically modified animals as disease models.

PCA and PLS-DA are the most widely used chemometric methods in the study of metabonomics, we have introduced these methods in pretty detail in this chapter. Principal component analysis (PCA) is a statistical procedure that uses an orthogonal transformation to convert a set of observations of possibly correlated variables into a set of values of linearly uncorrelated variables called principal components (PC) [29], and it's also a unsupervised techniques which can be easily mastered. PLS-DA is a widely used supervised method. This is a method which relates a data matrix containing independent variables from samples, such as spectral intensity values (X matrix), to a matrix containing dependent variables (e.g. measurementsof response, such as toxicity scores) for those samples (Y matrix) [30]. If a PLS-DA model is successfully fitted and tested valid, the model can provide a lot of information to researchers, such as VIP plot, scores, loadings and so on.

Metabonomics database is an easy way to get information from a large number of metabolites already identified. The widely used metabonomics databases include Wiley Registry of Mass Spectral Data and NIST/EPA/NIH Mass Spectral Database, The Golm Metabolome Database, MassBank, The METLIN metabolite database, The Fiehn GC-MS database, MoTo DB, The

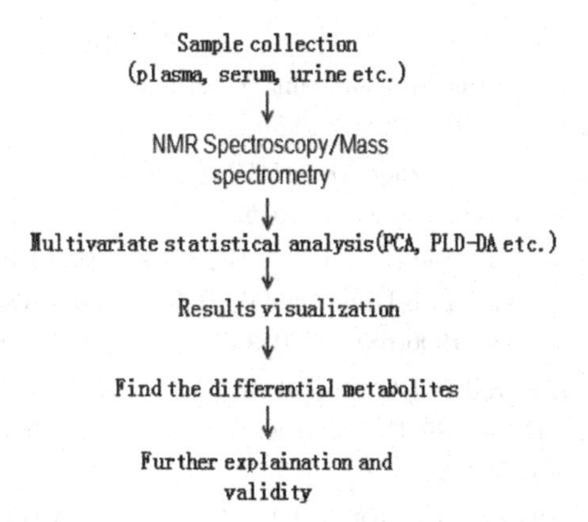

Fig. 9.18 Brief flow chart of metabonomics analysis

Kyoto Encyclopedia of Genes and Genomes, The MetaCyc and so on. These Mmetabonomics databases face to different researchers and data types.

At last, there is an universal flow chart (Figure 9.18) of metabonomics analysis which introduces the main processes briefly.

References

[1] Beger R D, et al. Metabolomics enables precision medicine: "A White Paper, Community Perspective". *Metabolomics*, 2016(12): 149 DOI 10.1007/s11306-016-1094-6.

[2] Lindon J C, Nicholson J K, Elaine Holmes. *The Handbook of Metabonomics and Metabolomics.* New York: Elsevier, 2007.

[3] Finishing C. The euchromatic sequence of the human genome. *Nature*, 2004(431): 931–945.

[4] Baldi P, and Hatfield G W. *DNA Microarrays and Gene Expression.* Cambridge: Cambridge University Press, 2002.

[5] Cacciatore S and Loda M. Innovation in Metabolomics to Improve Personalized Healthcare. *Ann. N. Y. Acad. Sci.*, 2015, 6, 1346(1): 57–62.

[6] Fiehn O. Metabolomics–the Link Between Genotypes and Phenotypes. *Plant. Mol. Biol.*, 2002(48): 155–171.

[7] Kaderbhai N N, Broadhurst D I, Ellis D I, and Kell D B. Functional Genomics via Metabolic Footprinting: Monitoring metabolite secretion by Escherichia coli tryptophan metabolism mutants using FT-IR and direct injection electrospray mass spectrometry. *Comp. Funct. Genom.*, 2003(4): 376–391.

[8] Gamache P H, Meyer D F, Granger M C, and Acworth I N. Metabolomic Applications of Electrochemistry/Mass Spectrometry. *J. Am. Soc. Mass Spectrom.*, 2004(15): 1717–1726.

[9] Palmnas S A and Vogel H J. The Future of NMR Metabolomics in Cancer Therapy: Towards Personalizing Treatment and Developing Targeted Drugs. *Metabolites*, 2013(3): 373-396.

[10] Claridge T D W. *High-Resolution NMR Techniques in Organic Chemistry.* 3rd. ed. Oxford: Elsevier Science, 2016.

[11] Liu M, Nicholson J K, and Lindon J C. High Resolution Diffusion and Relaxation Edited One and Two-Dimensional ^{1}H NMR Spectroscopy of Biological Fluids. *Anal. Chem.*, 1996(68): 3370–3376.

[12] Nicholson J K, Foxall P J D, Spraul M, Farrant R D, and Lindon J C. 750 MHz ^{1}H and ^{1}H-^{13}C NMR Spectroscopy of Human Blood Plasma. *Anal. Chem.*, 1995(67): 793–811.

[13] Keun H C, Beckonert O, Griffin J L, Richter C, Moskau D, Lindon J C, and Nicholson J K. Cryogenic Probe ^{13}C NMR Spectroscopy of Urine for

Metabonomic Studies. *Anal. Chem.*, 2002(74): 4588–4593.

[14] Price K E, Vandaveer S S, Lunte C E, and Larive C K. Tissue Targeted Metabonomics: Metabolic Profiling by Microdialysis Sampling and Microcoil NMR. J. Pharmaceut. *Biomed. Anal.*, 2005(38): 904–909.

[15] Tomlins A, Foxall P J D, Lindon J C, Lynch M J, Spraul M, Everett J R, and Nicholson J K. High Resolution Magic Angle Spinning ^{1}H Nuclear Magnetic Resonance Analysis of Intact Prostatic Hyperplastic and Tumour Tissues. *Anal. Commun.*, 1998(35): 113–115.

[16] Garrod S L, Humpfer E, Spraul M, Connor S C, Polley S, Connelly J, Lindon J C, Nicholson J K, and Holmes E. High-Resolution Magic Angle Spinning ^{1}H NMR Spectroscopic Studies on Intact Rat Renal Cortex and Medulla. *Magn. Reson. Med.*, 1999(41): 1108–1118.

[17] Cheng L L, Chang I W, Louis D N, and Gonzalez R G. Correlation of High-Resolution Magic Angle Spinning Proton Magnetic Resonance Spectroscopy with Histopathology of Intact Human Brain Tumor Specimens. *Cancer Res.*, 1998(58): 1825–1832.

[18] Krone N, Hughes B A, Lavery G G, et al. Gas Chromatogmphy/Mass Spectrometry (GC/MS) Remains a Preeminent Discovery Tool in Clinical Steroid Investigations Even in the Era of Fast Liquid Chromatography Tandem Mass Spectrometry (LC/MS/MS). *J. Steroid. Biochem. Mol. Biol.*, 2010, 121(3-5): 496-504.

[19] Drexler D M, Reily M D, Shipkova P A. Advances in Massspectrometry Applied to Pharmaceutical Metabolomics. *Anal. Bioanal. Chem.*, 2010, 399(8): 2645-2653.

[20] Yang J, Xu G, Zheng Y, Kong H, Pang T, Lv S, Yang Q. Diagnosis of Liver Cancer Using HPLC-Based Metabonomics Avoiding False-Positive Result From Hepatitis and Hepatocirrhosis Diseases. *J. Chromatogr.*, B, 2004(813): 59–65.

[21] Morris M, and Watkins S M. Focused Metabolomic Profiling in the Drug Development Process: Advances From Lipid Profiling. *Curr. Op. Chem. Biol.*, 2005(9): 407–412.

[22] Wilson I D, Nicholson J K, Castro-Perez J, Granger J H, Johnson K A, Smith B W, and Plumb R S. High resolution "ultra performance" liquid chromatography coupled to oa-TOF mass spectrometry as a tool for differential metabolic pathway profiling in functional genomic studies. *J. Proteome Res.*, 2005(4): 591–598.

[23] Soga T, Ohashi Y, Ueno Y, Nasaoka H, Tomita M, and Nishioka T. Quantitative Metabolome Analysis Using Capillary Electrophoresis Mass Spectrometry. *J. Proteome Res.*, 2003(2): 488–494.

[24] Lindon J C, Nicholson J K, and Wilson I D. Directly-Coupled HPLC-NMR and HPLC-NMRMS in Pharmaceutical Research and Development. *J. Chromatog.*, B, 2000(748): 233–258.

[25] Jordan Michael I, Bishop Christopher M. Neural Networks. In: Allen B. Tucker. *Computer Science Handbook*, Second Edition (Section VII: Intelligent Systems). Boca Raton: Chapman & Hall/CRC Press LLC, 2004.

[26] Mehryar Mohri, Afshin Rostamizadeh, Ameet Talwalkar. *Foundations of Machine Learning*. Cambridge: The MIT Press, 2012.

[27] Wold H. Partial Least Squares. In: Kotz S, and Johnson N L (Eds), *Encyclopedia of Statistical Sciences*. New York: John Wiley & Sons, 1985.

[28] Xiaoquan Qi, Xiaoya Chen, Yulan Wang. *Plant Metabolomics Methods and Applications*. Beijing: Chemical Industry Press, 2014.

[29] Jackson J E. *A User's Guide to Principal Components*. New York: Wiley, 1991.

[30] Wold S, Sjöström M, Eriksson L. PLS-Regression: A Basic Tool of Chemometrics. *Chemometrics and Intelligent Laboratory Systems*, 2001, 58(2): 109–130.

Chapter 10 Gene Ontology Database and KEGG Database

Hong Jiang[1]

The large-scale datasets generated by gene sequencing, proteomics, and other high-throughput experimental technologies are the bases for understanding life as a molecular system and for developing medical, industrial, and other practical applications. In order to facilitate bioinformatics analysis of such large-scale datasets, it is essential to organize our knowledge on higher levels of systemic functions in a computable form, so that it can be used as a reference for inferring molecular systems from the information contained in the building blocks. The Gene Ontology (GO) and Kyoto Encyclopedia of Genes and Genomes (KEGG) are the most used commonly and effectively databases.

10.1 Gene Ontology Database

10.1.1 Knowledge Representation

The GO project is a collaborative effort to address the need for consistent descriptions of gene products across databases. Founded in 1998, the project began as a collaboration between three model organism databases, FlyBase (Drosophila), the Saccharomyces Genome Database (SGD) and the Mouse Genome Database (MGD). The GO Consortium (GOC) has since grown to incorporate many databases, including several of the world's major repositories for plant, animal, and microbial genomes. The GO Contributors page lists all member organizations.

The GO project has developed three structured ontologies that describe gene products in terms of their associated biological processes, cellular com-

1. Jiang Hong, College of Life Sciences and Bioengineering, School of Science, Beijing Jiao Tong University, Beijing, China, 100044.

ponents and molecular functions in a species-independent manner. There are three separate aspects to this effort: first, the development and maintenance of the ontologies themselves; second, the annotation of gene products, which entails making associations between the ontologies and the genes and gene products in the collaborating databases; and third, the development of tools that facilitate the creation, maintenance and use of ontologies.

The use of GO terms by collaborating databases facilitates uniform queries across all of them. Controlled vocabularies are structured so they can be queried at different levels; for example, users may query GO to find all gene products in the mouse genome that are involved in signal transduction, or zoom in all receptor tyrosine kinases that have been annotated. This structure also allows annotators to assign properties to genes or gene products at different levels, depending on the depth of knowledge about that entity.

Shared vocabularies are an important step towards unifying biological databases, but additional work is still necessary as knowledge changes, updates lag behind, and individual curators evaluate data differently. The GO aims to serve as a platform where curators can agree on stating how and why a specific term is used, and how to consistently apply it, for example, to establish relationships between gene products.

The following areas are outside the scope of GO, and terms in these domains will not appear in the ontologies: Gene products (e.g. cytochrome C is not in the ontologies, but attributes of cytochrome C, such as oxidoreductase activity, is processes, functions or components that are unique to mutants or diseases (e.g. oncogenesis is not a valid GO term, as 'causing cancer' is the result of reprogrammed, not normal cells and thus it is not the normal function of a gene.); attributes of sequence such as 'intron' or 'exon' parameters belong in a separate sequence ontology; protein domains or structural features; protein-protein interactions; environment, evolution and expression; anatomical or histological features above the level of cellular components, including cell types.

10.1.2 GO Database

The GO database is a relational database comprised of the GO ontologies as well as the annotations of genes and gene products to terms in the those ontologies. Housing both the ontologies and the annotations in a single database allows powerful queries of the annotations using the ontology. The GO database is the source of all data available through the legacy AmiGO 1.8 browser and search engine.

The GO database is maintained as a MySQL database, built at regular intervals. GO database builds can be downloaded and installed on a local machine or can be queried remotely.

GO Database Guide: information about the GO database, including the database schema, accessing or downloading the database, and tips on usage.

GO Database schema: autogenerated schema documentation.

GO Database schema with views: autogenerated schema documentation, including database views. Database abbreviations are used in the GO Database.

Relational Databases for Biologists: an online introduction to databases in bioinformatics.

10.1.3 Annotation

Annotation is the process of assigning GO terms to gene products. The annotation data in the GO database is contributed by members of the GO Consortium, and the Consortium is continuously encouraging new groups to start contributing their annotations. The list of links below offer details on the GO annotation policies and the annotation process, as well as direct users to other pages of interest on GO annotation conventions, the standard operating procedures used by some consortium members, and the GO annotation file format guide.

1. Annotation policies and guidelines

An introduction to annotation and the rules and SOPs used for annotating gene products to GO terms. This GO annotation guide includes details about the annotation process. Collaborating databases annotate their gene products with GO terms, according to two general principles: first, annotations should be attributed to a source; second, each annotation should indicate the evidence on which it is based.

Long-term maintenance of annotation datasets. Annotation is carried out by curators in a range of bioinformatics database resource groups, such as Mouse Genome Informatics, Saccharomyces Genome Database and FlyBase. These groups then contribute their data to the central GO repository for storage and redistribution. After submission, the annotating groups may retain responsibility for updating the annotation data to take account of changes in annotation practices and in the structure of the ontologies. This is an ongoing responsibility. For groups who prefer not to maintain their annotation dataset in the long term, it is possible to submit data to the GO repository via another database group, which will undertake to maintain the data long-term.

Avoiding redundancy. Where two or more databases are submitting data on the same species we encourage the model whereby one database group collects all annotation data for that species, removes the redundant (duplicate) annotations, and then submits the total dataset to the central repository. This ensures that no redundant annotations will appear in the master dataset. Please see the list of species and relevant database groups for more details. We understand that annotating groups will also wish to make their full dataset available to the public. For this purpose, the GO Consortium makes all of the individual datasets available from the GO website, via the GO web CVS interface, or from the directory go/gene-associations/ in the GO CVS repository. All of the individual datasets are also listed in the annotation downloads table, and all individual groups will clearly be given credit for the work that they have done. The non-redundant set is only used as the master copy that appears in AmiGO and similar tools.

Credit for annotation work. Every annotation is marked with the name of the database that made the annotation as well as the name of the database that maintains and submits the annotation. This information is in two separate columns of the gene association file. This ensures that the database making the annotation, and the database maintaining the annotation, will both receive full credit for their work.

No single established database? Some model species research communities do not have an established database group with funding and time to commit to long-term maintenance of their datasets. Such groups can contribute annotations to the central repository via the UniProtKB GO Annotation (UniProtKB-GOA) multispecies annotation group. This is also a possible route for those groups just starting out in annotation who may wish to take up the responsibility for long-term maintenance of their datasets at a later date.

Sample annotation sets are available from a number of the databases in the Consortium; see the teaching resources page to download them. These sets comprise up to ten papers annotated by the database, with description of the GO terms and evidence codes used, along with reasons why these terms were chosen (where applicable).

For more information on annotation, please see the following resources in GO website, such as GO annotation conventions, GO annotation standard operating procedures, and Evidence codes used in GO annotation.

2. Guide to evidence code

Guide to evidence code explains, and indicates the nature of the evidence that supports a particular annotation.

A GO annotation consists of a GO term associated with a specific reference that describes the work or analysis upon which the association between a specific GO term and gene product is based. Each annotation must also include an evidence code to indicate how the annotation to a particular term is supported. Although evidence codes do reflect the type of work or analysis described in the cited reference which supports the GO term to gene product association, they are not necessarily a classification of types of experiments/analyses. Note that these evidence codes are intended for use in conjunction with GO terms, and should not be considered in isolation from the terms. If a reference describes multiple methods that each provide evidence to make a GO annotation to a particular term, then multiple annotations with identical GO identifiers and reference identifiers but different evidence codes may be made.

Out of all the evidence codes available, only Inferred from Electronic Annotation (IEA) is not assigned by a curator. Manually-assigned evidence codes fall into four general categories: experimental, computational analysis, author statements, and curatorial statements.

Use of an experimental evidence code in a GO annotation indicates that the cited paper displayed results from a physical characterization of a gene or gene product that has supported the association of a GO term. The Experimental Evidence codes are:

- Inferred from Experiment (EXP).
- Inferred from Direct Assay (IDA).
- Inferred from Physical Interaction (IPI).
- Inferred from Mutant Phenotype (IMP).
- Inferred from Genetic Interaction (IGI).
- Inferred from Expression Pattern (IEP).

Use of the computational analysis evidence codes indicates that the annotation is based on an in silico analysis of the gene sequence and/or other data as described in the cited reference. The evidence codes in this category also indicate a varying degree of curatorial input. The Computational Analysis evidence codes are:

- Inferred from Sequence or structural Similarity (ISS).
- Inferred from Sequence Orthology (ISO).
- Inferred from Sequence Alignment (ISA).
- Inferred from Sequence Model (ISM).
- Inferred from Genomic Context (IGC).
- Inferred from Biological aspect of Ancestor (IBA).
- Inferred from Biological aspect of Descendant (IBD).
- Inferred from Key Residues (IKR).
- Inferred from Rapid Divergence (IRD).
- Inferred from Reviewed Computational Analysis (RCA).

Author statement codes indicate that the annotation was made on the basis of a statement made by the author(s) in the reference cited. The Author Statement evidence codes used by GO are:

- Traceable Author Statement (TAS).

- Non-traceable Author Statement (NAS).

Use of the curatorial statement evidence codes indicates an annotation made on the basis of a curatorial judgement that does not fit into one of the other evidence code classifications. The Curatorial Statement codes are:

- Inferred by Curator (IC).
- No biological Data available (ND) evidence code.

All of the above evidence codes are assigned by curators. However, GO also used one evidence code that is assigned by automated methods, without curatorial judgement. The Automatically-Assigned evidence code is Inferred from Electronic Annotation (IEA).

Evidence codes are not statements of the quality of the annotation. Within each evidence code classification, some methods produce annotations of higher confidence or greater specificity than other methods, in addition the way in which a technique has been applied or interpreted in a paper will also affect the quality of the resulting annotation. Thus evidence codes cannot be used as a measure of the quality of the annotation.

3. Annotation convention

Annotation conventions are the guidelines which apply to all annotation methods and are particularly useful for manual literature-based annotation.

4. Annotation standard operating procedures

The annotation standard operating procedures are used by members of the GO Consortium during the process of annotation.

Database groups are responsible for collecting and submitting annotations for one or more species, as GO annotation reference collection, GO annotation (gene association) file format, GAF Quality Control/Validation, Annotation Quality Control checks, GAF 2.0 annotation file format, and GAF 1.0 annotation file format (deprecated).

10.2 KEGG Database

10.2.1 Knowledge Representation

KEGG is a database resource for understanding high-level functions and utilities of the biological system, such as the cell, the organism and the ecosystem, from genomic and molecular-level information, especially large-scale molecular datasets generated by genome sequencing and other high-throughput experimental technologies. It is a computer representation of the biological system, consisting of molecular building blocks of genes and proteins (genomic information) and chemical substances (chemical information) that are integrated with the knowledge on molecular wiring diagrams of interaction, reaction and relation networks (systems information). It also contains disease and drug information (health information) as perturbations to the biological system (Figure 10.1). The KEGG database has been in development by Kanehisa Laboratories since 1995, and is now a prominent reference knowledge base for integration and interpretation of large-scale molecular data sets generated by genome sequencing and other high-throughput experimental technologies. The current release is 83.2 published in September 1, 2017.

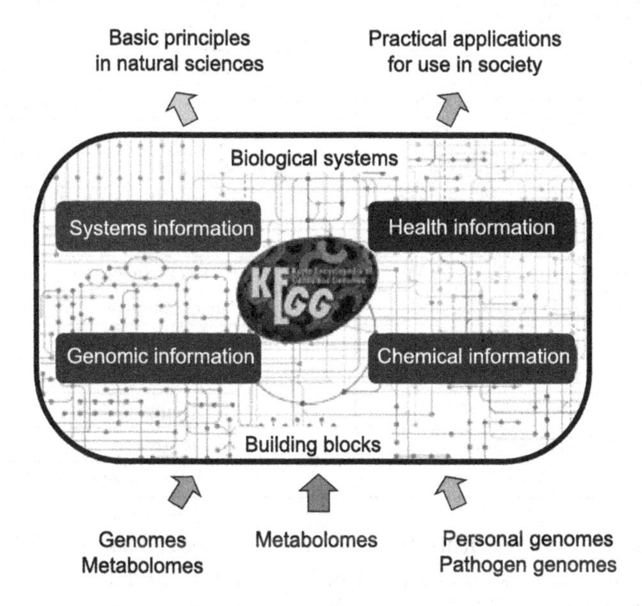

Fig. 10.1 The Framework of KEGG

10.2.2 Database in KEGG

KEGG is an integrated database resource consisting of sixteen databases shown below (Table 10.1). They are broadly categorized into systems information, genomic information, chemical information and health information, which are distinguished by color coding of web pages.

These database contain various data objects for computer representation of the biological systems. Thus, the database entry of each database is called the KEGG object, which is identified by the KEGG object identifier consisting of a database-dependent prefix and a five-digit number.

The most unique data object in KEGG is the molecular networks – molecular interaction, reaction and relation networks representing systemic functions of the cell and the organism. Experimental knowledge on such systemic functions is captured from literature and organized in the following forms: pathway map (in KEGG PATHWAY), brite hierarchy and table (in KEGG BRITE), membership of logical expression (in KEGG MODULE) and membership of simple list (in KEGG DISEASE).

Table 10.1 KEGG Database Resource

Category	Database	Content	Color
Systems information	KEGG PATHWAY	KEGG pathway maps	green
	KEGG BRITE	BRITE hierarchies and tables	
	KEGG MODULE	KEGG modules	
Genomic information	KEGG ORTHOLOGY (KO)	Functional orthologs	yellow
	KEGG GENOME	KEGG organisms (complete genomes)	red
	KEGG GENES	Genes and proteins	
	KEGG SSDB	GENES sequence similarity	
Chemical information	KEGG COMPOUND	Small molecules	blue
	KEGG GLYCAN	Glycans	
	KEGG REACTION	Biochemical reactions	
	KEGG RCLASS	Reaction class	
	KEGG ENZYME	Enzyme nomenclature	
Health information	KEGG DISEASE	Human diseases	purple
	KEGG DRUG	Drugs	
	KEGG DGROUP	Drug groups	
	KEGG ENVIRON	Health-related substances	

These databases constitute the reference knowledge base for biological interpretation of genomes and high-throughput molecular datasets through the process of KEGG mapping. The main component is the KEGG PATHWAY database, consisting of manually drawn graphical diagrams of molecular networks, called pathway maps, and representing various cellular processes and organism behaviors. KEGG PATHWAY is a reference database for pathway mapping, which is the process to match, for example, a genomic or transcriptomic content of genes against KEGG reference pathway maps to infer systemic functions of the cell or the organism. Pathway maps present detailed pictures of molecular networks. Often our knowledge is too fragmentary to be represented as pathway maps. Molecular lists are less detailed but more general representations. KEGG BRITE is an ontology database where molecular lists are hierarchically categorized, representing our knowledge on protein families, chemical compound families, and drug classifications, among others. KEGG MODULE is a supplement to KEGG PATHWAY indicating tighter functional units of pathways and complexes, represented simply as lists of molecules without specifying connection patterns. For human diseases our knowledge on genetic and molecular factors is organized in KEGG

DISEASE. More detailed molecular mechanisms, whenever known, are represented as disease pathway maps in KEGG PATHWAY. Each KEGG DISEASE entry is characterized by a list of known causative genes and other lists of molecules such as environmental factors, diagnostic markers, therapeutic drugs, etc. These lists are prepared such that they contain aspects of molecular systems behind the diseases, which may be used as a starting point of bioinformatics analysis for integration with other data and knowledge.

In 1995 the concept of mapping was first introduced in KEGG for linking genomes to metabolic pathways (metabolic reconstruction) using the EC number. Once the EC numbers were assigned to enzyme genes in the genome, organism-specific pathways could be generated automatically by matching against the enzyme (EC number) networks of the KEGG reference metabolic pathways. The EC number is no longer used as an identifier in KEGG after 1999. The KEGG Orthology (KO) system has been the basis for genome annotation and KEGG mapping since 2003.

From a different perspective, individual instances of genes are grouped into KO entries representing functional orthologs in the molecular networks. There are two more types of such generalization in KEGG, that is biochemical reactions into RC (reaction class) and drug interactions into DG (drug group).

10.2.3 Searching KEGG

The KEGG database is made accessible in two formats. One is the flat-file format for use in simple searches by the DBGET database retrieval system. The other is the relational database format for more sophisticated queries in selected databases.

DBGET has been developed as the backbone retrieval system for the GenomeNet and KEGG services (Table 10.2). It is based on a flat-file view of molecular biology databases, where the database is considered as a collection of entries. Thus, any entry of any database can be retrieved by the combination of the database name and the entry name in the form of: db: entry.

DBGET has two basic commands:

- bget - to retrieve ‘db:entry’.
- bfind - to find ‘db:entry’s by keyword search.

RDB Search interfaces in the following pages are direct SQL searches against the read-only copy of the internal KEGG database (Table 10.3).

KEGG MEDICUS search is an integrated search against drug labels (in Japan and/or the USA) together with relevant KEGG databases. The KEGG MEDICUS search is made against relevant KEGG databases including PATHWAY, DISEASE, DRUG, and ENVIRON, as well as outside databases for drug labels (package inserts). Currently, the Japanese drug labels obtained from JAPIC are fully integrated in the KEGG Relational Database (RDB), while only the search indices created from FDA's National Drug Code Directory (NDC) are stored in the KEGG RDB and actual contents are linked to NLM's DailyMed.

Table 10.2 KEGG database in DBGET

Database name			Abbreviation	Search fields
kegg	pathway		path	ENTRY, NAME
	brite		br	ENTRY, NAME, and entire text
	module		md	ENTRY, NAME
	orthology		ko	ENTRY, NAME, DEFINITION
	genomes	genome	gn	ENTRY, NAME, DEFINITION
		mgenome	mgnm	ENTRY, NAME, DEFINITION
	genes		org code	ENTRY, NAME, DEFINITION, ORTHOLOGY
	mgenes		T number	ENTRY, NAME, DEFINITION, ORTHOLOGY
	ligand	compound	cpd	ENTRY, NAME
		glycan	gl	ENTRY, NAME, COMPOSITION
		reaction	rn	ENTRY, NAME, DEFINITION
		rclass	rc	ENTRY, NAME, DEFINITION
		enzyme	ec	ENTRY, NAME
	disease		ds	ENTRY, NAME
	drug		dr	ENTRY, NAME
	dgroup		dg	ENTRY, NAME
	environ		ev	ENTRY, NAME

Table 10.3 The RDB search in KEGG

Database page		Search fields
PATHWAY		Entry, Name, Description, Map objects, Map legends
DISEASE		Entry, Name, Description, Category, Pathway, Gene
DRUG		Entry, Name, Product name, Component
ENVIRON		Entry, Name, Category, Source, Component
LIGAND	COMPOUND	Entry, Name, Formula, Exact mass, Reaction, Pathway, Enzyme, DBlinks
	GLYCAN	Entry, Name, Composition, Class, Reaction, Pathway, Enzyme, DBlinks
	REACTION	Entry, Name, RPair, Pathway, Enzyme
	RPAIR	Entry, Compound, RDM, Type

10.2.4 KEGG Mapping

KEGG mapping is the process to map elementary datasets (genes, proteins, small molecules, etc.) to network datasets (KEGG pathway maps, BRITE functional hierarchies, and KEGG modules). It is not simply an enrichment process, rather it is a set operation to generate a new set. From the beginning of the KEGG project, the basic idea was to automatically generate organism-specific pathways by the set operation between manually annotated genome data and manually created pathway maps. Thus, the KEGG mapping set operation has played a role to extend the KEGG knowledge base. In addition, it played another important role to assist integration and interpretation of users' datasets, especially large-scale datasets generated by high-throughput technologies. Currently there are four types of mapping operations available in KEGG: pathway mapping, brite mapping, module mapping and taxonomy mapping. The fourth type may involve molecular or non-molecular datasets (orthologs, modules, and organisms) and the network dataset (taxonomic tree).

Here is a summary of how the knowledge base of KEGG pathway maps and BRITE functional hierarchies is extended by the KEGG mapping set operations.

1. Organism-specific pathway/brite/module datasets

Organism-specific versions are created for KEGG pathway maps, BRITE functional hierarchies, and KEGG modules through the KEGG Orthology (KO) system, either as static files in the daily database update procedure (for the well-annotated genomes in KEGG GENES) or as temporary files on the fly (for GhostKOALA-annotated MGENES). The organism-specific pathway maps and module maps are colored in green, which is a KEGG convention.

2. Human pathway/brite datasets with disease genes and drug targets

On top of the human pathway maps and BRITE functional hierarchies, all known diseases genes accumulated in KEGG DISEASE and all known drug targets accumulated in KEGG DRUG are mapped and displayed in pink and light blue, respectively. The static pathway maps and BRITE hierarchies are identified by the special organism code 'hsadd' and the extension '_dd', respectively, enabling to be viewed and searched in a similar way as all other KEGG organisms.

3. Disease dataset with gene/genome information

On top of the BRITE hierarchy files for disease classifications, additional information is computationally included in an additional column, namely, human disease classification with known disease genes (extension '_gene') and infectious disease classification with known pathogen genomes (extension '_genome').

4. Drug dataset with molecular network information

On top of the BRITE hierarchy files for various drug classifications, additional information is computationally included either in an additional column or an additional hierarchy level. Examples include drug classifications with metabolizing enzyme data (extension '_enzyme') or target data (extension '_target').

5. Drug labels integrated into KEGG

The BRITE hierarchy files for drug classifications may also be used to integrated drug labels (package inserts), once they are properly linked to KEGG DRUG D numbers. Examples include drug labels from Japan (extension '_japic' or '_yj') and from the USA (extension '_ndc').

10.2.5 Identifier

The KEGG object identifer or simply the KEGG identifier (kid) is a unique identifier for each KEGG object, which is also the database entry identifier. Generally it takes the form of a prefix followed a five-digit number as shown below (Table 10.4).

Exceptions are GENES identifiers and EC numbers for enzymes, which follow the general form of the DBGET retrieval system (see About DBGET). Each entry in DBGET is identified by: <database>:<entry>, where <database> is the database name and <entry> is the entry name or the accession number.

The KEGG identifiers with database-dependent prefixes are often called D numbers, K numbers, C numbers, etc. Since these numbers are unique across the KEGG databases, <database> may be omitted in the DBGET system. For example, D01441, dr:D01441 and drug:D01441 are all equivalent. In addition to the T number shown above, an organism in KEGG is given a three- or four-letter KEGG organism code, which is treated like a database name. Therefore, individual genes in an organism can be identified in the following way: <org>:<gene>, where <org> is the KEGG organism code and <gene> is the KEGG GENES entry name. The KEGG organism code is also used as a prefix to identify organism-specific pathway maps or BRITE functional hierarchies.

Table 10.4 The KEGG object identifer

Database	Object	Prefix		Example
pathway	KEGG pathway map	map, ko ec, rn	⟨org⟩	map00010 hsa04930
brite	BRITE functional hierarchy	br, jp ko	⟨org⟩	br08303 ko01003
module	KEGG module	M	⟨org⟩_M	M00010
ko	KO functional ortholog	K		K04527
genome	KEGG organism (complete genome)	T		T01001 (hsa)
genes ⟨org⟩ vg ag	Gene/protein			hsa:3643 vg:155971 ag:CAA76703
compound	Small molecule	C		C00031
glycan	Glycan	G		G00109
reaction	Reaction	R		R00259
rclass	Reaction class	RC		RC00046
enzyme	Enzyme			ec:2.7.10.1
disease	Human disease	H		H00004
drug	Drug	D		D01441
dgroup	Drug group	DG		DG00710
environ	Health-related substance	E		E00048

⟨org⟩ represents three- or four-letter organism code

10.2.6 Samples

Major efforts have been undertaken to manually create a knowledge base for such systemic functions by capturing and summarizing experimental knowledge in computable forms; namely, in the forms of molecular networks called KEGG pathway maps, BRITE functional hierarchies, and KEGG modules. Continuous efforts have also been made to develop and improve the cross-species (ortholog-based) annotation procedure for linking genomes to the molecular networks. As the result, KEGG is widely used as a reference knowledge base for integration and interpretation of large-scale datasets generated by genome sequencing and other high-throughput experimental technologies.

References

[1] Kanehisa Laboratories. KEGG. [2017-08-30]. http://www.genome.jp/kegg/.

[2] Minoru Kanehisa, Vachiranee Limviphuvadh, and Mao Tanabe. Knowledge-

Based Analysis of Protein Interaction Networks in Neurodegenerative Diseases. In: Alzate O, editor. *Neuroproteomics*. Boca Raton (FL): CRC Press/Taylor & Francis, 2010.

[3] The Gene Ontology (CC-B1/40). [2017-08-30]. http://www.geneontology.org.

PART III
STATISTICS AND PROGRAMMING

Chapter 11 Basic Algorithms for Bioinformatics

Xiyu Cheng[1], Shangyuan Tang[2], Peng Ye[3]

Algorithm is regarded as the key to open the door of bioinformatics. We will see how popular bioinformatics algorithms work and we will see what principles are behind their design. It is important to understand how an algorithm works in order to be confident in its results; it is even more important to understand an algorithm's design methodology in order to find its weaknesses and improve them. In this chapter, principles of basic algorithms such as Graph Theory, Dynamic Programming, Hidden Markov Model, Neural Network and Clustering Analysis will be introduced together with some applications to show their power that solve biological problems.

11.1 Algorithms

11.1.1 Introduction to Algorithms

An algorithm, which is a sequence of instructions, is designed to solve a well-formulated problem. Problems will be specified in terms of their inputs and their outputs, and the algorithm will be the method of translating the inputs into the outputs. A well-formulated problem is unambiguous and precise, leaving no room for misinterpretation [1].

In order to solve a problem, some entity needs to perform the steps specified by the algorithm. A human with a pen and paper would be able to do this task, but humans are generally slow, make mistakes, and prefer not to

1. Cheng Xiyu, College of Life Sciences and Bioengineering, School of Science, Beijing Jiao Tong University, Beijing, China, 100044.

2. Tang Shangyuan, College of Life Sciences and Bioengineering, School of Science, Beijing Jiao Tong University, Beijing, China, 100044.

3. Ye Peng, College of Life Sciences and Bioengineering, School of Science, Beijing Jiao Tong University, Beijing, China, 100044.

perform repetitive work. A computer is less intelligent but can perform simple steps quickly and reliably. Algorithms must be designed in a programming language (i.e. C or Java) in order to give specific instructions to the processor because that a computer cannot understand our language such as English. All details, including some trifling details that a person would naturally understand, must be specified to the computer in exactly the right format. For example, if a computer were to put on shoes, one would need to tell it to find a pair that both matches and fits, to put the left shoe on the left foot, the right shoe on the right, and to tie the laces [1].

11.1.2 Biological Algorithms

Nature uses algorithm-like procedures to solve biological problems such as the process of DNA replication [1]. A cell must make a complete copy of all its genetic material before it can divide. Replication of DNA proceeds in phases, each of which requires an elaborate cooperation between different types of molecules. For the sake of simplicity, we describe the replication process as it occurs in bacteria, rather than the replication process in humans, which is quite a bit more complicated. James Watson and Francis Crick proposed the basic mechanism in this process in the early 1950s. It could only be verified by the ingenious Meselson-Stahl experiment of 1957. The replication process starts from a pair of complementary strands of DNA and ends up with two pairs of complementary strands. A huge amount of molecular logistics is required to ensure completely accurate DNA replication. For example, DNA helicase separates strands, and DNA polymerase ensures proper complementarity. However, in terms of the logic of the process, none of this complicated molecular machinery actually matters—to mimic this process in an algorithm we simply need to take a string which represents the DNA and return a copy of it.

The number of operations that a real computer performs to copy a string is really large. For one particular computer architecture, we may end up issuing thousands of instructions to a computer processor. Computer scientists release themselves from this complexity by inventing programming languages that allow one to ignore many of these boring details. Biologists have not yet invented a similar 'language' to illustrate biological algorithms working in the cell. The amount of 'intelligence' that the simplest organism, such as a bacterium, shows to perform any routine task—including replication—is really amazing. Unlike STRINGCOPY, which only performs abstract operations, the bacterium really builds new DNA using materials from outsides. What would happen if it ran out? To prevent this, a bacterium examines the surroundings, absorbs new materials from outside, or moves off to forage for food. Moreover, it waits to begin its DNA copy until materials are sufficient. Actually, it may be an impossible mission even for the most sophisticated computer programs to carry out the complicated behavior displayed by even a single-celled organism [1].

11.2 Graph Theory

11.2.1 Introduction to Graph Theory

Many bioinformatics algorithms may be formulated in the language of graph theory. The use of the word 'graph' here is different than in many physical science contexts: we do not mean a chart of data in a Cartesian coordinate system. In order to work with graphs, we will need to define a few concepts that may not appear at first to be particularly well motivated by biological examples, but after introducing some of the mathematical theory we will find how powerful they can be in such bioinformatics applications as DNA sequencing and protein identification, etc [1-3].

1. Graph

Diagrams with collections of points connected by lines are examples of graphs. The points are called vertices and lines are called edges. A simple graph shown in Figure 11.1, consists of five vertices and six edges. A graph is denoted by $G = G(V, E)$ and is described by its set of vertices V and set of edges E (every edge can be shown as a pair of vertices). The graph in Figure 11.1 is described by the vertex set $V = \{a, b, c, d, e\}$ and the edge set $E = \{(a, b), (a, c), (b, c), (b, d), (c, d), (c, e)\}$. The way the graph is actually drawn is irrelevant; two graphs with the same vertex and edge sets are equivalent, even if the special pictures that represent the graph appear different (see Figure 11.1). The only important feature of a graph is which vertices are connected and which are

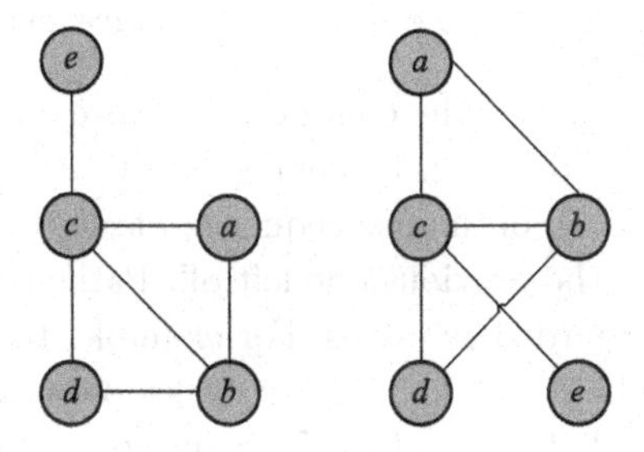

Fig. 11.1 Two equivalent representations of a simple graph with five vertices and six edges

not [1].

The number of edges incident to a given vertex v is called the degree of the vertex and is denoted $d(v)$. For example, vertex 8 in Figure 11.2 has degree 2 while vertex 12 has degree 3. The sum of degrees of all 12 vertices in Figure 11.2 is 32 (8 vertices of degree 3 and 4 vertices of degree 2), twice the number of edges in the graph. This is not a coincidence: for every graph G with vertex set V and edge set E, $\sum_{v \in V} d(v) = 2 \cdot |E|$. Indeed, an edge connecting vertices v and w is counted in the sum $\sum_{v \in V} d(v)$ twice: first in the term $d(v)$ and again in the term $d(w)$. You may wonder why I cannot connect fifteen phones such that each is connected to exactly seven others, and why a country with exactly three roads out of every city cannot have precisely 1,000 roads. This can be explained by the equality $\sum_{v \in V} d(v) = 2 \cdot |E|$.

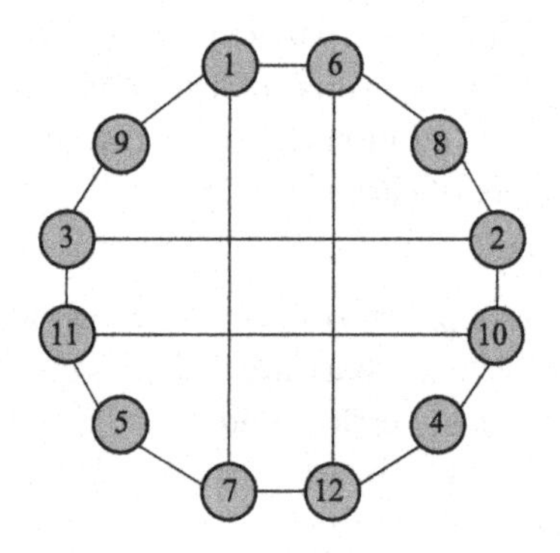

Fig. 11.2 A graph with twelve vertices and sixteen edges

Directed graphs, in which every edge is directed from one vertex to another, are used to solve many bioinformatics problems. Every vertex v in a directed graph is characterized by indegree (v) (the number of incoming edges) and outdegree (v) (the number of outgoing edges). For every directed graph $G(V, E)$, $\sum_{v \in V} \text{indegree}(v) = \sum_{v \in V} \text{outdegree}(v)$, since every edge is counted once on the right-hand side of the equation and once on the left-hand side.

When all pairs of vertices can be connected by a path, a graph is called connected. The path is a continuous sequence of edges, where each successive edge begins where the previous one left off. Paths that start and end at the same vertex are regarded as cycles. For example, the paths (1-6-12-7-1) and (11-10-4-12-7-5-11) in Figure 11.2 are cycles. Graphs that are not connected are disconnected (Figure 11.3). Disconnected graphs can be devided into connected components. One can think of a graph as a map showing cities (vertices) and the freeways (edges) that connect them. Not all cities are connected by freeways: for example, you cannot drive from Miami to Honolulu. These two cities belong to two different connected components of the graph. A graph is called complete if there is an edge between every two vertices.

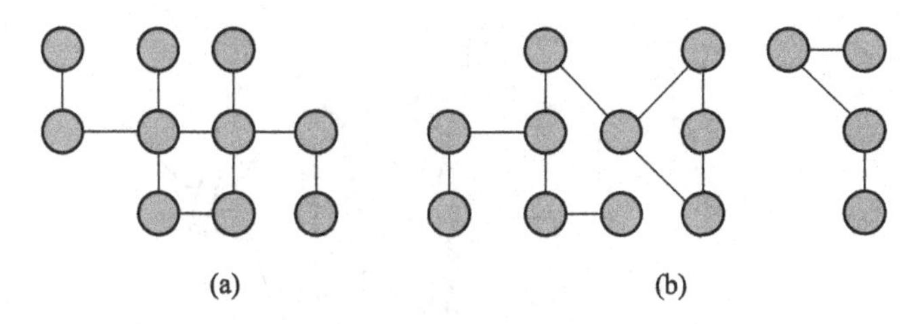

Fig. 11.3 A connected (a) and a disconnected (b) graph

2. Euler graph

Graph theory was born in the eighteenth century when Leonhard Euler solved the famous Königsberg Bridge problem as below.

Bridge Obsession Problem: *Find a tour through a city (located on n islands connected by m bridges) that starts on one of the islands, visits every bridge exactly once, and returns to the originating island.*

Input: A map of the city with n islands and m bridges.

Output: A tour through the city that visits every bridge exactly once and returns to the starting island.

Königsberg is located on the banks of the Pregel River, with a small island in the middle. Serveral bridges connected the different parts of the city (Figure 11.4). Euler was interested in whether he could arrange a tour of the city in such a way that the tour visits each bridge exactly once. For Königsberg this turned out to be impossible, but Euler basically invented an algorithm to solve this problem for any city [1].

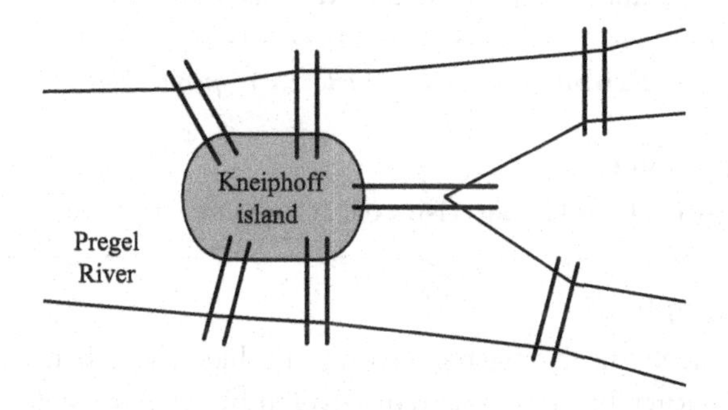

Fig. 11.4 Bridges of the Pregel River in Königsberg

Figure 11.5(a) shows a more complicated city map with ten islands and sixteen bridges.

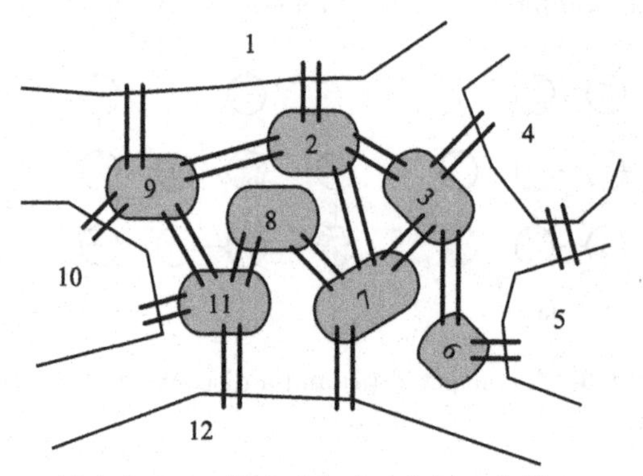

(a) A city map with ten islands and sixteen bridges

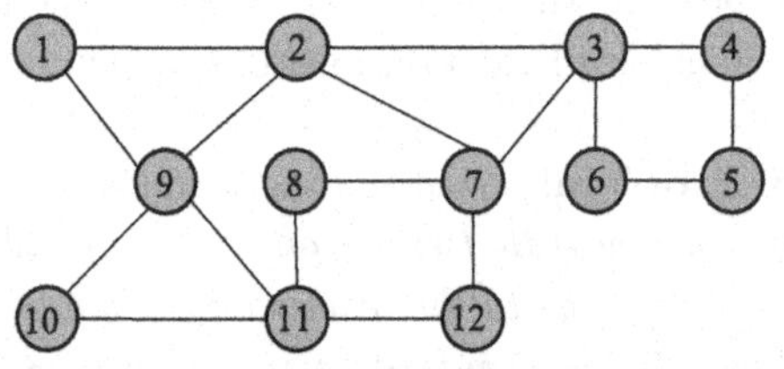

(b) Euler transformed the map into a graph and obtained an Eulerian Cycle

Fig. 11.5 A more complicated version of Bridge Obsession problem

Euler transformed the map into a graph with ten vertices and sixteen edges (every island corresponds to a vertex and every bridge corresponds to an edge) and obtained an Eulerian cycle. The path that runs through vertices 1-2-3-4-5-6-3-7-2-9-11-8-7-12-11-10-9-1 is an Eulerian cycle (Figure 11.5) [1]. After this transformation, the Bridge Obsession problem turns into the Eulerian Cycle problem as below. This problem was solved by Euler and later found thousands of applications in various areas.

Eulerian Cycle Problem: *Find a cycle in a graph that visits every edge exactly once.*

Input: A graph G.

Output: A cycle in G that visits every edge exactly once.

---

3. Tree graph

After Euler solved the Königsberg Bridge problem, graph theory was forgotten for a century before it was rediscovered by Arthur Cayley [1]. Arthur Cayley studied the chemical structures of (noncyclic) saturated hydrocarbons

C_nH_{2n+2} (Figure 11.6).

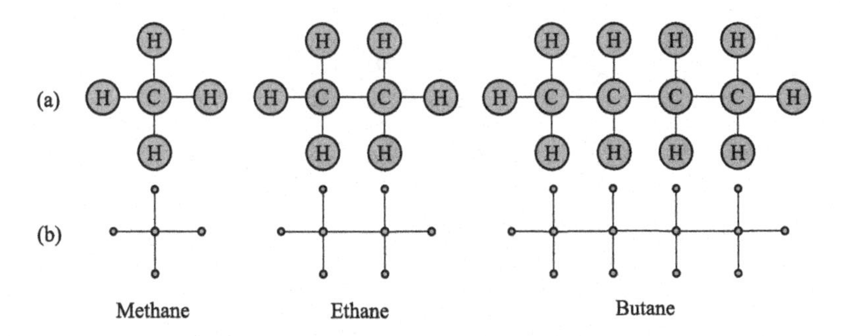

Fig. 11.6 Hydrocarbons as chemists see them (a), and their graph representation (b)

Structures of this type of hydrocarbon are examples of trees, which are simply connected graphs with no cycles. It is not hard to indicate that every tree has at least one vertex with degree 1, or leaf. This observation immediately implies that every tree on n vertices has $n - 1$ edges, regardless of the structure of the tree. Indeed, since every tree has a leaf, we can delete it and its attached edge, resulting in another tree. So far, we have removed one edge and one vertex. In this smaller tree there exists a leaf that we, again, remove. So far, we have removed two vertices and two edges. We keep this up until we are left with a graph with a single vertex and no edges. Since we have removed $n - 1$ vertices and $n - 1$ edges, the number of edges in every tree is $n - 1$ (Figure 11.7).

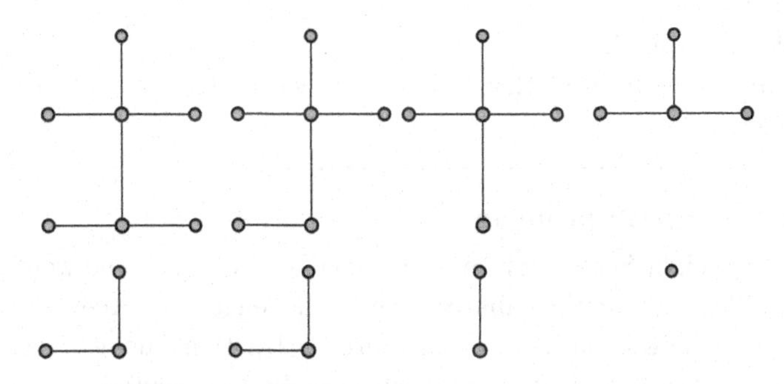

Fig. 11.7 A tree with n vertices has $n - 1$ edges

4. Hamilton graph

A game, which is corresponding to a graph whose twenty vertices were labeled with the names of twenty famous cities, is invented by William Hamilton after Cayley's work on tree enumeration (Figure 11.8) [1].

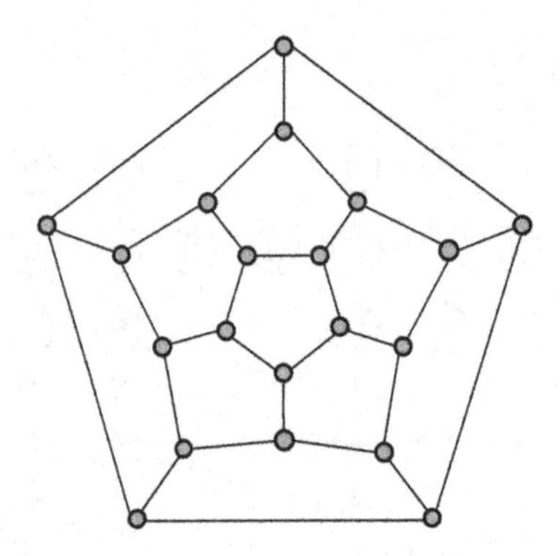

Fig. 11.8 A case of Hamiltonian Cycle

The aim of the game is to visit all twenty cities in such a way that every city is visited exactly once before returning back to the city where the tour started. As the story goes, Hamilton tried to sell the game for 25 pounds to a London game dealer and it failed miserably. Despite the commercial failure of this great idea, the more general problem of finding 'Hamiltonian Cycles' in arbitrary graphs is of great importance to many scientific and engineering disciplines. The problem of finding Hamiltonian Cycles looks deceivingly simple and somehow similar to the Eulerian Cycle problem. However, it turns out to be NP-complete while the Eulerian Cycle problem can be solved in linear time.

Hamiltonian Cycle Problem: *Find a cycle in a graph that visits every vertex exactly once.*

Input: A graph G.

Output: A cycle in G that visits every vertex exactly once (if such a cycle exists).

---

5. Shortest path problem

Some sort of weight is often given to every edge of graphs. The weight of an edge may reflect different attributes such as the length of a freeway segment connecting two cities, the number of tourist attractions along a city block, and so on [1]. Weighted graphs are often formally represented as an ordered triple, $G = (V, E, w)$, where V is the set of vertices in the graph, E is the set of edges, and w is a weight function defined for every edge e in E (i.e., $w(e)$ is a number reflecting the weight of edge e). Given a weighted graph, one may be interested in finding the shortest path between two vertices (e.g., the shortest path between Beijing and Shenzhen). If you were given a complicated road map, this problem may sound quite difficult. However, it turns out that there exist useful algorithms to answer this question.

Shortest Path Problem: *Given a weighted graph and two vertices, find the shortest distance between them.*

Input: A weighted graph, G = (V, E, w), and two distinguished vertices s and t.

Output: The shortest path between s and t in graph G.

Graphs are powerful methods in many applied researches and their role goes well beyond analysis of maps of freeways. For example, at first glance, the following 'basketball' problem has nothing to do with graph theory: *Fifteen teams play a basketball tournament in which each team play with each other team. Prove that the teams can always be numbered from 1 to 15 in such a way that team 1 defeated team 2, team 2 defeated team 3, ..., team 14 defeated team 15.* A careful analysis reduces this problem to finding a Hamiltonian path in a directed graph on fifteen vertices.

11.2.2 Applications of Graph Theory

Graph theory flourished in the twentieth century to become a critical component of discrete mathematics based on achievements of Euler, Cayley, and Hamilton etc. In the 1950s, it was known that genes behaved as functional units of DNA, much like pearls on a necklace, but the chemical structure and organization of the genes was unknown. Were genes broken into still smaller components? If so, how were they organized? It seemed a reasonable hypothesis that the DNA content of genes was branched, or even looped, rather than linear. These two organizations have very different topological implications. Prior to Watson and Crick's elucidation of the DNA double helix, Seymour Benzer used graph theory to indicate that genes are linear. Nowdays, graph theoretical tools are widely used to the study of protein interaction networks, protein function, protein folding, gene coregulation, gene coexpression, and so on [2-5].

1. Protein interaction network analysis

In most of conditions, graphs are not required to have any type of regularity. This makes them very flexible combinatorial objects and it can be used to represent complex and diverse relationships [4]. Graph theory is used in bioinformatics to study the relationships between biological entities. For example, a graph, the so-called protein-protein interaction network, where proteins are nodes and every pair of interacting proteins is connected by an edge, is used to represent experimentally determined protein interactions. Such a representation may not show all the complexity of protein interactions in underlying biological processes. The study of the topological properties of these networks has still become a potential tool in searching for general principles that govern the organization of molecular networks [4]. For example, it was observed that in protein interaction networks some types of small-size subnetworks are much more abundant than would be expected by chance [6]. The discovery of these overrepresented subnetworks or network motifs has led to investigation of their information processing properties [7] and network evolution mechanisms that could account for their emergence [8].

Graph theory is used to study stability of protein complex interaction networks [9]. It is known that protein complexes can interact with other complexes to form protein complex interaction network (PCIN) that involves in important cellular processes. Huang et al. [9] employed graph theoretical approach to reveal hidden properties and features of four species PCINs. Two main issues are addressed, (i) the global and local network topological properties, and (ii) the stability of the networks under 12 types of perturbations. According to the topological parameter classification, they identified some critical protein complexes and validated that the topological analysis approach could provide meaningful biological interpretations of the protein complex systems. Local topological parameters are good indicators to characterize the structure of PCINs. They also found that the degree-based, betweenness-based and brokering-coefficient-based perturbations have the largest effect on network stability.

Network reconstruction based on proteomic data and prior knowledge of protein connectivity was carried out by Stavrakas et al. [10] using graph theory. Modeling of signal transduction pathways is instrumental for understanding cells' function. Stavrakas et al. [10] propose a method to interrogate signal transduction pathways in order to produce cell-specific signaling models. They integrate available prior knowledge of protein connectivity, in a form of a Prior Knowledge Network (PKN) with phosphoproteomic data to construct predictive models of the protein connectivity of the interrogated cell type. The pathways are combined through a heuristic formulation to produce final topology handling inconsistencies between the PKN and the experimental scenarios. Their results show that the algorithm is efficient and accurate for the construction of medium and large scale signaling networks. They demonstrate the applicability of the proposed approach by interrogating a manually curated interaction graph model of EGF/TNFA stimulation against made up experimental data.

The integrative and cooperative nature of protein structure contains the assessment of topological and global features of constituent parts. High compatibility to structural concepts or physicochemical properties in addition to exploiting a remarkable simplification in the system has made network an ideal tool to study biological systems. Varied protein structural and functional characteristics have been illustrated based on graph theory. Niknam et al. [5] present an interactive and user-friendly Matlab-based toolbox, PDB2Graph, devoted to protein structure network construction, visualization, and analysis. PDB2Graph also provides a promising tool for identifying critical nodes involved in protein structural robustness and function based on centrality indices. It maps critical amino acids in protein networks and can greatly aid structural biologists in selecting proper amino acid candidates for manipulating protein structures in a more reasonable and rational manner [5]. The capability and efficiency of PDB2Graph was shown by the case of the structural modification of Calmodulin through allosteric binding of Ca^{2+}.

2. Protein function analysis

Developments of sequencing technology result in constant expansion of the gap between the number of the known sequences and their functions. It is indispensable to develop a computational method for the annotation of protein function. A novel method is proposed by Li et al. [3] to identify protein function based on the weighted protein-protein interaction network and graph theory. The network topology features with local and global information are presented to characterise proteins. The minimum redundancy maximum relevance algorithm is used to select 227 optimized feature subsets and support vector machine technique is utilized to build the prediction models. The performance of their method is assessed through 10-fold cross-validation test, and the range of accuracies is from 67.63% to 100%. Generally, such network topology features provide insights into the relationship between protein functions and network architectures [3].

3. Protein folding analysis

In another case, graph theory is also used for studying protein folding. How many structurally different microscopic routes are accessible to a protein molecule while folding? This has been a challenging question to address experimentally. Atomistic simulations, on the other hand, are restricted by sampling and the inability to reproduce thermodynamic observables directly. Gopi et al. [2] overcome these bottlenecks and provide a quantitative description of folding pathway heterogeneity by developing a comprehensive, scalable and yet experimentally consistent approach combining concepts from statistical mechanics, physical kinetics and graph theory. The predictive methodology developed in their study reveals the presence of rich ensembles of folding mechanisms that are generally invisible in experiments, reconciles the contradictory observations from experiments and simulations and provides an experimentally consistent avenue to quantify folding heterogeneity.

4. Identifing non-random somatic mutations

A graph theoretic approach to utilizing protein structure to identify non-random somatic mutations has been successfully developed by Ryslik et al. [11]. It is well known that the development of cancer is caused by the accumulation of somatic mutations within the genome. A variety of approaches have been developed to identify potential driver mutations using methods such as machine learning and mutational clustering. Ryslik et al. [11] have designed and implemented GraphPAC (Graph Protein Amino acid Clustering), which increases their power to identify mutational clusters by taking into account protein tertiary structure via a graph theoretical approach. Using GraphPAC, you are able to detect novel clusters in proteins that are known to exhibit mutation clustering as well as identify clusters in proteins without evidence of prior clustering based on current methods. GraphPAC provides an alternative to iPAC and an extension to current methodology when identifying potential activating driver mutations by utilizing a graph theoretic approach when considering protein tertiary structure.

11.3 Dynamic Programming

11.3.1 Introduction to Dynamic Programming

A problem can be break into smaller subproblems using some algorithms and the solution of the larger one is therefore constructed based on the solutions of the subproblems. During this process, the number of subproblems may be huge, and some algorithms solve the same subproblem repeatedly, needlessly increasing the running time. Dynamic programming organizes computations to avoid recomputing values that you already know, and a great deal of time can be saved. We use the following problem to illustrate the technique.

Suppose that instead of answering the phone you decide to play the 'Rocks' game from the previous book with two piles of rocks, say ten in each [1]. The reader will be reminded that in each turn, one player may take either one rock (from either pile) or two rocks (one from each pile). Once the rocks are taken, they are removed from play. The player that takes the last rock wins the game. You make the first move. To find the winning strategy for the 10+10 game, we can get a table, which we can call R, shown below. Instead of solving a problem with 10 rocks in each pile, a more general problem will be solved with n rocks in one pile and m rocks in another (the $n + m$ game) where n and m are arbitrary. If Player 1 can always win the game of $5 + 6$, then we would say $R_{5,6} = W$, but if Player 1 has no winning strategy against a player that always makes the right moves, we would write

$R_{5,6} = L$. Computing $R_{n,m}$ for an arbitrary n and m seems very difficult, but we can build on smaller values. Some games, notably $R_{0,1}$, $R_{1,0}$, and $R_{1,1}$, are clearly winning propositions for Player 1 since in the first move, Player 1 can win. Thus, we fill in entries (1, 1), (0, 1) and (1, 0) as W as follows Figure 11.9 [1].

	0	1	2	3	4	5	6	7	8	9	10
0		W									
1	W	W									
2											
3											
4											
5											
6											
7											
8											
9											
10											

Fig. 11.9 'Rocks' game No.1

After the entries (0, 1), (1, 0), and (1, 1) are finished, we fill other entries. In the (2, 0) case, the only move that Player 1 can make leads to the (1, 0) case that is a winning position for his opponent. A analysis applies to the (0, 2) case is similar, producing the result in Figure 11.10.

	0	1	2	3	4	5	6	7	8	9	10
0		W	L								
1	W	W									
2	L										
3											
4											
5											
6											
7											
8											
9											
10											

Fig. 11.10 'Rocks' game No.2

In the (2, 1) case, Player 1 have three choices that lead respectively to the games of (1, 1), (2, 0), or (1, 0). The (2, 0) case gives a losing position for his opponent and therefore (2, 1) is a winning position. The (1, 2) case is symmetric to (2, 1), and the following Figure 11.11 is thus obtained.

	0	1	2	3	4	5	6	7	8	9	10
0		W	L								
1	W	W	W								
2	L	W									
3											
4											
5											
6											
7											
8											
9											
10											

Fig. 11.11 'Rocks' game No.3

Now we can fill in $R_{2,2}$. In the (2, 2) case, Player 1 can make three various moves that lead to entries (2, 1), (1, 2), and (1, 1). All of these entries are winning positions for his opponent and therefore $R_{2,2} = L$ as shown in Figure 11.12.

	0	1	2	3	4	5	6	7	8	9	10
0		W	L								
1	W	W	W								
2	L	W	L								
3											
4											
5											
6											
7											
8											
9											
10											

Fig. 11.12 'Rocks' game No.4

We can proceed filling in R at the same way by noticing that for the entry (i, j) to be L, the entries above, diagonally to the left and directly to the left, must be W. These entries ($(i-1, j)$, $(i-1, j-1)$, and $(i, j-1)$) correspond to the three possible moves that Player 1 can make.

The ROCKS algorithm determines if Player 1 wins or loses. If Player 1 wins in an $n+m$ game, ROCKS returns W. If Player 1 loses, ROCKS returns L. The ROCKS algorithm introduces an artificial initial condition, $R_{0,0} = L$

	0	1	2	3	4	5	6	7	8	9	10
0		*W*	*L*	*W*	*L*	*W*	*L*	*W*	*L*	*W*	*L*
1	*W*	*W*	*W*	*W*	*W*	*W*	*W*	*W*	*W*	*W*	*W*
2	*L*	*W*	*L*	*W*	*L*	*W*	*L*	*W*	*L*	*W*	*L*
3	*W*	*W*	*W*	*W*	*W*	*W*	*W*	*W*	*W*	*W*	*W*
4	*L*	*W*	*L*	*W*	*L*	*W*	*L*	*W*	*L*	*W*	*L*
5	*W*	*W*	*W*	*W*	*W*	*W*	*W*	*W*	*W*	*W*	*W*
6	*L*	*W*	*L*	*W*	*L*	*W*	*L*	*W*	*L*	*W*	*L*
7	*W*	*W*	*W*	*W*	*W*	*W*	*W*	*W*	*W*	*W*	*W*
8	*L*	*W*	*L*	*W*	*L*	*W*	*L*	*W*	*L*	*W*	*L*
9	*W*	*W*	*W*	*W*	*W*	*W*	*W*	*W*	*W*	*W*	*W*
10	*L*	*W*	*L*	*W*	*L*	*W*	*L*	*W*	*L*	*W*	*L*

Fig. 11.13 'Rocks' game No.5

to simplify the pseudocode.

```
ROCKS(n, m)
1  R0,0 = L
2  for i ← 1 to n
3     if Ri−1,0 = W
4        Ri,0 ← L
5     else
6        Ri,0 ← W
7  for j ← 1 to m
8     if R0,j−1 = W
9        R0,j ← L
10      else
11         R0,j ← W
12 for  i ← 1 to n
13      for j ← 1 to m
14         if Ri−1,j−1 = W and Ri,j−1 = W and Ri−1,j = W
15            Ri,j ← L
16         else
17            Ri,j ← W
18 return  Rn,m
```

In point of fact, a faster algorithm to solve the Rocks puzzle depends on the simply pattern in R, and checks to see if n and m are both even, in which case the player loses. However, though the following FASTROCKS is more efficient than ROCKS, it may be difficult to modify it for other games, for example a game in which each player can move up to three rocks at a time from the piles. This is one example where the slower algorithm is more instructive than a faster one. But obviously, it is often better to use the faster one when you really need to solve the corresponding problem.

```
FASTROCKS(n, m)
1  if n and m are both even
2      return L
3  else
4      return W
```

11.3.2 Applications of Dynamic Programming

In 1953, dynamic programming algorithm was first introduced by Richard Bellman to study multistage decision problems, but he may not anticipate its broad applications in current computer programming and bioinformatics. Dynamic programming algorithm gives polynomial time solutions to a class of optimization problems that have an optimal substructure [4]. The optimal solution of the overall problem can be solved from the optimal solutions of many overlapping subproblems that can be computed independently and memorized for repeated use. Dynamic programming is one of the early algorithms introduced in bioinformatics and it has been broadly applied since then [12-15].

1. Comparing sequence

A basic framework for understanding DNA sequence comparison algorithms is carried out by dynamic programming [1]. And many of these algorithms have been used by scientists to make important inferences about gene function and evolutionary mechanism. Similar to the shortest path problem in graph theory [16], dynamic programming can solve some problems that are often associated to the objects with a similar optimal substructure. A string, with naturally ordered letters, is a typical example of such objects. Hence, many computational problems related to strings can be solved using dynamic programming. Interestingly, the primary structures of deoxyribonucleic acids (DNAs) and proteins, which are both linear molecules, thus can be represented by plain sequences, although on two different alphabets with limited size (4 nucleotides and 20 amino acids, respectively). Life is complicated in many conditions, but it is also simple, in this perspective. Dynamic programming provided a powerful tool to compare their sequences [4].

After we found a new gene, we often have no idea about its function. A common way to studying a newly sequenced gene's function is to check similarities with genes of known function [1]. The discovery of the gene of cystic fibrosis, which is a fatal disease associated with abnormal secretion, and is diagnosed in children at a rate of 1 in 3,900, is a good example of a successful similarity search. In 1989 the study for the cystic fibrosis gene was narrowed to a region of 1 million nucleotides on the chromosome 7. However, the exact location of the gene was not revealed yet. When the area around the cystic fibrosis gene was sequenced, researchers compared the region against a database of all known genes, and discovered similarities between some segment within this region and a gene that had already been discovered, and was known to code for adenosine triphosphate (ATP) binding proteins. These proteins span the cell membrane multiple times as part of the ion transport channel; this seemed a plausible function for a cystic fibrosis gene, given the fact that the disease involves sweat secretions with abnormally high sodium content. Finally, the similarity search illustrated a damaged mechanism in faulty cystic fibrosis genes [1]. Establishing a link between cancer-causing genes and normal growth genes and illustrating the nature of cystic fibrosis were only the first success study in sequence comparison. Many applications of sequence comparison algorithms were quickly found, and today bioinformatics approaches are among the dominant techniques for the discovery of gene function.

Nearly fifty years ago, Needleman and Wunsch [17] used bottom-up dynamic programming to study an optimal pairwise alignment between two protein sequences. Although this algorithm gives a similar assessment of a pair of sequences, it assumes the similarity between two input sequences is across the entire sequences. Smith and Waterman [18] carried out a simple yet important modification to this algorithm to perform local alignments, in which similar parts of input sequences were aligned. The advantage of local alignments in identifying common functional domains or motifs is obvious. It has attracted considerable interests and gave us several powerful tools in bioinformatics nowadays, such as FASTA and BLAST. Similar to DNA and protein, ribonucleic acids (RNAs) are also linear. They fold into stable secondary structures to perform their biological functions. So, they are often represented by sequences of four letters, similar to DNAs, but with annotated *arcs*, where each arc represents a base pair. Interestingly, the base pairs in native secondary structure of an RNA usually do not form pseudoknots, that is, the arcs are not crossing. As a result, RNA sequences with annotated arcs can also be sorted into partial ordered trees (instead of sequences) [19]. Therefore, many bioinformatics problems related to RNAs such as RNA secondary structure prediction, RNA structure comparison, RNA consensus folding and so on, can be carried out by dynamic program algorithms [4,19].

2. Protein structure study

Unlike RNAs, the native three-dimensional (3D) structures of proteins are difficult to be predicted from their primary sequences and are determined mainly by experimental methods including crystallography and nuclear magnetic resonance (NMR). It has been observed that proteins sharing similar 3D structures may have unrelated primary sequences. With more and more protein structures being solved experimentally, it is important to automatically identify proteins with similar structure but lacking obvious sequence similarity [20]. Although it is not straightforward to represent the protein 3D structures as partially ordered sequences, several commonly used methods for protein structure comparison are also designed based on dynamic programming algorithms [4].

It is known that the determination of secondary structure topology is a critical step in deriving the atomic structures from the protein density maps obtained from electron cryomicroscopy technique [12]. This step often relies on matching the secondary structure traces detected from the protein density map to the secondary structure sequence segments predicted from the amino acid sequence. A pool of possible secondary structure positions needs to be sampled because of inaccuracies in both sources of information. Biswas et al. [12] developed a dynamic programming method to find the optimal placement for a secondary structure topology and found that the algorithm requires significantly less computational time than the brute force method.

One of the main tasks towards the prediction of protein β-sheet structure is to predict the native alignment of β-strands. The alignment of two β-strands defines similar regions that may reflect functional, structural, or evolutionary relationships between them. Therefore, any improvement in β-strands alignment not only reduces the computational search space but also improves β-sheet structure prediction accuracy. Sabzekar et al. [14] developed a dynamic programming algorithm for protein β-strand alignment. They utilized both β-residues and β-structure information in alignment of β-strands. The structure of dynamic programming of the alignment algorithm is changed in order to work with their prior knowledge. Moreover, the Four Russians method is applied to the proposed alignment algorithm in order to reduce the time complexity of the problem. The modified algorithm was applied to the β-sheet structure prediction. The experimental results on the BetaSheet 916 data set showed significant improvements in the execution time, the accuracy of β-strands' alignment and consequently β-sheet structure prediction accuracy.

As mentioned above, predicting the β-sheet structure of a protein is one of the most important intermediate steps towards the identification of its tertiary structure. To achieve reliable long-range interactions, a promising approach is to enumerate and rank all β-sheet conformations for a given protein and find the one with the highest score. The problem is that the search space of the problem grows exponentially with respect to the number of β-strands. Additionally, brute-force calculation in this conformational space leads to dealing with a combinatorial explosion problem with intractable computational complexity. Sabzekar et al. [15] have developed an efficient dynamic programming algorithm to generate and search the space of the problem efficiently to reduce the time complexity of the problem. More accurate β-sheet structures are found by searching all possible conformations, and the time complexity of the problem is reduced by searching the space of the problem efficiently which makes the proposed method applicable to predict β-sheet structures with high number of β-strands. Experimental results on the BetaSheet 916 dataset showed significant improvements of the proposed method in both execution time and the prediction accuracy in comparison with the state-of-the-art β-sheet structure prediction methods [15].

3. Genome study

The exemplar breakpoint distance problem is motivated by finding conserved sets of genes between two genomes. It asks to find respective exemplars in two genomes to minimize the breakpoint distance between them. If one genome has no repeated gene (called trivial genome) and the other has genes repeating at most twice, it is referred to as the (1, 2)-exemplar breakpoint distance problem, EBD(1, 2) for short. Wei et al. [21] proposed a parameter to describe the maximum physical span between two copies of a gene in a genome, and based on it, designed a fixed-parameter algorithm for EBD(1, 2). Using a dynamic programming approach, this algorithm can take $O(4^s n^2)$ time and $O(4^s n)$ space to solve an EBD(1, 2) instance that has two genomes of n genes where the second genome has each two copies of a gene spanning at most s copies of the genes. This algorithm can also be used to compute the maximum adjacencies between two genomes.

4. Algebraic dynamic programming over general data structures

Dynamic programming algorithms provide exact solutions to many problems in computational biology, such as sequence alignment, RNA folding, hidden Markov models (HMMs), and scoring of phylogenetic trees [22]. The ideas of algebraic dynamic programming (ADP) are generalized to a much wider scope of data structures by relaxing the concept of parsing. This allows us to formalize the conceptual complementarity of inside and outside variables in a natural way. It is demonstrated that outside recursions are generically derivable from inside decomposition schemes. In addition to rephrasing the well-known algorithms for HMMs, pairwise sequence alignment, and RNA folding they show how the TSP and the shortest Hamiltonian path problem can be implemented efficiently in the extended ADP framework. The ancient evolution of HOX gene clusters in terms of shortest Hamiltonian paths was successfully investigated as a show case application. The generalized ADP framework greatly facilitates the development and implementation of dynamic programming algorithms for a wide spectrum of applications [22].

11.4 Bayesian Statistics

11.4.1 Introduction to Bayesian Statistics

Given many advantages of Bayesian methods over frequentist-based methods, they have become widely adopted in bioinformatics in recent years. First of all, they provide an intuitive method to incorporating prior biological knowledge in data analysis and interpretation. Secondly, different kinds of genomics and proteomics data can be integrated in a principled fashion under a consistent modeling framework. Thirdly, many Bayesian computational tools that have been developed over the past 30 years are readily available to complex models for genomics and proteomics data that are difficult to deal with from a frequentist perspective [23].

In addition to numerous journal articles published on Bayesian methods for computational biology problems, several books dedicated to Bayesian models in bioinformatics have appeared, for example Alterovitz et al. [23], Dey et al. [24], and Mallick et al. [25], as well as tutorials [26]. In this part, we start with a discussion of the fundamental Bayes theorem and several simple examples of how it can be applied to study some biological problems. Several more complex problems that have benefited from Bayesian modeling and analysis are then covered.

In most probability and statistics textbooks, the Bayes theorem is introduced in the following form: $P(B|A) = P(A \text{ and } B)/P(A)$, where A and B represent two events, $P(B|A)$ is the conditional probability that B occurs given that A has already happened, $P(A \text{ and } B)$ is the joint probability that both A and B occur, and $P(A)$ is the unconditional (marginal) probability that A occurs [23]. If events A and B are independent, that is, $P(A \text{ and } B) = P(A)P(B)$, then $P(B|A) = P(B)$. The value of $P(B|A)/P(B)$ characterizes the degree of dependency of event B on event A. Let us first deal with a simple example using the Bayes theorem. When we are interested in the dependency of the neighboring nucleotides in a genome, event A may correspond to a base being a specific nucleotide, say 'C,' and event B may correspond to the next base being a specific nucleotide, say also 'C.' Suppose that the proportion of bases in this genome being 'C' is 0.25, that is, $P(A) = 0.25$, and the proportion of two consecutive bases being 'CC' is 0.10, that is, $P(A$ and $B) = 0.10$. Then $P(B|A) = P(A \text{ and } B)/P(A) = 0.10/0.25 = 0.4$. Therefore, $P(B|A)/P(B) = 0.4/0.25 = 1.6$; that is, having a base being 'C' increases the probability of the next base being 'C' by 60%. In this case, knowing that the previous base is 'C' (prior information) affects the chance that the current base is 'C'.

Another example is the inference of the ethnic origin of an individual based on genetic marker information, which is commonly happened in genetics or forensic studies. For example, genetic markers may be collected at a number of single nucleotide polymorphisms from an individual and is used to infer if this person is a Caucasian or an African. In this case, event B and A are the ethnic origin of this individual and the genetic marker data for this person, respectively. This problem can be done based on the Bayes theorem as follows:

P(an individual is a Caucasian|marker data)
$= P$(an individual is a Caucasian and marker data)|P(marker data)
$= P$(an individual is a Caucasian)P(marker data|an individual is a Caucasian)|P(marker data).

It is to be noted that we have used the following results:

$$P(B|A) = P(A \text{ and } B)|P(A) = P(B)P(A|B)/P(A)$$

above; in another word, the Bayesian theorem was used twice with one conditional on A and the other conditional on B. We have three quantities that need to be calculated in this formulation: P(an individual is a Caucasian), P(marker data | an individual is a Caucasian), and P(marker data). The information about the value of P(an individual is a Caucasian) often comes from the knowledge on the general population where this individual is sampled from, therefore some injection of prior information in this calculation. To calculate P(marker data|an individual is a Caucasian), information on the properties of the markers in a Caucasian population, such as allele and genotype frequencies, is required. Again, some relevant knowledge about the markers is also needed in this process. Lastly, to calculate P(marker data), all possible ethnic origins for this individual need be considered. In the case of two origins, for instance Caucasian and African, we can calculate P(marker data) as

P(marker data)= P(an individual is a Caucasian and marker data) + P(an individual is an African and marker data)
$= P$(an individual is a Caucasian)P(maker data | an individual is a Caucasian) + P(an individual is an African)P(marker data| an individual is an African).

We may consider P(an individual is a Caucasian) as the prior information about the ethnicity of a sampled individual in the population being selected, and P(an individual is a Caucasian|marker data) as the posterior probability for this person's ethnicity after the relevant information (i.e. genetic marker data) about this individual are collected. Therefore, the Bayes theorem provides an intuitive way to combine prior information and data to carry out our inferential objective. Although A and B show events in the above notion, the equation still holds if they are replaced by more general entities, such as data sets or parameters. For example, consider A as the probability that a coin lands on heads; therefore A can be replaced by a parameter h, the chance of seeing a head. Now consider B as the results from tossing this coin N times; therefore B can be replaced by a set of observations $Y = (Y_1, \ldots, Y_N)$, where $Y_i = 1$ or 0 corresponds to a head or tail from the ith tossing. Then $P(B|A) = P(Y|h) = hH(1-h)T$, where H is the total number of heads and T is the total number of tails observed from N experiments. In this case, the Bayes theorem can be used to infer the probability that this coin lands on heads based on the observations Y, $P(h|Y)$, through the Bayes theorem as follows. It is to be noted that the roles of A and B can be exchanged so that $P(A|B) = P(A \text{ and } B)/P(B)$. Therefore, $P(h|Y) = P(h \text{ and } Y)/P(Y) = P(h)P(Y|h)/P(Y)$. It is to be noted that the posterior distribution of h depends on three quantities: the prior distribution for h, $P(h)$; the probability distribution for the observations Y conditional on the parameter h, $P(Y|h)$; and the probability for Y integrated over all possible values of h, $P(Y) = \int P(Y|h)P(h)dh$, which is independent of h. Depending on the specific choices of the prior distribution for h, different posterior inference conditional on the observed data Y will be observed. This example is simple, and it domonstrates the general approach for inferring parameters of interest, denoted by θ, from a prior distribution $\pi(\theta)$ and observations Y, through the following equation:

$$\pi(\theta|Y) = \pi(\theta)P(Y|\theta)/P(Y) \propto \pi(\theta)P(Y|\theta).$$

A population genetics problem is similar to the one discussed above. Our goal is not to infer one individual's ethnic background. We aim to infer the allele frequencies of a set of genetic markers from the collection of a set of individual samples. For a single marker, the collected data are the genotypes from a set of individuals at this marker, and the parameter is the allele frequency of this marker. Then we have P(allele frequency|marker data) = P(marker data|allele frequency) P(allele frequency)|P(marker data). P(marker data|allele frequency) can be calculated based on the Hardy–Weinberg equilibrium. The prior information comes in the form of P(allele frequency), which is usually assumed to have the Dirichlet distribution, and an integration is required to calculate P(marker data) where the integration is over the prior distribution for the allele frequencies.

These examples indicate that the simple Bayes theorem can be applied to a variety of problems where the events A and B can represent different entities. For most Bayesian methods, A and B usually represent the observed data and parameters of interest, respectively, which can be generically denoted by Y and θ, and the goal is to infer the posterior distribution of $\pi(\theta|Y) = \pi(\theta)P(Y|\theta)/P(Y)$. The following elements are key in applying the Bayesian theorem to a specific problem: ① model specifications, which are needed to evaluate $P(Y|\theta)$; ② prior specifications, which are needed to define $\pi(\theta)$; and ③ computational methods needed to infer the posterior distributions, because it is usually not easy to directly evaluate $P(Y)$. Statistical inference is usually achieved through sampling from the posterior distribution $\pi(\theta|Y)$. We need pay attention to specify model forms and prior distributions such that, ① the models are comprehensive enough to appropriately model the obtained data; ② the degree of knowledge about the model parameters can be reflected by the prior distributions; and ③ it is feasible to infer the posterior distributions using suitable computational procedures. It is worth to note that the development and applications of Markov chain Monte Carlo methods for posterior inferences provided intensive driving force that has made Bayesian methods widely employed in recent years [27]. When there are a large number of parameters involved in a model, the Gibbs sampler is the most commonly employed one among different Markov chain Monte Carlo methods. Let θ denote the collection of all the model parameters. It may be tough to sample from the joint posterior distribution of $\pi(\theta|Y)$ when θ has a large dimension. On the other hand, if θ can be partitioned into C subsets in the form of $\theta = (\theta_1, \theta_2, \ldots, \theta_C)$, and if it is relatively easy to sample from $\pi(\theta_1|\{Y, \theta_2, \theta_3, \ldots, \theta_C\}), \pi(\theta_2|\{Y, \theta_1, \theta_3, \ldots, \theta_C\}), \ldots, \pi(\theta_C|\{Y, \theta_1, \theta_2, \ldots,$ $\theta_{C-1}\})$, the Gibbs sampler proceeds by repeatedly sampling $\theta_1, \theta_2, \ldots$, and θ_C from these distributions to arrive at an empirical distribution of the joint posterior distribution of $\pi(\theta|Y)$.

11.4.2 Applications of Bayesian Statistics

1. Analysis of transcriptional regulatory networks

Bayesian methods can also be used to integrate data from multiple sources. Researchers aim to infer the regulatory targets of transcription factors based on joint analysis of gene expression data and protein–DNA binding data. It is easy to gather gene expression data from microarrays having probes targeting transcripts, whereas protein–DNA interaction data can be collected from chromatin immunoprecipitation experiments coupled with microarrays with probes targeting regulatory regions. These data shows different aspects of the gene regulation process, with the protein–DNA binding data indicating the potential targets of transcription factors, and the gene expression data suggestting the results of the complex regulation process. If there is no experimental noise and all binding targets are functional, the protein–DNA interaction data would be sufficient to deduce the regulatory network. However, as with any high-throughout data, substantial noises are often observed in protein interaction measurements. More importantly, the observed physical binding between a transcription factor and the regulatory region of a gene does not necessarily prove that the binding is functional. Therefore, the work to resolve some of the ambiguities and errors in the protein–DNA interaction data will be benefited from the incorporation of gene expression data.

Differential networks allow us to better understand the changes in cellular processes that are exhibited in conditions of interest, identifying variations in gene regulation or protein interaction between, for example, cases and controls, or in response to external stimuli. Thorne [28] reported a novel methodology for the inference of differential gene regulatory networks from gene expression microarray data. A Bayesian model selection approach was applied to compare models of conserved and varying network structure. Gaussian graphical models were used to represent the network structures. They applied a variational inference approach to the learning of Gaussian graphical models of gene regulatory networks, that enables us to perform Bayesian model selection that is significantly more computationally efficient than Markov chain Monte Carlo approaches. This method is demonstrated to be more robust than independent analysis of data from multiple conditions when applied to synthetic network data, generating fewer false positive predictions of differential edges.

The general Bayesian framework based on computational tools provide a powerful approach for modeling and analyzing complex genomics and proteomics data. These methods have been applied to many problems [23, 29-31]. Statistical inference for Bayesian models is usually based on the Gibbs sampling scheme where the posterior distributions are inferred from iteratively sampling from a set of conditional distributions. Aside from significant computational demands, it is important to ensure that the samples thus obtained can be used to represent the correct posterior distributions. Therefore, it is needed to monitor the convergence of the samples, and multiple runs are sometimes needed to check for the consistency across many runs. As for any Bayesian method, the choices of prior distributions may affect the conclusions drawn based on the posterior distributions. Despite that prior distributions may be dominated by the observed data when there are sufficient data, there is no guarantee this would always be the case, especially when the number of parameters is huge in some cases such as the analysis of genomics and proteomics data. It may be needed to carry out some types of sensitivity analysis to evaluate the impacts of the changes in the prior distributions.

The Bayesian methods offer a consistent framework for knowledge and data integration in the analysis of genomics and proteomics data, which is very important when many types of data need to be analyzed to extract the key information. For example, if the regulatory relationship among a set of transcription factors and genes is already known, this knowledge can be easily brought into the analysis by fixing these relationships. However, such prior knowledge may be difficult to incorporate under a frequentist approach. In addition, Bayesian methods are useful for complex models where the number of parameters reach as high as hundreds or thousands in the model. Although it may be possible to infer these parameters in a frequentist setting, such as through the expectation-maximization algorithm to maximize the likelihood, the likelihood surface may have many modes, making the inference unstable and difficult. On the other hand, Bayesian methods through sampling from the posterior distributions may produce more stable results. Sparsity constraints can also be easily incorporated through sparse priors. Notably, several excellent books are available to provide the foundations on Bayesian analysis [32] as well as computational implementations [33].

2. Predicting motifs in nucleotide sequences

Position weight matrices (PWMs) are the standard model for DNA and RNA regulatory motifs. In PWMs nucleotide probabilities are independent of nucleotides at other positions. Models that account for dependencies need many parameters and are prone to overfitting. Siebert and Söding [34] have developed a Bayesian approach for motif discovery using Markov models. This Bayesian Markov model (BaMM) training automatically adapts model complexity to the amount of available data. They also derive an EM algorithm for de-novo discovery of enriched motifs. For transcription factor binding, BaMMs achieve significantly ($P = 1/16$) higher cross-validated partial AUC than PWMs in 97% of 446 ChIP-seq ENCODE datasets and improve performance by 36% on average. BaMMs also learn complex multipartite motifs, improving predictions of transcription start sites, polyadenylation sites, bacterial pause sites, and RNA binding sites by 26%$\sim$101% [34].

3. Dfferential methylation detection

The advent of high-throughput DNA methylation profiling techniques has enabled the possibility of accurate identification of differentially methylated genes for cancer research. The large number of measured loci facilitates whole genome methylation study, yet posing great challenges for differential methylation detection due to the high variability in tumor samples. Wang et al. [35] have developed a novel probabilistic approach, differential methylation detection using a hierarchical Bayesian model exploiting local dependency, to detect differentially methylated genes based on a Bayesian framework.

This approach features a joint model to capture both the local dependency of measured loci and the dependency of methylation change in samples. A hierarchical Bayesian model is developed to fully take into account the local dependency for differential analysis, in which differential states are embedded as hidden variables. Simulation studies demonstrate that this method outperforms existing methods for differential methylation detection, particularly when the methylation change is moderate and the variability of methylation in samples is high. It has been applied to breast cancer data to identify important methylated genes (such as polycomb target genes and genes involved in transcription factor activity) associated with breast cancer recurrence [35].

4. Genomic prediction

Wang et al. [36] applied a new method, HyB_BR (for Hybrid BayesR), which implements a mixture model of normal distributions and hybridizes an Expectation-Maximization (EM) algorithm followed by Markov chain Monte Carlo (MCMC) sampling, to genomic prediction in a large dairy cattle population with imputed whole genome sequence data. The imputed whole genome sequence data included 994,019 variant genotypes of 16,214 Holstein and Jersey bulls and cows. Traits included fat yield, milk volume, protein kg, fat% and protein% in milk, as well as fertility and heat tolerance. HyB_BR achieved genomic prediction accuracies as high as the full MCMC implementation of BayesR, both for predicting a validation set of Holstein and Jersey bulls (multi-breed prediction) and a validation set of Australian Red bulls (across-breed prediction). HyB_BR had a ten fold reduction in compute time, compared with the MCMC implementation of BayesR (48 hours versus 594 hours). In many cases, HyB_BR identified sequence variants with a high posterior probability of affecting the milk production or fertility traits that were similar to those identified in BayesR. The results demonstrate that HyB_BR is a feasible method for simultaneous genomic prediction and QTL mapping with whole genome sequence in large reference populations.

5. Protein structure and protein interaction network analysis

Bayesian comparison of protein structures was carried out using partial Procrustes distance. It is known that an important topic in bioinformatics is the protein structure alignment. Most of previous statistical methods align two protein structures based on the global geometric information without considering the effect of neighbourhood in the structures. Ejlali et al. [37] provided a Bayesian model to align protein structures, by considering the effect of both local and global geometric information of protein structures. Parameters are estimated using a Markov chain Monte Carlo approach. This model was applied to a real dataset and the accuracy and convergence rate was assessed. Results showed that this model is more efficient than previous approaches.

Comparative analysis of protein-protein interaction (PPI) networks provides an effective means of detecting conserved functional network modules across different species. Jeong et al. [38] reported a novel probabilistic framework for comparing PPI networks and effectively predicting the correspondence between proteins, represented as network nodes, that belong to conserved functional modules across the given PPI networks. The basic idea is to estimate the steady-state network flow between nodes that belong to different PPI networks based on a Markov random walk model. The random walker is designed to make random moves to adjacent nodes within a PPI network as well as cross-network moves between potential orthologous nodes with high sequence similarity. Through evaluations based on multiple real PPI networks, the proposed scheme led to improved alignment results that are biologically more meaningful at reduced computational cost, outperforming the current state-of-the-art algorithms [38].

Recently, Bayesian network model was successfully used for identification of pathways by integrating protein interaction with genetic interaction data [39]. It is known that molecular interaction data at proteomic and genetic levels provide physical and functional insights into a molecular biosystem and are helpful for the construction of pathway structures complementarily. Despite advances in inferring biological pathways using genetic interaction data, there still exists weakness in developed models, such as activity pathway networks (APN), when integrating the data from proteomic and genetic levels. Fu et al. [39] utilized probabilistic graphical model to develop a new method that integrates genetic interaction and protein interaction data and infers exquisitely detailed pathway structure. The pathway network was modeled as Bayesian network. They applied this model to infer pathways for the coherent subsets of the global genetic interaction profiles, and the available data set of endoplasmic reticulum genes. This method can accurately reconstruct known cellular pathway structures such as SWR complex, ER-Associated Degradation pathway, Elongator complex and so on. This method is able to overcome its weakness (certain edges are inexplicable). The results indicate that the developed method based on Bayesian network performs better in predicting signaling pathways than previously described models.

11.5 Markov Models

11.5.1 Introduction to Markov Models

A Markov chain, also named as Markov model, describes a sequence of events that happen one after another in a chain. Each event determines the probability of the next event. A Markov model can be regarded as a process that moves in one direction from one state to the next with a certain probability, which is known as transition probability. A good example of a Markov model is the signal change of traffic lights in which the state of the current signal depends on the state of the previous signal (e.g., green light switches on after red light, which switches on after yellow light). Markov chains can also describe biological sequences written as strings of letters; each letter showing a state is linked together with transitional probability values [40]. The description of biological sequences using Markov chains allows the calculation of probability values for a given residue based on the unique distribution frequencies of nucleotides or amino acids.

Several types of Markov models are used to describe data sets of different complexities. In each type of Markov model, different mathematical solutions are derived. A zero-order Markov model describes the probability of the current state independent of the previous state. This is typical for a random sequence, in which every residue occurs with an equal frequency. A first-order Markov model describes the probability of the current state being determined by the previous state. This corresponds to the unique frequencies of two linked residues (dimer) occurring simultaneously. Similarly, a second-order Markov model describes the situation in which the probability of the current state is determined by the previous two states [40]. This corresponds to the unique trimer frequencies in biological sequences. For example, in a protein-coding sequence, the frequency of unique trimers should be various from that in a noncoding or random sequence. This discrepancy can be demonstrated by the second-order Markov model. In addition, there are even higher orders of Markov models for biological sequence analysis.

A stochastic process is a family of functions of a variable t, $\{X(t,\omega), t \in T, \omega \in \Omega\}$ (t is usually understood as a time), parametrized by random outcomes ω. For any fixed outcome ω, $X(.,\omega)$ is a function; for any fixed time t, $X(t,.)$ is a random variable. A Markov process, which is a special case of a stochastic process in that it has a limited memory, constitute the best-known and useful class of stochastic processes. Limited memory represents that for a process $X(t,\omega)$ which has been running in the past ($t \leqslant t_0$), the future $\{X(t,\omega),\ t > t_0\}$ is characterized by the present, i.e., $X(t_0,\omega)$. This latter property is known as the Markov property [41].

A Markov chain is a Markov process for which $X(t,\omega) \in S$, where S is a discrete set. Usually the state space S is a subset of the integers. That is, a Markov chain exhibits random transitions between discrete states. The theory shown here is focused on the case of a finite number of states, N, numbered $1, 2, \ldots, N$. Also, we discuss most systematically the case of discrete times $0, 1, 2, \ldots, k, \ldots$. However, we also increase some facts about the case of continuous time. Most frequently, we write $X_k(\omega)$ or X_k instead of $X(k,\omega)$. The defining property of a Markov chain is that the future of the chain is determined by the present, i.e., X_k. This can be expressed by the following equation:

$$P(X_{k+1}=j|X_k=i,X_{k-1}=i_1,X_{k-2}=i_2,\ldots)=P(X_{k+1}=j|X_k=i). \tag{11.1}$$

The conditional probability $P(X_{k+1}=j|X_k=i)$ is called the transition probability from $X_k=i$ to $X_{k+1}=j$ and is denoted by p_{ij}, where

$$p_{ij}=P(X_{k+1}=j|X_k=i). \tag{11.2}$$

Time homogeneity of the Markov chains is their important property, which means that their transition probabilities p_{ij} do not depend on time. The Markov property (11.1) has most important consequences for the analysis of Markov chains and allows us to derive recursive relations for probabilities related to X_k. In particular, the probability of the occurrence of the sequence of states $i_0,i_1,\ldots,i_K$ is given by the product of transition probabilities

$$P[i_0,i_1,\ldots,i_K]=\pi_{i_0}p_{i_0i_1}\cdots p_{i_{K-1}i_K}, \tag{11.3}$$

where $\pi_{i_0}=P[X_0=i_0]$. The above equation can be derived by using the chain rule [41] and the Markov property (11.1).

1. Reversible Markov models

Here we consider a Markov model in the reversed order, $\{X_k,X_{k-1},X_{k-2},\ldots\}$ [41]. It is not difficult to prove that the process X_k, X_{k-1}, X_{k-2}, $\ldots$ again has the Markov property. By using the Bayes' rule, we can compute the transition probability from state i to state j in reversed time,

$$\begin{aligned} p_{ij}^{\text{reservsed}} &= P[X_{k-1}=j|X_k=i]=\frac{P[X_{k-1}=j|P[X_k=i|X_{k-1}=j]}{P[X_k=i]} \\ &= \frac{\pi_j(k-1)p_{ji}}{\pi_i(k)} \end{aligned} \tag{11.4}$$

There is an inconsistency in the notation in expression (11.4), since $p_{ij}^{\text{reservsed}}$ depends on the time instant k. For simplicity of notation, index k is suppressed. Nevertheless, we learn from (11.4) that the Markov model with reversed time becomes inhomogeneous. In most applications, analysis of the reversed Markov model under the additional assumption of stationarity is very important. In that case the Markov model with reversed time becomes homogeneous. We have $P[X_{k-1}=j]=\pi_{Sj}$ and $P[X_k=i]=\pi_{Si}$, and (11.4) becomes

$$p_{ij}^{\text{reservsed}}=\frac{\pi_{Sj}p_{ji}}{\pi_{Si}} \tag{11.5}$$

We call a Markov model reversible if it satisfies

$$p_{ij}^{\text{reservsed}} = p_{ji} \tag{11.6}$$

It is interesting that reversibility implies stationarity of both the forward and the reversed chain. Indeed, if $p_{ij} = \dfrac{\pi_j(k-1)p_{ji}}{\pi_i(k)}$ for all i, j then if we set $i = j$ we have $\pi_i(k-1)/\pi_i(k) = 1$.

From the definition (11.6), we can understand that when we observe (or record) states of a reversible Markov model, we cannot tell whether it is proceeding forward or backward. Combining (11.5) and (11.6), we obtain the following condition for the reversibility of a Markov model:

$$p_{ij}\pi_{Si} = \pi_{Sj}p_{ji}. \tag{11.7}$$

It is also named the local balance condition, or the detailed balance condition, owing to the following interpretation. Assume that we are recording events in a Markov model. The average number of transitions from state i to j, per recorded event, is $p_{ij}\pi_{Si}$. Analogously, for transitions from state j to i the average number per event is $\pi_{Sj}p_{ji}$. By the condition (11.7), in a reversible Markov model these numbers are equal.

2. Time-continuous Markov models

In the above we have assumed that transitions between states could only happen at discrete times $0, 1, 2, \ldots, k, \ldots$. Now, we assume that transitions between discrete states $1, 2, \ldots, N$ can occur at any time t, which is a real number [41]. We denote the resulting stochastic process by $X(t)$ and introduce the $N \times N$ transition matrix $P(t-s)$, with entries

$$p_{ij}(t-s) = P[X(t) = j | X(s) = i] \tag{11.8}$$

The Markov property of $X(t)$ is equivalent to the Chapman–Kolmogorov equation

$$p_{ij}(s+t) = \sum_{n=1}^{N} p_{in}(s)p_{nj}(t) \tag{11.9}$$

Using the matrix notation $P(t)$, Equ. (11.9) can be written as

$$P(s+t) = P(s)P(t). \tag{11.10}$$

In the above, $s \geqslant 0, t \geqslant 0$, and

$$P(0) = I, \tag{11.11}$$

where I means the identity matrix. $P(t)$ is differentiable, and by computing the derivative, from (11.10) we obtain

$$\frac{\mathrm{d}P(t)}{\mathrm{d}t} = P'(t) = QP(t) \tag{11.12}$$

where the matrix Q called the intensity matrix of the time – continuous Markov model $X(t)$, is given by the limit of the derivative at zero,

$$Q = \lim_{t \to 0^+} \frac{\mathrm{d}P(t)}{\mathrm{d}t}$$

The constructions of Markov processes $X(t)$ applied in practical studies, for example in the nucleotide substitution models, start with defining the intensity matrix Q. Such a method is the most natural. Given intensity matrix Q, state transition matrix $P(t)$ can be obtained by solving (11.12) with the initial condition (11.11). The solution is

$$P(t) = \exp(Q_t) = \sum_{m=1}^{\infty} \frac{(Q_t)^m}{m!} \tag{11.13}$$

For each $t \geqslant 0$, $P(t)$ is a stochastic matrix, and given an initial probability distribution $\pi(0)$ of states $1, 2, \ldots, N$, we can compute the distribution at time t, from

$$\pi(t) = \pi(0)P(t). \tag{11.14}$$

The construction of the process using intensities indicates that for any state i, the probability of a transition $i \to j$ in the interval $(t, t + \Delta t)$ is equal to $q_{ij}\Delta t + o(\Delta t)$, i.e.,

$$P[X(t + \Delta t) = j | X(t) = i] = q_{ij}\Delta t + o(\Delta t) \tag{11.15}$$

For the diagonal elements of the intensity matrix Q, we define

$$q_{ii} = -\sum_{j \neq i} q_{ij} \tag{11.16}$$

It is possible to derive (11.12), (11.13) from (11.15) and (11.16) [41].

3. Markov model Monte Carlo methods

It is well known that Monte Carlo methods, based on random number generators, allow one to perform a variety of tasks, including optimizing functions, stochastic simulations, and computing integrals in high dimension. The Markov model Monte Carlo approach is performing these tasks using Markov models. The Metropolis–Hastings algorithm is regarded as an important tool in Markov model Monte Carlo methods. It was originally developed for computing (or estimating) integrals in high-dimensional state spaces in molecular physics, but subsequently found many different applications. The Metropolis–Hastings method provides an avenue to the following problem: construct an ergodic Markov model with states $1, 2, \ldots, N$ and with a prescribed stationary distribution, given by a vector $\boldsymbol{\pi}_S$. By building a Markov model, we mean defining its state transition probabilities. Clearly, there are an infinite number of Markov models with a stationary distribution $\boldsymbol{\pi}_S$. Given transition probabilities we can get the stationary distribution $\boldsymbol{\pi}_S$, but there is no explicit formula for the inverse relation. Metropolis–Hastings method gives a solution to this problem by starting from any ergodic Markov model with states $1, 2, \ldots, N$ and then modifying its transition probabilities in such a way that the local balance condition (11.7) is enforced. Therefore the modified Markov model becomes reversible and has the desired stationary distribution $\boldsymbol{\pi}_S$ [41].

Employing this idea, let us assume that we have defined an irreducible, aperiodic Markov model with states $1, 2, \ldots, N$ and transition probabilities q_{ij}. In the next step, these probabilities are modified by multiplying them by factors a_{ij}, which leads to a new Markov model with transition probabilities

$$p_{ij} = a_{ij} q_{ij}. \tag{11.17}$$

We want to choose the factors a_{ij} such that transition probabilities p_{ij} satisfy the local balance condition (11.17). Substituting (11.17) in (11.7), we obtain

$$a_{ij} q_{ij} \pi_{Si} = a_{ji} q_{ji} \pi_{Sj}. \tag{11.18}$$

There are two variables and one equation here, so again an infinite number of solutions is possible. A simple solution is to assume that one of the factors a_{ij} and a_{ji} is equal to one. There are two possibilities. However, it should be taken into account the condition that multiplying factors should satisfy $a_{ij} \leqslant 1$ for all i, j. This condition stems from the fact that the scaling in (11.17) must not lead to probabilities out of the range $(0, 1]$. This, finally, leads to the solution

$$a_{ij} = \min(1, \frac{q_{ji} \pi_{Sj}}{q_{ij} \pi_{Si}}). \tag{11.19}$$

Equation (11.17), with a_{ij} specified in (11.19), allows us to compute the transition probabilities p_{ij} for all $i \neq j$. For the probabilities p_{ii}, we use the formula

$$p_{ii} = 1 - \sum_{j \neq i} p_{ij}, \tag{11.20}$$

As seen from the rule (11.19), the expression for a_{ij} does not rely on the absolute values of $\boldsymbol{\pi}_{Si}$ but only on their ratios. This indicates that it is enough to know $\boldsymbol{\pi}_S$ up to proportionality constant. This is an important feature, which allows one to simulate distributions for which a norming constant is difficult to find. The details of Markov model Monte Carlo methods was described before [41].

11.5.2 Applications of Markov models

1. Protein structure study

Large-scale conformational changes in proteins are implicated in many important biological functions. These structural transitions can often be rationalized in terms of relative movements of rigid domains. A probabilistic model was designed for detecting rigid-body movements in protein structures [42]. This model aims to approximate alternative conformational states by a few structural parts that are rigidly transformed under the action of a rotation and a translation. By using Bayesian inference and Markov chain Monte Carlo sampling, they estimated all parameters of the model, including a segmentation of the protein into rigid domains, the structures of the domains themselves, and the rigid transformations that generate the observed structures. The results showed that Gibbs sampling algorithm can also estimate the optimal number of rigid domains with high efficiency and accuracy.

The structural transition of prion proteins from a native α-helix (PrPC) to a misfolded β-sheet-rich conformation (PrPSc) is believed to be the main cause of a number of prion diseases in humans and animals. Understanding the molecular basis of misfolding and aggregation of prion proteins will be valuable for unveiling the etiology of prion diseases. However, due to the limitation of conventional experimental techniques and the heterogeneous property of oligomers, little is known about the molecular architecture of misfolded PrPSc and the mechanism of structural transition from PrPC to PrPSc [43]. The prion fragment 127-147 (PrP127-147) has been reported to be a critical region for PrPSc formation in Gerstmann-Straussler-Scheinker (GSS) syndrome and thus has been used as a model for the study of prion aggregation. Zhou et al. [43] used molecular dynamics (MD) simulation techniques to study the conformational change of this fragment that could be relevant to the PrPC-PrPSc transition. Employing extensive replica exchange molecular dynamics and conventional MD simulations, a huge number of conformations of PrP127-147 was sampled. Using the Markov state model (MSM), Zhou et al. [43] identified the metastable conformational states of this fragment and the kinetic network of transitions between the states. The resulting MSM reveals that disordered random-coiled conformations are the dominant structures. A key metastable folded state with typical extended β-sheet structures is identified with Pro137 being located in a turn region, consistent with a previous experimental report. Conformational analysis reveals that intrapeptide hydrophobic interaction and two key residue interactions, including Arg136-His140 and Pro137-His140, contribute a lot to the formation of ordered extended β-sheet states. This study provides insights into the molecular details of the early stage of prion aggregation [43].

2. Protein-protein association analysis

Protein-protein association is fundamental to many biological processes. However, a microscopic model describing the structures and kinetics during association and dissociation is lacking on account of the long lifetimes of associated states, which have prevented efficient sampling by direct molecular dynamics simulations. Plattner et al. [44] demonstrated protein-protein association and dissociation in atomistic resolution for the ribonuclease barnase and its inhibitor barstar by combining adaptive high-throughput molecular dynamics simulations and Markov modelling. The model reveals experimentally consistent intermediate structures, energetics and kinetics on timescales from microseconds to hours. A variety of flexibly attached intermediates and misbound states funnel down to a transition state and a native basin consisting of the loosely bound near-native state and the tightly bound crystallographic state. These results offer a deeper level of insight into macromolecular recognition.

3. Mechanism study on channel gating regulation

The P2X4 receptor (P2X4R) is a member of a family of purinergic channels activated by extracellular ATP through three orthosteric binding sites and allosterically regulated by ivermectin (IVM), a broad-spectrum antiparasitic agent [45]. Mackay et al. [45] developed a Markov model based on the presence of one IVM binding site, which described some effects of IVM on rat P2X4R. They further reported two novel models, both with three IVM binding sites. The simpler one-layer model can reproduce many of the observed time series of evoked currents, but does not capture well the short time scales of activation, desensitization, and deactivation. A more complex two-layer model can reproduce the transient changes in desensitization observed upon IVM application, the significant increase in ATP-induced current amplitudes at low IVM concentrations, and the modest increase in the unitary conductance. The two-layer model also suggests that this receptor can exist in a deeply inactivated state, not responsive to ATP, and that its desensitization rate can be altered by each of the three IVM binding sites. Their results provide a detailed analysis of P2X4R kinetics and elucidate the orthosteric and allosteric mechanisms regulating its channel gating [45].

4. Genetics and evolution study

We assume that we have some knowledge about the state of a population at two known times, when the dynamics is governed by a Markov chain such as a Wright-Fisher model. Such knowledge could be obtained, for example, from observations made on ancient and contemporary DNA, or during laboratory experiments involving long term evolution. A natural assumption is that the behaviour of the population, between observations, is related to (or constrained by) what was actually observed. Zhao et al. [46] reported that this assumption had limited validity. When the time interval between observations is larger than a characteristic value, which is a property of the population under consideration, there is a range of intermediate times where the behaviour of the population has reduced or no dependence on what was observed and an equilibrium-like distribution applies. Thus, for example, if the frequency of an allele is observed at two different times, then for a large enough time interval between observations, the population has reduced or no dependence on the two observed frequencies for a range of intermediate times. Given observations of a population at two times, they provided a general theoretical analysis of the behaviour of the population at all intermediate times, and determine an expression for the characteristic time interval, beyond which the observations do not constrain the population's behaviour over a range of intermediate times. The findings of this study relate to what can be meaningfully inferred about a population at intermediate times, given knowledge of terminal states [46].

11.6 Hidden Markov Model

Hidden Markov Model (HMM) is a popular machine learning method in bioinformatics. Machine learning algorithms are presented with training data, which are used to derive important insights about the (often hidden) parameters. Once an algorithm has been suitably trained, it can apply these insights to the analysis of a test sample. As the number of training data is increased, the accuracy of the machine learning algorithm is typically improved as well. The parameters that are learned during training are knowledge; and application of the algorithm with those parameters to new data means the algorithm's use of that knowledge [47]. The Hidden Markov Models approach, considered in this section, learn some unknown probabilistic parameters from training samples and utilizes these parameters in the framework of some algorithmic techniques such as dynamic programming to find the best explanation for the experimental data.

11.6.1 Introduction to Hidden Markov Models

In the 1990s, only about a third of the newly predicted protein sequences using pairwise comparisons indicate convincing similarity to other known sequences [48,49]. This situation is even tougher in the case of new, incomplete sequences or fragments. Large databases of fragments are becoming available as a result of large genome, cDNA, and other sequencing projects, especially those producing expressed sequences tags [50]. At the beginning of 1997, approximately half of GenBank consisted of fragment data. Such data cover a substantial fraction, if not all, of the expressed human genome. It is of course of great importance to recognize and classify such fragments, and recover any additional useful information. A potential method to improve the sensitivity and speed of current database-searching techniques has been to use consensus models built from multiple alignments of protein families [47,51]. Unlike conventional pairwise comparisons, a consensus model can in principle study additional information, such as the position and identity of residues that are more or less conserved throughout the family, as well as variable insertion and deletion probabilities. All descriptions of sequence consensus, such as flexible patterns and blocks, can be regarded as special cases of the HMM approach [51]. Over the past few decades, HMMs provide another useful class of probabilistic graphical models used to model a variety of time series, especially in speech recognition but also in a number of other areas, such as ion channel recordings and optical character recognition [48]. Some other computational biology problems, including the modeling of coding/noncoding regions in DNA, of protein binding sites in DNA, and of protein superfamilies, have also been studied using HMMs.

1. HMM Definition

A first-order discrete HMM is a stochastic generative model for time series defined by a finite set S of states, a discrete alphabet A of symbols, a probability transition matrix $T = (t_{ji})$, and a probability emission matrix $E = (e_{iX})$ [48]. The system randomly evolves from state to state while emitting symbols from the alphabet. When the system is in a given state i, it has a probability t_{ji} of moving to state j and a probability e_{iX} of emitting symbol X. Therefore, an HMM can be visualized by imagining that two different dice are associated with each state: an emission die and a transition die. The essential first-order Markov assumption of course is that the emissions and transitions depend on the current state only, and not on the past. Only the symbols emitted by the system are observable, not the underlying random walk between states; hence the qualification 'hidden.' The hidden random walks can be viewed as hidden or latent variables underlying the observations.

As in the case of neural networks, the directed graph associated with nonzero t_{ji} connections is also called the architecture of the HMM. It is always be considered that there are two special states, the start state and the end state. At time 0, the system is always in the start state. Alternatively, one can use a distribution over all states at time 0. The transition and emission probabilities are the parameters of the model. An equivalent theory can be developed by associating emissions with transitions, rather than with states.

A very simple example of an HMM is shown in Figure 11.14. It can be imagined that there are two 'DNA dice'. They have emission probability vectors of ($e_{1A} = 0.25, e_{1C} = 0.25, e_{1G} = 0.25, e_{1T} = 0.25$) and ($e_{2A} = 0.1, e_{2C} = 0.1, e_{2G} = 0.1, e_{2T} = 0.7$), respectively. The transition probabilities are given in the figure. Suppose that we now find a sequence such as ATCCTTTTTTTCA. There are at least three questions that one can ask immediately as follows. How likely is this sequence for this particular HMM? (This is the likelihood question.) What is the most probable sequence of transitions and emissions through the HMM underlying the production of this particular sequence? (This is the decoding question.) And assuming that the transition and emission parameters are not known with certainty, how should their values be changed in light of the observed sequence? (This is the learning question.) Precise algorithmic answers for all three problems in the general case will be given in the following sections. Now different types of HMM architectures for biological applications are considered.

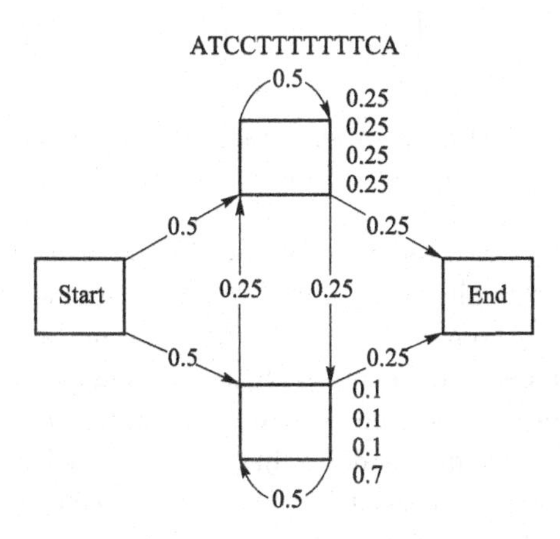

Fig. 11.14 A simple example of an HMM, with two states in addition to the start and end states

2. HMMs for Biological Sequences

The main HMM alphabets for biological sequence applications are the 20-letter amino acid alphabet and the four-letter nucleotide alphabet for proteins and DNA/RNA problems, respectively [48]. Moreover, a number of other alphabets can also be used, such as a 64-letter alphabet of triplets, a three-letter alphabet (α, β, coil) for secondary structure, and Cartesian products of alphabets. A space symbol can be added to any of these alphabets when it is necessary. In the simple HMM example above, there are only two hidden states, with a fully interconnected architecture between them. In real applications it is very required to consider more complex HMM architectures, with many more states and typically sparser connectivity. The selection of architecture is highly problem-dependent. In biological sequences, as in speech recognition, the linear aspects of the sequences are often well obtained by the so-called left–right architectures. Architecture is left–right if it prevents returning to any state once a transition from that state to any other state has occurred. Let us first review the most basic and widely used left–right architecture for biological sequences, the standard linear architecture (Figure 11.15).

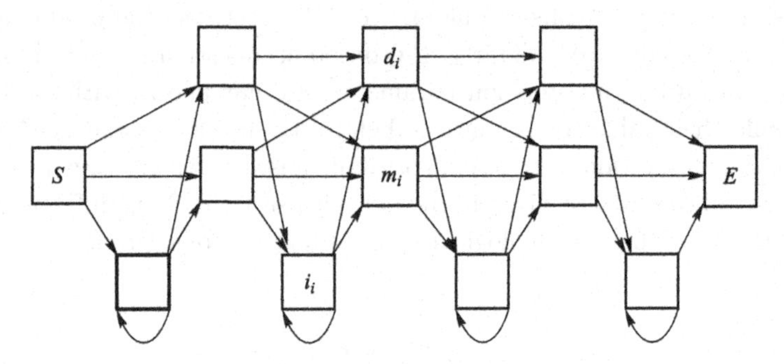

Fig. 11.15 The Standard HMM Architecture. S: start state, E: end state, d_i, : delete, m_i: main, i_i: insert states

At the beginning, we consider the problem of modeling a family of related sequences, such as a family of proteins. As in the application of HMMs to speech recognition, a family of proteins can be found as a set of different utterances of the same word, generated by a common underlying HMM. The standard architecture can be seen as a very simple variation of the multiple-die model. The multiple-die model is in fact a trivial HMM with a linear sequence of states, one for each die. Transition probabilities from one state to the next are all set to 1. The emission probability of each die is associated with the composition of the family in the corresponding column. The main problem with such a model is that there are insertions and deletions: the sequences in the family in general do not have the same length N. Even if a gap symbol is added to the die alphabet, a preexisting multiple alignment is needed to calculate the emission probabilities of each die. The standard architecture is a simple but fundamental variation of the simple die model, where special states for insertions and deletions are added at all possible positions. In the standard architecture, there are three other classes of states besides start and end: the main states, the delete states, and the insert states, with $S = \{\text{start}, m_1, \ldots, m_N, i_1, \ldots, i_{N+1}, d_1, \ldots, d_N, \text{end}\}$. Delete states are also called gap or skip states. N is the length of the model, typically equal to the average length of the sequences in the family. The main and insert states always emit an amino acid symbol, whereas the delete states are mute.

This is of course equivalent to adding a space symbol to the alphabet and forcing the emission of the delete states to be concentrated on this symbol [48]. The linear sequence of state transitions, start $\rightarrow m_1 \rightarrow m_2 \rightarrow \cdots \rightarrow m_N \rightarrow$ end, is the backbone of the model. These are the states corresponding to a multipledie model. For each main state, corresponding insert and delete states are needed to model insertions and deletions. More precisely, there is a 1:1 correspondence between main states and delete states, and a 1:1 correspondence between backbone transitions and insert states. The self-loop on the insert states allows for multiple insertions at a given site. With an alphabet of size $|A|$, the standard architecture has approximately $2N|A|$ emission parameters and $9N$ transition parameters, without taking into account small boundary effects (the exact numbers are $(2N+1)|A|$ emissions and $9N+3$ transitions). Thus, for large N, the number of parameters is of the order of $49N$ for protein models and $17N$ for DNA models. Of course, neglecting boundary effects, there are also $2N$ normalization emission constraints and $3N$ normalization transition constraints.

11.6.2 Applications of Hidden Markov Models

A number of real-world systems have common underlying patterns among them and deducing these patterns is important for us in order to understand the environment around us. These patterns in some instances are apparent upon observation while in many others especially those found in nature are well hidden. Moreover, the inherent stochasticity in these systems introduces sufficient noise that we need models capable to handling it in order to decipher the underlying pattern. Hidden Markov Model is a probabilistic model that is frequently used for studying the hidden patterns in an observed sequence or sets of observed sequences. Since its conception in the late 1960s it has been extensively applied in biology to capture patterns in various disciplines ranging from small DNA and protein molecules, their structure and architecture that forms the basis of life to multicellular levels such as movement analysis in humans [52].

Regardless of the design and training method, once an HMM has been successfully derived from a family of sequences, it can be used in a number of different tasks as follows: ① multiple alignments, ② database mining and classification of sequences and fragments, and ③ structural analysis and pattern discovery. All these tasks are carried out based on the computation, for any given sequence, of its probability according to the model as well as its most likely associated path, and on the analysis of the model structure itself. In most cases, HMMs have functioned well on all tasks. HMM libraries of models can also be combined in a hierarchical and modular fashion to yield increasingly refined probabilistic models of sequence space regions. HMMs could in principle be used in generative mode also to produce de novo sequences having a high likelihood with respect to a target family, although this property has not been exploited [48].

HMMs are very useful in bioinformatics. Although an HMM has to be trained based on multiple sequence alignment, once it is trained, it can in turn be used for the construction of multiple alignment of related sequences. We can use HMMs not only for database searching to detect distant sequence homologs, but also in protein family classification through motif and pattern identification. We can also employ HMMs for advanced gene and promoter prediction, transmembrane protein prediction, as well as protein fold recognition [40]. Several examples of their applications will be given in the following section.

1. Protein classification study

In the case of proteins, HMMs have been successfully applied to many families, such as globins, immunoglobulins, kinases, and G-protein-coupled receptors [47]. HMMs have also been studied to model secondary structure elements, such as alpha-helices, as well as secondary structure consensus patterns of protein superfamilies [53]. In fact, by the end of 1997, HMM data bases of protein families (Pfam) and protein family secondary structures (FORESST) became available [53, 54].

Informative phylogenetic analysis is dependent on the presence of curated and annotated sequences. This may be complemented by the simultaneous availability of empirical data pertaining to their in vivo function. Confounding sequences, with their similarity to more than one functional cluster, can therefore, render any categorization ambiguous, subjective, and imprecise. Kundu [55] analyzed the development of a mathematical expression that can characterize a potential confounding protein sequence. Specifically, statistical descriptors of combinatorially arranged profile HMM scores are computed and evaluated. The resultant data is then incorporated into an index of sequence suitability. The sequence may then be recommended as either suitable for inclusion or be excluded all together. The index is independent of experimental data, and can be computed from the primary structure of the protein sequence. This can be utilized to trim previously grouped sequences and can either finalize the composition of training set or reduce the search space of sequences to be tested [55].

The limitation of most HMMs is their inherent high dimensionality. Lampros et al. [56] developed several variations of low complexity models that can be applied even to protein families with a few members. These variations include the use of a Hidden Markov Model, with a small number of states (called reduced state-space HMM), which is trained with both amino acid sequence and secondary structure of proteins whose 3D structure is known and it is used for protein fold classification. They used data from Protein Data Bank and annotation from SCOP database for training and evaluation of the proposed HMM variations for a number of protein folds that belong to major structural classes. Results indicate that the variations have similar performance, or even better in some cases, on classifying proteins than SAM, which is a widely used HMM-based method for protein classification. The major advantage of the proposed variations is that they employed a small number of states and the algorithms used for training and scoring are of low complexity and thus relatively fast. The main variations examined include a version of the reduced state-space HMM with seven states (7-HMM), a version of the reduced state-space HMM with three states (3-HMM) and an optimized version of the reduced state-space HMM with three states, where an optimization process is applied to its scores (optimized 3-HMM) [56].

2. Protein structure study

Protein structure prediction and analysis are more significant for living organs to perfect asses the living organ functionalities. Several protein structure prediction methods use neural network (NN). However, the HMM is more interpretable and effective for more biological data analysis compared to the NN. It employs statistical data analysis to enhance the prediction accuracy. Kamal et al. [57] proposed a protein prediction approach from protein images based on HMM and Chapman Kolmogrov equation. Initially, a preprocessing stage was applied for protein images' binarization using Otsu technique in order to convert the protein image into binary matrix. Subsequently, two counting algorithms, namely the Flood fill and Warshall are employed to classify the protein structures. Finally, HMM and Chapman Kolmogrov equation are applied on the classified structures for predicting the protein structure. The execution time and algorithmic performances are measured to evaluate the primary, secondary and tertiary protein structure prediction.

Protein domain identification and analysis are cornerstones of modern proteomics. The tools available to protein domain researchers avail a variety of approaches to understanding large protein domain families. HMMs form the basis for identifying and categorizing evolutionarily linked protein domains. Jablonowski [58] reported the use of HMM models for predicting and identifying Src Homology 2 (SH2) domains within the proteome.

3. Protein prediction and discovery of novel enzymes

Alpha helical transmembrane (TM) proteins constitute an important structural class of membrane proteins involved in a wide variety of cellular functions. The prediction of their transmembrane topology, as well as their discrimination in newly sequenced genomes, is of great importance for the elucidation of their structure and function. Several methods based on various algorithmic techniques have been applied for the prediction of the transmembrane segments and the topology of alpha helical transmembrane proteins. HMMs have been efficiently used in the development of several computational methods used for this task. Tsaousis et al. [59] give a brief review of different available prediction methods for alpha helical transmembrane proteins pointing out sequence and structural features that should be incorporated in a prediction method. The procedure, which involves in the design and development of a HMM capable of predicting the transmembrane alpha helices in proteins and discriminating them from globular proteins, is also available in this paper [59].

HMMs are also usded for finding new enzymes. Renewable biopolymers, such as cellulose, starch and chitin are highly resistance to enzymatic degradation. Therefore, it is quite important to upgrade current degradation processes by including novel enzymes. Lytic polysaccharide mono-oxygenases (LPMOs) can disrupt recalcitrant biopolymers, thereby enhancing hydrolysis by conventional enzymes. However, novel LPMO families are difficult to identify using existing methods. Voshol et al. [60] developed a novel profile HMM and used it to mine genomes of ascomycetous fungi for novel LPMOs. Results verified that the alignment was correct. In the alignment, several known conserved features, such as the histidine brace and the N/Q/E-X-F/Y motif and previously unidentified conserved proline and glycine residues were identified. These residues are distal from the active site, suggesting a role in structure rather than activity. The multiple protein alignment was subsequently used to build a profile HMM. The profile HMM was successfully used to mine fungal genomes for a novel family of LPMOs. The model is not limited to bacterial and fungal genomes. This is illustrated by the fact that the model was also able to identify another new LPMO family in *Drosophila melanogaster*. Furthermore, the HMM was used to verify the more distant blast hits from the new fungal family of LPMOs, which belong to the Bivalves, Stony corals and Sea anemones. As a result, the HMM will help the broader scientific community in identifying other yet unknown LPMOs [60].

4. Finding RNA-protein interaction sites

RNA-binding proteins play important roles in the various stages of RNA maturation through binding to its target RNAs. Cross-linking immunoprecipitation coupled with high-throughput sequencing (CLIP-Seq) has made it possible to identify the targeting sites of RNA-binding proteins in various cell culture systems and tissue types on a genome-wide scale. Several HMM-based approaches have been suggested to identify protein-RNA binding sites from CLIP-Seq datasets. Wang et al. [61] well summarized how HMM can be applied to analyze CLIP-Seq datasets, including the bioinformatics preprocessing steps to extract count information from the sequencing data before HMM and the downstream analysis steps following peak-calling.

5. DNA and RNA applications

Multiple alignments of nucleotide sequences are relatively harder to make than alignments of protein sequences. One reason is that parameters in amino acid substitution matrices can be estimated by means of evolutionary and biochemical analysis, while it is hard to obtain good measures of general mutation and deletion costs of individual nucleotides in nucleic acids. The 'twilight zone' of dubious alignment significance is reached faster for sequences from a shorter alphabet, and fewer evolutionary events are therefore required to get into the twilight zone when aligning DNA. HMMs do not a priori require an explicit definition of the substitution costs. The computationally hard many-to-many multiple sequence alignment problem is avoided based on the HMM method by recasting it as a many-to-one sequence-to- HMM alignment problem. The different positions in a model can in practice have individual implicit substitution costs associated with them. These features have contributed to the fact that in several cases HMMs applied to nucleic acids have led to the discovery of new patterns not previously reported by other methods. In protein-related applications, HMMs have more often produce improvements of earlier other methods [48].

(1) Gene finding and gene site identifciation

A number of powerful HMMs and other probabilistic models for gene finding in eukaryotes had been developed [48]. Eukaryotic gene models are typically built by assembling a number of components, such as submodels for splice sites, exons, and introns to use the corresponding weak consensus signals and compositional differences. The individual submodels must be relatively small if the goal is to scan entire genomes in reasonable time. Other key elements are the use of three exon submodels in parallel in order to take into account the three possible ways introns may interrupt the reading frame, as well as features to incorporate exon and intron length distributions, promoters, poly-adenylation signals, intergenic sequences, and strand asymmetry. It is better to train the entire recognition system at once, rather than each of its components separately. In addition, the standard HMM algorithms can be modified to optimize the global gene parse produced by the system rather than the sequence likelihoods.

It is known that integrons are genetic elements that facilitate the horizontal gene transfer in bacteria and are known to harbor genes associated with antibiotic resistance. The gene mobility in the integrons is governed by the presence of *attC* sites, which are 55 to 141-nucleotide-long imperfect inverted repeats. Pereira et al. [62] reported HattCI, a new method for fast and accurate identification of *attC* sites in large DNA data sets. The method is based on a generalized hidden Markov model that describes each core component of an *attC* site individually. Using twofold cross-validation experiments on a manually curated reference data set of 231 *attC* sites from class 1 and 2 integrons, HattCI showed high sensitivities of up to 91.9% while maintaining satisfactory false-positive rates. When applied to a metagenomic data set of 35 microbial communities from different environments, HattCI found a substantially higher number of *attC* sites in the samples that are known to contain more horizontally transferred elements. HattCI will significantly increase the ability to identify *attC* sites and thus integron-mediated genes in genomic and metagenomic data.

(2) Detecting genome-wide Copy Number Variation

Association of Copy Number Variation (CNV) with schizophrenia, autism, developmental disabilities and fatal diseases such as cancer is verified. Next Generation Sequencing have provided powerful tool for the CNV studies. However, many of the current CNV detection tools are not capable of discriminating tandem duplication from non-tandem duplications. Malekpour et al. [63] proposed MGP-HMM as a tool which besides detecting genome-wide deletions discriminates tandem duplications from non-tandem duplications. MGP-HMM takes mate pair abnormalities into account and predicts the digitized number of tandem or non-tandem copies. Abnormalities in the mate pair directions and insertion sizes, after being mapped to the reference genome, are elucidated using a HMM. For this purpose, a Mixture Gaussian density with time-dependent parameters is applied for emitting mate pair insertion sizes from HMM states. MGP-HMM also uses a Poisson distribution for modeling read depth data. Hidden state of the proposed HMM is the digitized copy number of a genomic segment and states correspond to the multipliers of the mixture Gaussian components. A set of next generation sequencing real and simulated data have validated the accuracy of our model [63].

(3) Sequence clustering and cluster visualization

Ferles et al. [64] devised mapping methodologies and projection techniques that visualize and demonstrate biological sequence data clustering results. The Sequence Data Density Display (SDDD) and Sequence Likelihood Projection (SLP) visualizations represent the input symbolical sequences in a lower-dimensional space in such a way that the clusters and relations of data elements are depicted graphically. Both operate in combination/synergy with the self-organizing hidden Markov model map. The resulting unified framework is in position to analyze automatically and directly raw sequence data. This analysis is carried out with little, or even completes absence of, prior information/domain knowledge.

(4) HMMs of human splice sites, exons, and introns

A number of studies have been directed to illustrate the molecular mechanism responsible for intron splicing ever since it was discovered that eukaryotic genes contain intervening sequences that are removed from the mRNA molecules before they leave the nucleus to be translated. It is known that the necessary and sufficient sequence determinants for proper splicing are still largely unrevealed. And we have used probabilistic models in the form of HMMs to study the splicing signals found experimentally. It is not similar to the case of protein families. In this case, it is important to remark that all exons and their associated splice site junctions are neither directly nor closely related by evolution. It should be noted that they still form a 'family' in the sense of sharing certain general characteristics. For example, in a multiple alignment of a set of flanked exons, the consensus sequences of the splice sites should stand out as highly conserved regions in the model, exactly like a protein motif in the case of a protein family. As a result, we had better be particularly careful to regard insertions and deletions in the HMM as formal string operations rather than evolutionary events. A model with the standard architecture, which was trained on 1000 randomly selected flanked donor and acceptor sites, was used to check if an HMM would pick up easily known features of human acceptor and donor sites [48].

(5) HMMs of human promoter regions

A number of HMMs have also been trained using DNA sequences from human promoter regions. Promoter data were obtained from the GenBank. All human sequences, which contained at least 250 nucleotides upstream and downstream, were extracted from an experimentally determined transcriptional start point. Sequences containing non-nucleotide symbols were excluded.

To sum up, the numerous advantages of HMMs in computational molecular biology are very clear. An advantage of HMMs over profiles is that the probability modeling in HMMs has more predictive power. Because the handling of insertions and deletions is a major problem in recognizing highly divergent sequences, HMMs are therefore more robust in describing subtle patterns of a sequence family than standard profile analysis. They come with a solid statistical foundation and with efficient learning algorithms. A consistent treatment of insertions and deletion penalties, which is in the form of locally learnable probabilities, is allowed in these methods. Learning can take place directly from raw sequence data. Unlike conventional supervised NNs, HMMs can accommodate inputs of variable length and a teacher is unnecessary for this process. They are the most flexible generalization of sequence profiles. They can be used efficiently in a number of tasks ranging from multiple alignments, to data mining and classification, to structural analysis and pattern discovery. They are easily combined into libraries and in modular and hierarchical ways [48].

However, HMMs can also suffer in particular from two weaknesses. First, they often have a large number of unstructured parameters. This can be a problem when only a few sequences are available in a family, not an uncommon situation in early stages of genome projects. Second, first-order Markov property of HMMs limited first-order HMMs: they can not express dependencies between hidden states. Proteins fold into complex 3D shapes determining their function. Subtle long-range correlations in their polypeptide chains may exist that are not accessible to a single HMM. However, it is worth to note that HMMs can easily capture long-range correlations that are expressed in a constant way across a family of sequences, even when such correlations are the result of 3D interactions. This is the case, for example, for two linearly distant regions in a protein family that must share the same hydropathy as a result of 3D closeness. The same hydropathy pattern will be observed in all the members of the family and is likely to be reflected in the corresponding HMM emission parameters with training.

11.7 Neural Networks

11.7.1 Introduction to Neural Networks

1. General introduction

Artificial neural networks (NNs) were originally designed with the goal of modeling information processing and learning in the brain. The brain metaphor is a useful source of inspiration. And it is clear today that the artificial neurons used in most NNs are quite remote from biological neurons. However, it is worth to mention that the development of NNs has produced a number of practical applications in various fields such as computational molecular biology [48]. NNs is now a powerful tool in the arsenal of machine-learning techniques that can be applied to sequence analysis and pattern recognition problems. At the most basic level, NNs can be viewed as a broad class of parameterized graphical models consisting of networks with interconnected units evolving in time. We can not only use pairwise connections but also more elaborate connections associated with the interaction of more than two units, leading to the 'higher-order' or 'sigma-pi' networks. The connection from unit j to unit i usually comes with a weight denoted by w_{ij}. Thus we can represent an NN with a weight-directed graph or 'architecture'. For simplicity, any self-interactions are not used, so that we can assume that $w_{ii} = 0$ for all the units. It is customary to distinguish a number of important architectures, including recurrent, feed-forward, and layered. A recurrent architecture is an architecture that contains directed loops. An architecture devoid of directed loops is said to be feed-forward. Recurrent architectures are more complex with richer dynamics. An architecture is layered if the units are partitioned into classes, also called layers, and the connectivity patterns are defined between the classes. It is not necessary for a feed-forward architecture to be layered.

As shown in Figure 11.16, the architectures used are layered feed-forward architectures in many current applications of NNs to molecular biology [48]. The units are often devideded into visible units and hidden units. The visible units are those in contact with the external world, including input and output units. In most conditions, the input units and the output units in simple architectures are grouped in layers, forming the input layer and the output layer. A layer containing only hidden units is named as a hidden layer. The number of layers is often referred to as the 'depth' of a network. Naturally NNs can be assembled in modular and hierarchical fashion to create complex overall architectures. The design of the visible part of an NN depends on the input representation chosen to encode the sequence data and the output that may typically represent structural or functional features.

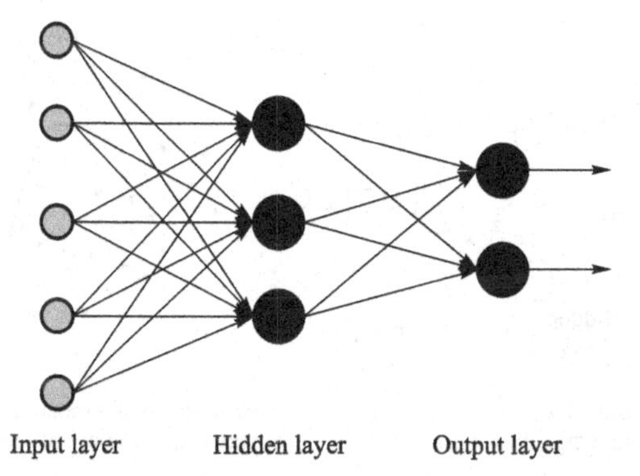

Fig. 11.16 Feedforward Neural Networks. Layers may contain different numbers of units and connectivity patterns between layers may different

2. Artificial Neural Networks

Linear discrimination can be carried out by linear classifiers or single neurons. In other words, the separation can only be done by means of lines, planes, or hyperplanes. An example of a linear separation of feature vectors belonging to two classes, depicted by circles and squares [23]. However, we may wish to develop discriminant systems which allow more complicated boundaries between classes. We can reach this goal by combining several neurons into a network, as shown in Figure 11.17. The neural network shown in the left part of Figure 11.17 is called a multilayer perceptron, or hidden-layer perceptron. This is a simple example, where the input vector $\boldsymbol{x}$ has two components x_1, x_2 and the total number of neurons in the network is three. We organized this neural network into three layers. The first, input layer is built from the input signals x_1, x_2. The second, hidden layer contains two neurons, which, as their inputs, take sums of the input signals with different weights, w_{11}^1 and w_{12}^1 with an offset w_{10}^1 for the first neuron, and w_{11}^1 and w_{12}^1 with an offset w_{20}^1 for the second neuron. The superscript 1 represents the first layer. The outputs from the neurons in the second layer are fed into the last, third layer, which has only one neuron, with an output signal y. In the right plot in Figure 11.17 we present the shape of a separation line which can be obtained with the use of the neural net in the left plot. This line separates two different classes determined by states of the experiment, marked by circles and squares. Such a shape of the separation line cannot be obtained with a single-neuron classifier. Artificial neural networks of the type shown in the left part of Figure 11.17 can have more than one hidden layer, as well as more neurons in each of the layers. The crucial task is the training algorithms for artificial neural networks. Back propagation is a well-known recursive algorithm for adjusting the values of the weights.

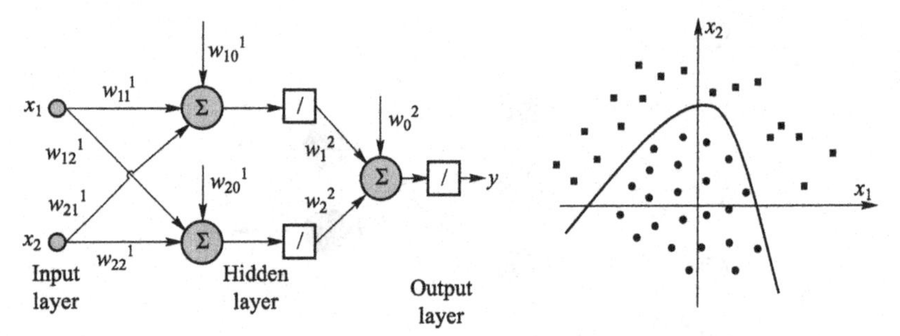

(a) A neural net with three layers, two input signals x_1 and x_2, and one output signal y

(b) The shape of a separation line (Note: it canbe obtained with the use of the left neural net)

Fig. 11.17 A case of neural net

11.7.2 Applications of Neural Networks

When compared to the age of the whole field, the history of application of neural network algorithms to problems within the field of biological sequence analysis is fairly long. In 1982 the perceptron was used to predict ribosome binding sites using amino acid sequence input [65]. In 1986, the backpropagation training algorithm for the multilayer perceptron was commonly used, which produced the real boost in the application of neural network techniques [66]. A study on prediction of protein secondary structure was then published in 1988. This and other papers that quickly followed were based on an adaptation of the NetTalk multilayer perceptron architecture, which from its input of letters in English text predicted the associated phonemes needed for speech synthesis and for reading the text aloud [48]. By changing the input alphabet into alphabets of the amino acids or nucleotides, this approach could immediately be adapted to tasks within the field of sequence analysis. Likewise, the encoding of the phonemes could easily be transformed into structural classes, like those commonly used for the assignment of protein secondary structure, or functional categories representing binding sites, cleavage sites, or residues being posttranslationally modified. In this chapter, some of the work such as those within the application areas of nucleic acids and proteins etc will be introduced.

1. Protein interaction network analysis

Protein-protein interactions (PPIs) play an important role in most biological processes [67]. How to correctly and efficiently detect protein interaction is a problem that is worth studying. Although high-throughput technologies provide the possibility to detect large-scale PPIs, these cannot be used to detect whole PPIs, and unreliable data may be generated. Wang et al. [67] reported a novel computational method to effectively predict the PPIs using the information of a protein sequence. The method adopts Zernike moments to extract the protein sequence feature from a position specific scoring matrix (PSSM). Then, these extracted features were reconstructed using the stacked autoencoder. Finally, a novel probabilistic classification vector machine (PCVM) classifier was employed to predict the protein-protein interactions. When performed on the PPIs datasets of *Yeast* and *H. pylori*, the proposed method could achieve average accuracies of 96.60% and 91.19%, respectively. The promising result shows that the proposed method has a better ability to detect PPIs than other detection methods. The method was also applied to predict PPIs on other species, and promising results were obtained. To evaluate the ability of the method, Wang et al. [67] compared it with the-state-of-the-art support vector machine (SVM) classifier for the Yeast dataset. The results obtained via multiple experiments prove that the method is powerful, efficient and feasible.

Despite the importance suggested by many proteins variants statistically associated with human disease, nearly all such variants have unknown mechanisms, for example, protein-protein interactions (PPIs). Du et al. [68] addressed this challenge using a recent machine learning advance-deep neural networks (DNNs). They proposed a method called DeepPPI (Deep neural networks for Protein-Protein Interactions prediction) for improving the performance of PPIs prediction. This method employs deep neural networks to learn effectively the representations of proteins from common protein descriptors. The experimental results indicated that DeepPPI achieves superior performance on the test data set with an accuracy of 92.50%, precision of 94.38%, recall of 90.56%, specificity of 94.49%, Matthews correlation coefficient of 85.08% and area under the curve of 97.43%, respectively. Extensive experiments showed that DeepPPI can learn useful features of proteins pairs by a layer-wise abstraction, and thus achieves better prediction performance than existing methods.

The ability to engineer zinc finger proteins binding to a DNA sequence of choice is essential for targeted genome editing to be possible. Experimental techniques and molecular docking have been successful in predicting protein-DNA interactions, however, they are highly time and resource intensive. Dutta et al. [69] reported a novel algorithm designed for high throughput prediction of optimal zinc finger protein for 9 bp DNA sequences of choice. In accordance with the principles of information theory, a subset identified by using k-means clustering was used as a representative for the space of all possible 9 bp DNA sequences. The modeling and simulation results assuming synergistic mode of binding obtained from this subset were used to train an ensemble micro neural network. Synergistic mode of binding is the closest to the DNA-protein binding seen in nature, and gives much higher quality predictions, while the time and resources increase exponentially in the trade off. This algorithm is inspired from an ensemble machine learning approach, and incorporates the predictions made by 100 parallel neural networks, each with a different hidden layer architecture designed to pick up different features from the training dataset to predict optimal zinc finger proteins for any 9 bp target DNA. The model gave an accuracy of an average 83% sequence identity for the testing dataset. For final validation of approach, they compared their predictions against optimal ZFPs reported in literature for a set of experimentally studied DNA sequences. The accuracy, as measured by the average string identity between our predictions and the optimal zinc finger protein reported in literature for a 9 bp DNA target was found to be as high as 81% for DNA targets with a consensus sequence GCNGNNGCN reported in literature. Moreover, the average string identity of our predictions for a catalogue of over 100 9 bp DNA for which the optimal zinc finger protein has been reported in literature was found to be 71%. Validation with experimental data shows that this tool is capable of domain adaptation and thus scales well to datasets other than the training set with high accuracy. This algorithm allows designing zinc finger proteins for DNA targets of the user's choice, opening up new frontiers in the field of targeted genome editing [69].

2. Identification of proteins

In several years, deep learning is a modern machine learning technique using in a variety of fields with state-of-the-art performance. Therefore, utilization of deep learning to enhance performance is also an important solution for current bioinformatics field. Le et al. [70] tried to use deep learning via convolutional neural networks and position specific scoring matrices to identify electron transport proteins, which is an important molecular function in transmembrane proteins. The deep learning method can approach a precise model for identifying of electron transport proteins with achieved sensitivity of 80.3%, specificity of 94.4%, and accuracy of 92.3%, with MCC of 0.71 for independent dataset. The proposed technique can serve as a powerful tool for identifying electron transport proteins and can help biologists understand the function of the electron transport proteins. Moreover, this study provides a basis for further research that can enrich a field of applying deep learning in bioinformatics.

3. Prediction of protein secondary structure

When we inspect graphical visualizations of protein backbones on a computer screen, local folding regularities in the form of repeated structures are immediately visible. Two such types of secondary structures are maintained by backbone hydrogen bonds. And they were actually suggested by theoretical considerations before they were found in the first structures to be solved by X-ray crystallography. There is no canonical definition of classes of secondary structure, but Ramachandran plots representing pairs of dihedral angles for each amino acid residue show that certain angular regions tend to be heavily overrepresented in real proteins. One region corresponds to alpha-helices, where backbone hydrogen bonds link residues i and $i+4$; another, to betasheets, where hydrogen bonds link two sequence segments in either a parallel or antiparallel fashion.

One of the classic problems in computational molecular biology is secondary structure prediction in which the sequence preferences and correlations involved. Many different neural network architectures have been used to carry out this task [48]. The assignment of the secondary structure categories to the experimentally determined 3D structure is nontrivial, and has in most of the work been performed by the widely used DSSP and STRIDE program. DSSP works by analysis of the repetitive pattern of potential hydrogen bonds from the 3D coordinates of the backbone atoms, while STRIDE uses both hydrogen bond energy and backbone dihedral angles rather than hydrogen bonds alone.

None of these programs can be said to be perfect. The ability to assign what visually appears as a helix or a sheet, in a situation where the coordinate data have limited precision, is not a trivial algorithmic task. Quantum chemistry does not give a nice analytical expression for the strength of a hydrogen bond, which is another factor contributing to the difficulty. In the prediction context it would be ideal not to focus solely on the visual, or topological, aspects of the assignment problem, but also to try to produce a more predictable assignment scheme. A reduced assignment scheme, which would leave out some of the helices and sheets and thereby make it possible to obtain close to perfect prediction, could be very useful, for example in tertiary structure prediction, which often uses a predicted secondary structure as starting point.

4. Biomolecular property predictions

Although deep learning approaches have had tremendous success in image, video and audio processing, computer vision, and speech recognition, their applications to three-dimensional (3D) biomolecular structural data sets have been hindered by the geometric and biological complexity. To address this problem, Cang and Wei [71] introduced the element-specific persistent homology (ESPH) method. ESPH represents 3D complex geometry by one-dimensional (1D) topological invariants and retains important biological information via a multichannel image-like representation. This representation reveals hidden structure-function relationships in biomolecules. They further integrated ESPH and deep convolutional neural networks to construct a multichannel topological neural network (TopologyNet) for the predictions of protein-ligand binding affinities and protein stability changes upon mutation. To overcome the deep learning limitations from small and noisy training sets, a multi-task multichannel topological convolutional neural network (MM-TCNN) was proposed. Results demonstrated that TopologyNet outperforms the latest methods in the prediction of protein-ligand binding affinities, mutation induced globular protein folding free energy changes, and mutation induced membrane protein folding free energy changes.

5. DNA and RNA applications

Nearly forty years ago, a highly diverse set of methods has been proposed for the problem of identifying protein-coding regions in newly sequenced eukaryotic DNA. Correct exon assignments may be obtained by two independent approaches: prediction of the location of the alternating sequence of donor and acceptor sites, or classification of nucleotides— or continuous segments of nucleotides—as belonging to either the coding or the noncoding category. While intron splice sites have a relatively confined local pattern with lengths of 15–60 nucleotides; protein-coding regions (exons) are often much larger, having typical lengths of 100–150 nucleotides, an interval that is quite stable across a wide range of eukaryotic species. For both types of objects, the pattern strength or regularity is the major factor influencing the potential accuracy of their detection.

Some intron splice site sequences are very close to the 'center of gravity' in a sequence space, while others deviate considerably from the consensus pattern. Likewise, exon sequence may conform strongly or weakly to the prevailing reading frame pattern in a particular organism. The prediction of intron splice sites was studied using neural networks and the results indicated that there is a kind of compensating relationship between the strength of the donor and acceptor site pattern and the strength of the pattern present in the associated coding region [72]. Easily detectable exons may allow weaker splice sites, and vice versa. In particular, very short exons, which usually carry a weak signal as coding regions, are associated with strong splice sites. This relation is also moderated by the distribution for the intron length, which shows a considerable difference among different organisms. The correlation between splice site strength and exonness has been studied in the artificial neural network-based prediction scheme known as Net-Gene, where two local splice site networks are used jointly with an exon prediction network equipped with a large window of 301 nucleotides. This method markedly reduces the number of false positive predictions and, at the same time, enhances the detection of weak splice sites by lowering the prediction threshold when the signal from the exon prediction network is sharp in the transition region between coding and noncoding sequence segments (Details can be seen from Chapter 6 of Baldi and Brunak [48]).

Gene structure prediction can also be carried out by sensor integration. A combination of sensors for detection of various signals related to a complex object has been used for a long time in the theory of pattern recognition. Several schemes have been developed in which NN components play a key role, the earliest GRAIL and GeneParser systems. The performance of the GRAIL system has evolved over the years primarily by the development of more complex sensor indicators, and not by more sophisticated neural networks. The MLP with one hidden layer trained by backpropagation is the same. The GRAIL system can not only be used for recognition of coding-region candidates, but also gene modeling (exon assembly), detection of indel errors and suggestion of likely corrections, detection of CpG islands, and recognition of PolII promoters and polyadenylation sites. In the GeneParser scheme intron/exon and splice site indicators are weighted by a neural network to approximate the log-likelihood that a sequence segment exactly means an intron or exon (first, internal, or last) [73]. A dynamic programming algorithm is then used to this data to check the combination of introns and exons that maximizes the likelihood function. When tested against the training data, GeneParser precisely identifies 75% of the exons and correctly predicts 86% of coding nucleotides as coding, while only 13% of non-exon bps were predicted to be coding. This corresponds to a correlation coefficient for exon prediction of 0.85. Generalized performance is nearly as good as with the training set based on the simplicity of the network weighting scheme [48].

Accurate computational identification of promoters remains a challenge as these key DNA regulatory regions have variable structures composed of functional motifs that provide gene-specific initiation of transcription. Umarov and Solovyev [74] utilized Convolutional Neural Networks (CNN) to analyze sequence characteristics of prokaryotic and eukaryotic promoters and build their predictive models. A similar CNN architecture, which is on promoters of five distant organisms: human, mouse, plant (*Arabidopsis*), and two bacteria, was trained. Results showed that CNN trained on sigma70 subclass of *Escherichia coli* promoter gives an excellent classification of promoters and non-promoter sequences (Sn=0.90, Sp=0.96, CC=0.84). The *Bacillus* subtilis promoters identification CNN model achieves Sn=0.91, Sp=0.95, and CC=0.86. For human, mouse and *Arabidopsis* promoters we employed CNNs for identification of two well-known promoter classes (TATA and non-TATA promoters). CNN models nicely recognize these complex functional regions. The developed CNN models, implemented in CNNProm program, demonstrated the ability of deep learning approach to grasp complex promoter sequence characteristics and achieve significantly higher accuracy compared to the previously developed promoter prediction programs. Umarov and Solovyev [74] also proposed random substitution procedure to discover positionally conserved promoter functional elements. As the suggested approach does not require knowledge of any specific promoter features, it can be easily extended to identify promoters and other complex functional regions in sequences of many other and especially newly sequenced genomes.

11.8 Clustering Analysis

11.8.1 Introduction to Clustering Analysis

One important problem in bioinformatics is to partition a set of experimental data into groups (clusters) in such a way that the data points within the same cluster are highly similar while data points in various clusters are very various [1]. Therefore, clustering has become the most widespread exploratory technique used in genomic data analysis. In this section, several algorithms that perform different types of clustering will be introduced. There is no simple recipe for choosing one particular method over another for a particular clustering problem, just as there is no universal notion of what constitutes a 'good cluster'. However, these algorithms provide significant insight into data and allow us to obtain clusters of genes with similar functions even when it is unclear what particular role these genes play. Studies of evolutionary tree reconstruction, which is closely related to clustering will also be introduced.

1. Gene expression analysis

Sequence comparison often provides a promising avenue to discover the function of a newly sequenced gene by checking similarities between the new gene and previously sequenced genes with known functions. However, it is noted that the sequence similarity of genes in a functional family is often so weak that we cannot reliably derive the function of the newly sequenced gene based on sequence alone. Moreover, genes with the same function sometimes have no sequence similarity at all. Therefore, the functions of more than 40% of the genes in sequenced genomes remain unknown. Fortunately, DNA array, which is used to analyze gene functions, has been developed. DNA arrays provide a powerful tool to analyze the expression levels of many genes and to reveal which genes are switched on and switched off in the cell. The outcome of this kind of study is an $n \times m$ expression matrix I, with the n rows corresponding to genes, and the m columns corresponding to different time points and different conditions [1]. The expression matrix I shows intensities of hybridization signals as provided by a DNA array.

The element $I_{i,j}$ of the expression matrix means the expression level of gene i in experiment j; the entire ith row of the expression matrix is called the expression pattern of gene i. It is possible to look for pairs of genes in an expression matrix with similar expression patterns, which would be manifested as two similar rows. Therefore, if the expression pattern of one gene is similar to another gene, there is a good chance that these genes are somehow related. In other word, these genes either have similar functions or are involved in the same biological process. If the expression pattern of a newly sequenced gene is similar to the expression pattern of a gene with known function, we may have reason to suspect that these genes perform similar or related functions. The deciphering of regulatory pathways is also important application of expression analysis. Similar expression patterns usually imply coregulation. At the same time, it should be mentioned that DNA arrays typically produce noisy data with high error rates. Therefore, expression analysis should be done very carefully.

Genes with similar expression patterns can be grouped into clusters by clustering algorithms. These clusters correspond to groups of functionally related genes. To cluster the expression data, the $n \times m$ expression matrix is usually transformed into an $n \times n$ distance matrix $d = (d_{i,j})$ where $d_{i,j}$ represents how similar the expression patterns of genes i and j are (see Figure 11.19). Clustering aims to group genes into clusters satisfying the following two conditions: ① Homogeneity. Genes within a cluster should be highly similar to each other. ② Separation. Genes from different clusters should be quite different.

A good example of clustering is indicated in Figure 11.18. Based on the above two properties, a good partition is shown in Figure 11.18(a), while a bad one is shown in Figure 11.18(b). Clustering algorithms aims to find a good partition. A good clustering of data is one that adheres to these goals. While we hope that a better clustering of genes gives rise to a better grouping of genes on a functional level, the final analysis of resulting clusters is left to biologists. Varied tissues express different genes, and there are more than 10,000 genes expressed in any one tissue. Given the big number of over 100 different tissue types as well as many time points for measurements of expression levels, gene expression experiments will generate vast amounts of data which can be hard to interpret. Moreover, expression levels of related genes may vary by several orders of magnitude, thus creating the problem of achieving accurate measurements over a large range of expression levels. Genes with low expression levels may be related to genes with high expression levels.

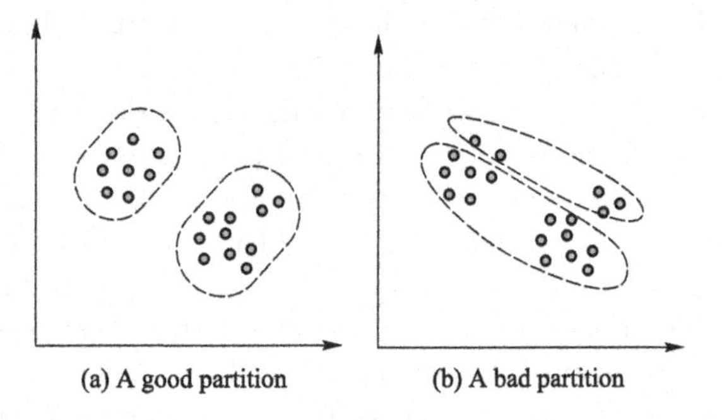

Fig. 11.18 Data can be grouped into different clusters

2. Hierarchical clustering

It is not very difficult to construct a clustering from any data set. As we know, clusters may have subclusters; these may have subsubclusters, and so on. For example, mammals can be broken down into primates, carnivora, bats, marsupials, and many other orders. Then, the order carnivora can be broken down into cats, hyenas, bears, seals, etc. Moreover, cats can be broken into thirty seven species. Hierarchical clustering, which organizes elements into a tree, rather than forming an explicit partitioning of the elements into clusters, is considered as a powerful technique [1]. In this technique, the genes are represented as the leaves of a tree. The edges of the trees are assigned lengths and the distances between leaves—that is, the length of the path in the tree that connects two leaves—correlate with entries in the distance matrix. Such trees are often applied in both the analysis of expression data and in studies of molecular evolution. As shown in Figure 11.19, clustering of the data in previous study [1] is represented by a tree. This tree actually indicates a family of different partitions into clusters, each with a different number of clusters (one for each value from 1 to n). We can see what these partitions by drawing a horizontal line through the tree. Each line crosses the tree at i points ($1 \leqslant i \leqslant k$) and correspond to i clusters.

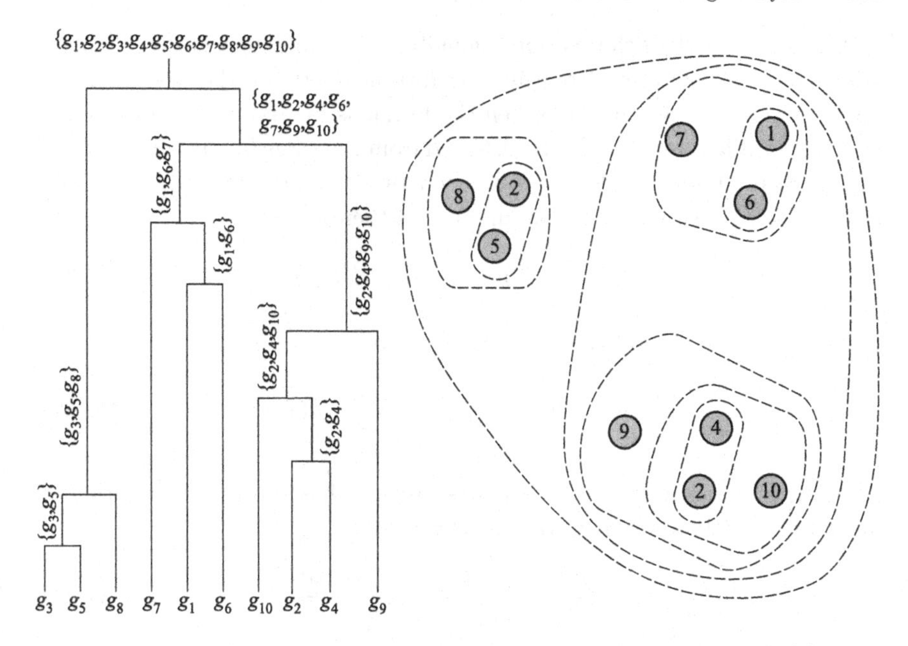

Fig. 11.19 A hierarchical clustering of the data in previous study [1]

The HIERARCHICALCLUSTERING algorithm below uses an $n \times n$ distance matrix d as an input, and progressively produces n different partitions of the data as the tree it outputs. The largest partition has n single-element clusters, with every element forming its own cluster. The second-largest partition, by which the two closest clusters from the largest partition are combined, thus has $n - 1$ clusters. Generally speaking, the ith partition, by which the two closest clusters from the $(i - 1)$th partition are combined, has $n - i + 1$ clusters [1].

```
HIERARCHICALCLUSTERING(d, n)
1 Form n clusters, each with 1 element
2 Construct a graph T by assigning an isolated vertex to each cluster
3 while there is more than 1 cluster
4    Find the two closest clusters C1 and C2
5    Merge C1 and C2 into new cluster C with |C1| + |C2| elements
6    Compute distance from C to all other clusters
7    Add a new vertex C to T and connect to vertices C1 and C2
8    Remove rows and columns of d corresponding to C1 and C2
9    Add a row and column to d for the new cluster C
10 return T
```

Line 6 in the algorithm is (intentionally) left ambiguous; clustering algorithms vary in how they determine the distance between the newly formed cluster and any other cluster. When the formulas for recomputing distances are various, different answers are observed from the same hierarchical clustering algorithm. For example, the distance between two clusters can be defined as the smallest distance between any pair of their elements

$$d_{\min}(C^*, C) = \min_{x \in C^*, y \in C} d(x, y) \tag{11.21}$$

or the average distance between their elements

$$d_{\text{avg}}(C^*, C) = \frac{1}{|C^*||C|} \sum_{x \in C^*, y \in C} d(x, y) \tag{11.22}$$

Another distance function estimates distance based on the separation of C_1 and C_2 in HIERARCHICALCLUSTERING:

$$d(C^*, C) = \frac{d(C^*, C_1) + d(C^*, C_2) + d(C_1, C_2)}{2} \tag{11.23}$$

Hierarchical clustering has been used to analyze the expression profiles of 8,600 genes over thirteen time points to find the genes responsible for the growth response of starved human cells. The HIERARCHICALCLUSTERING produced a tree consisting of five main subtrees and many smaller subtrees. The genes within these five clusters had similar functions. These results indicate that the resulting clusters are biologically sensible.

3. *k*-Means Clustering

If we pretend that k—the number of clusters—is known in advance, we can view the n rows of the $n \times m$ expression matrix as a set of n points in m-dimensional space and partition them into k subsets. One of the most popular clustering methods for points in multidimensional spaces is called k-means clustering [1]. Given a set of n data points in m-dimensional space and an integer k, the problem is to find a set of k points, or centers, in m-dimensional space that minimize the squared error distortion defined below. Given a data point v and a set of k centers $X = \{x_1, \ldots, x_k\}$, define the distance from v to the centers X as the distance from v to the closest point in X, that is, $d(v, X) = \min_{1 \leqslant i \leqslant k} d(v, x_i)$. The $d(v, x_i)$ is assumed as the Euclidean distance in m dimensions. The squared error distortion for a set of n points $V = \{v_1, \ldots, v_n\}$, and a set of k centers $X = \{x_1, \ldots, x_k\}$, is defined as the mean squared distance from each data point to its nearest center:

$$d(V, X) = \frac{\sum_{i=1}^{n} d(v_i, X)^2}{n} \tag{11.24}$$

---

k-means clustering problem:
Given n data points, find k center points minimizing the squared error distortion.

Input: A set, V, consisting of n points and a parameter k.

Output: A set X consisting of k points (called centers) that minimizes d(V, X) over all possible choices of X.

If the above formulation does not explicitly address clustering n points into k clusters, a clustering can be done by simply assigning each point to its closest center. The k-means clustering problem looks relatively simple. However, there are no efficient algorithms known for it. The Lloyd k-means clustering algorithm is known as one of the most popular clustering heuristics that often generates good solutions in gene expression analysis. The Lloyd algorithm randomly chooses an arbitrary partition of points into k clusters and aims to improve this partition by moving some points between clusters. In the beginning we can select arbitrary k points as cluster representatives. The algorithm iteratively carries out the following two steps until either it converges or until the fluctuations become very small: ① assign each data point to the cluster C_i corresponding to the closest cluster representative x_i $(1 \leqslant i \leqslant k)$; and ② after the assignments of all n data points, determine new cluster representatives based on the center of gravity of each cluster, that is, the new cluster representative is $\frac{\sum_{v \in C} v}{|C|}$ for every cluster C.

The Lloyd algorithm usually converges to a local minimum of the squared error distortion function rather than the global minimum. Unfortunately, interesting objective functions other than the squared error distortion produce similarly difficult problems. If you try to minimize $\sum_{i-1}^{n} d(v_i, X)$ (k-median problem) or $\max_{1 \leqslant i \leqslant n} d(v_i, X)$ (k-center problem) instead of the squared error distortion ($\sum_{i-1}^{n} d(v_i, X)^2$), it can be quite difficult to find a good clustering. It is worth to remark that all of these definitions of clustering cost emphasize the homogeneity condition and more or less ignore the other important goal of clustering, the separation condition. Moreover, in some unlucky instances of the k-means clustering problem, the algorithm may focus on a local minimum that is arbitrarily bad compared to an optimal solution. The Lloyd algorithm is very fast. It can significantly rearrange every cluster in every iteration. A more conservative avenue is to move only one element between clusters in each iteration. It is assumed that every partition P of the n-element set into k

clusters has an associated clustering cost, denoted cost (P), that determines the quality of the partition P: the smaller the clustering cost of a partition, the better that clustering is. The squared error distortion is one particular choice of cost (P) and assumes that each center point is the center of gravity of its cluster. The pseudocode is shown below. It implicitly assumes that cost (P) can be efficiently determined based either on the distance matrix or on the expression matrix. Given a partition P, a cluster C within this partition, and an element i outside C, $P_{i \to C}$ denotes the partition obtained from P by moving the element i from its cluster to C. This move enhances the clustering cost only if $\Delta(i \to C) = \text{cost}(P) - \text{cost}(P_{i \to C}) > 0$, and the PROGRESSIVEGREEDYK-MEANS algorithm searches for the 'best' move in each step (i.e., a move that maximizes $\Delta(i \to C)$ for all C and for all $i \notin C$).

```
PROGRESSIVEGREEDYK-MEANS(k)
1 Select an arbitrary partition P into k clusters.
2 while forever
3   bestChange←0
4   for every cluster C
5     for every element i ∉ C
6       if moving i to cluster C reduces the clustering cost
7         if Δ(i → C) > bestChange
8           bestChange← Δ(i → C)
9           i* ← i
10          C* ← C
11    if bestChange > 0
12      change partition P by moving i* to C*
13    else
14      return P
```

Even though this algorithm may loop endlessly as revealed by line 2, the return statement on line 14 saves us from an infinitely long wait. Iterating is stoped when no move allows for an improvement in the score; this eventually has to happen.

11.8.2 Applications of Clustering Analysis

1. Evolutionary trees

Many years ago, biologists constructed evolutionary trees based on morphological features, like beak shapes or the presence or absence of fins. Now, DNA sequences are often used for the reconstruction of evolutionary trees. A

DNA-based evolutionary tree of bears and raccoons is shown in Figure 11.20 [1]. It helped biologists to decide whether the giant panda belongs to the bear family or the raccoon family. This question is not easy since bears and raccoons diverged just 35 million years ago and many morphological features are observed from both of them. For over a hundred years biologists could not agree on whether the giant panda should be classified in the bear family or in the raccoon family. In 1870 an amateur naturalist, Père Armand David, returned to Paris from China with the bones of the mysterious creature which he called 'black and white bear'. The bones are examined. And it was concluded that they more closely resembled the bones of a red panda than those of bears. Red pandas were part of the raccoon family. As a reslut, giant pandas were also classified as raccoons. Giant pandas look like bears. However, they have features that are unusual for bears and typical of raccoons: they do not hibernate in the winter like other bears do, their male genitalia are tiny and backward-pointing (like raccoons' genitalia), and they do not roar like bears but bleat like raccoons.

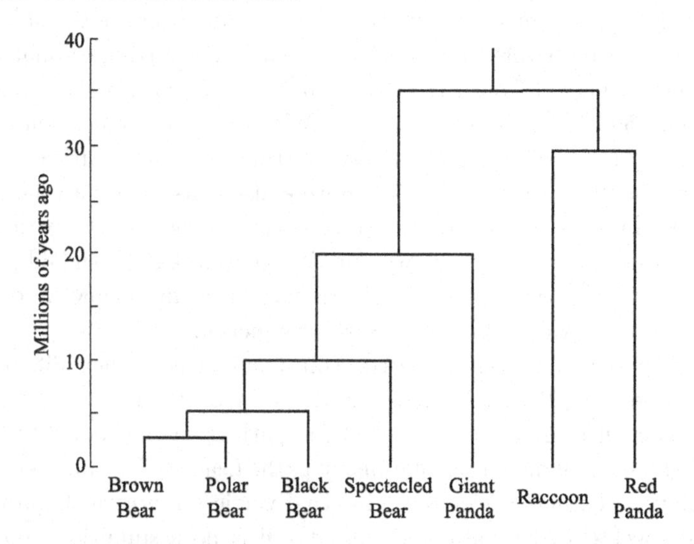

Fig. 11.20 An evolutionary tree showing the divergence of raccoons and bears. Despite their difference in size and shape, these two families are closely related

The problem of giant panda classification was finally solved in 1985 using DNA sequences and algorithms. The final analysis indicated that DNA sequences provide an important source of information to test evolutionary hypotheses. About 500,000 nucleotides were used to construct the evolutionary tree of bears and raccoons. When the giant panda controversy was resolved by Steven O'Brien, Rebecca Cann, Mark Stoneking and Allan Wilson constructed an evolutionary tree of humans and instantly created a new controversy. This tree produces the Out of Africa hypothesis, which thinks that humans have a common ancestor who lived in Africa 200,000 years ago. This study made the question of human origins into an algorithmic puzzle [1].

Mitochondrial DNA (mtDNA) sequences of people of different races and nationalities were used to construct the tree. Sequences of mtDNA from people representing African, Asian, Australian, Caucasian, and New Guinean ethnic groups were compared and 133 variants of mtDNA was found by Wilson and his colleagues. Next, the evolutionary tree was constructed for these DNA sequences that showed a trunk splitting into two major branches. One branch consisted only of Africans, the other included some modern Africans and some people from everywhere else. It is concluded that a population of Africans, the first modern humans, forms the trunk and the first branch of the tree while the second branch represents a subgroup that left Africa and later spread out to the rest of the world. All of the mtDNA, even samples from regions of the world far away from Africa, were significantly similar. These results suggested that our species is relatively young. The African samples had the most mutations, thus implying that the African lineage is the oldest. All modern humans trace their roots back to Africa. It was further estimated that modern man emerged from Africa 200,000 years ago with racial differences arising only 50,000 years ago [1]. Allan Wilson and colleagues constructed the human mtDNA evolutionary tree supporting the Out of Africa hypothesis. After that, 100 distinct trees, which were also consistent with data, were constructed by Alan Templeton to provide evidence against the African origin hypothesis. This is a cautionary tale. It suggests that we should proceed carefully when constructing large evolutionary trees and below we describe some algorithms for evolutionary tree reconstruction.

We use either unrooted or rooted evolutionary trees. Their difference is shown in Figure 11.21. In a rooted evolutionary tree, the root represents the most ancient ancestor in the tree, and the path from the root to a leaf in the rooted tree is named an evolutionary path. Leaves of evolutionary trees represent the existing species while internal vertices represent hypothetical ancestral species. In the unrooted case, there is no assumption about the position of an evolutionary ancestor (root) in the tree. It is also remarked that rooted trees can be viewed as directed graphs if one directs the edges of the rooted tree from the root to the leaves. We often work with binary weighted trees where every internal vertex has degree equal to 3 and every edge has an assigned positive weight (sometimes referred to as the length). The weight of an edge (v, w) may represent the number of mutations on the evolutionary path from v to w or a time estimate for the evolution of species v into species w. It is sometimes assumed the existence of a molecular clock that assigns a time $t(v)$ to every internal vertex v in the tree and a length of $t(w) - t(v)$ to an edge (v, w). Here, time means the 'moment' when the species v produced its descendants; every leaf species means time 0 and every internal vertex presumably means some negative time.

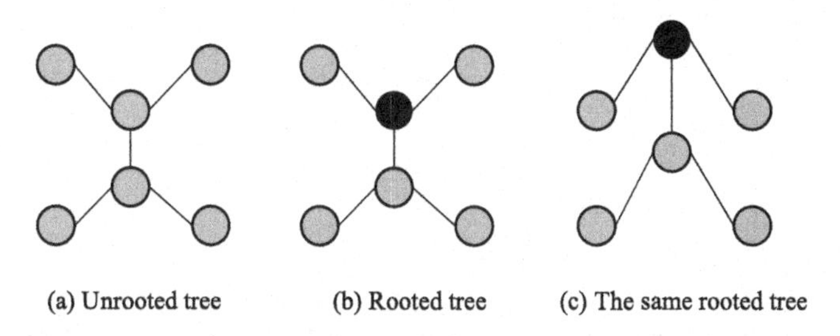

Fig. 11.21 The difference of trees

2. Tree reconstruction based on distance and additive matrices
Given a weighted tree T with n leaves, the length of the path between any two leaves i and j, $d_{i,j}(T)$ can be calculated as the sum of the weights of the edges in the path between them (Figure 11.22 a) [1]. For example, $d(1,5) = 12 + 13 + 14 + 17 + 12 = 68$ (Figure 11.22 a).

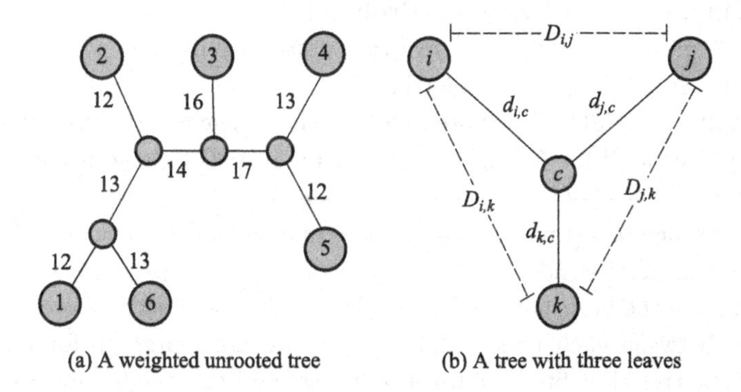

Fig. 11.22 Tree cases

One usually faces the opposite problem: you measure the $n \times n$ distance matrix $(D_{i,j})$, and then must find a tree T that has n leaves and fits the data. In other word, $d_{i,j}(T) = D_{i,j}$ for every two leaves i and j. A number of ways can be used to generate distance matrices. For example, one can sequence a particular gene in n species and define $D_{i,j}$ as the edit distance between this gene in species i and species j. It is easy to construct a tree that fits any given 3×3 matrix D. This binary unrooted tree has four vertices i, j, k as leaves and vertex c as the center. In the tree, the lengths of each edge are defined by the following three equations with three variables $d_{i,c}, d_{j,c}$, and

$d_{k,c}$ (Figure 11.22 b):

$$d_{i,c} + d_{j,c} = D_{i,j} d_{i,c} + d_{k,c} = D_{i,k} d_{j,c} + d_{k,c} = D_{j,k} \tag{11.25}$$

The solution is given by,

$$\begin{aligned} d_{i,c} &= \frac{D_{i,j} + D_{i,k} - D_{j,k}}{2} d_{j,c} \\ &= \frac{D_{j,i} + D_{j,k} - D_{i,k}}{2} d_{k,c} \\ &= \frac{D_{k,i} + D_{k,j} - D_{i,j}}{2} \end{aligned} \tag{11.26}$$

An unrooted binary tree with n leaves has $2n - 3$ edges, so fitting a given tree to an $n \times n$ distance matrix D leads to solving a system of $\binom{n}{2}$ equations with $2n - 3$ variables. For $n = 4$ this amounts to solving six equations with only five variables. It should be mentioned that it is not always possible to solve this system, making it hard or impossible to construct a tree from D. A matrix $(D_{i,j})$ is called additive if there exists a tree T with $d_{i,j}(T) = D_{i,j}$, and nonadditive otherwise.

Distance-Based Phylogeny Problem

Reconstruct an evolutionary tree from a distance matrix.

Input: An n × n distance matrix $(D_{i,j})$.

Output: A weighted unrooted tree T with n leaves fitting D, that is, a tree such that $d_{i,j}(T) = D_{i,j}$ for all $1 \leqslant i < j \leqslant n$ if $(D_{i,j})$ is additive.

The distance-based phylogeny problem may not have a solution, but if it does—that is, if D is additive—there exists a simple algorithm to solve it. We emphasize the fact that we are somehow given the matrix of evolutionary distances between each pair of species, and we are searching for both the shape of the tree that fits this distance matrix and the weights for each edge in the tree [1].

A 'simple' way to carry out the distance-based phylogeny problem for additive trees is to search for a pair of neighboring leaves, that is, leaves that have the same parent vertex. Reconstructing trees from additive matrices can be done using recursive algorithm such as ADDITIVEPHYLOGENY algorithm. This algorithm provides a way to check if the matrix D is additive. While this algorithm is intuitive and simple, it is not the most efficient way to construct additive trees. Another way to check additivity is by using the following 'four-point condition'. Let $1 \leqslant i, j, k, l \leqslant n$ be four distinct indices. Compute 3 sums: $D_{i,j} + D_{k,l}$, $D_{i,k} + D_{j,l}$, and $D_{i,l} + D_{j,k}$. If D is an additive matrix then these three sums can be represented by a tree with four leaves.

Moreover, two of these sums represent the same number while the third sum represents another smaller number. It is considered that elements $1 \leqslant i, j, k, l \leqslant n$ satisfy the fourpoint condition if two of the sums $D_{i,j} + D_{k,l}$, $D_{i,k} + D_{j,l}$, and $D_{i,l} + D_{j,k}$ are the same, and the third one is smaller than these two.

Theorem: *An $n \times n$ matrix D is additive if and only if the four point condition holds for every 4 distinct elements $1 \leqslant i, j, k, l \leqslant n$.*

If the distance matrix D is not additive, one might want instead to find a tree that approximates D using the sum of squared errors $\sum_{i,j} (d_{i,j}(T) - D_{i,j})^2$ as a measure of the quality of the approximation. This leads to the (NP-hard) Least Squares Distance-Based Phylogeny problem [1].

Least Squares Distance-Based Phylogeny Problem

Given a distance matrix, find the evolutionary tree that minimizes squared error.

Input: An $n \times n$ distance matrix $(D_{i,j})$

Output: A weighted tree T with n leaves minimizing $\sum_{i,j} (d_{i,j}(T) - D_{i,j})^2$ over all weighted trees with n leaves.

---

3. Evolutionary trees and hierarchical clustering

Variants of hierarchical clustering are often used to construct evolutionary trees. Unweighted pair group method with arithmetic mean (UPGMA) algorithm, which is a variant of HIERARCHICALCLUSTERING that uses a different way to compute the distance between clusters, and assigns heights to vertices of the constructed tree, is considered as a extremely simple clustering algorithm [1]. Thus, the length of an edge (u, v) represents the difference in heights of the vertices v and u. The height plays the role of the molecular clock, and lets one to 'date' the divergence point for every vertex in the evolutionary tree. Given clusters C_1 and C_2 in UPGMA, the distance between them is defined to be the average pairwise distance:

$$D(C_1, C_2) = \frac{1}{|C_1||C_2|} \sum_{i \in C_1} \sum_{j \in C_2} D(i, j) \tag{11.27}$$

UPGMA algorithm is shown as follows.

UPGMA(D, n)

1 Form n clusters, each with a single element

2 Construct a graph T by assigning an isolated vertex to each cluster

3 Assign height h(v)= 0 to every vertex v in this graph

4 **while** there is more than one cluster

5 Find the two closest clusters C_1 and C_2
6 Merge C_1 and C_2 into a new cluster C with $|C_1| + |C_2|$ elements
7 for every cluster $C^* \neq C$
8 $D(C, C^*) = \frac{1}{|C||C^*|} \sum_{i \in C} \sum_{j \in C^*} D(i, j)$
9 Add a new vertex C to T and connect to vertices C_1 and C_2
10 $h(C) \leftarrow \frac{D(C_1, C_2)}{2}$
11 Assign length $h(C) - h(C_1)$ to the edge (C_1, C)
12 Assign length $h(C) - h(C_2)$ to the edge (C_2, C)
13 Remove rows and columns of D corresponding to C_1 and C_2
14 Add a row and column to D for the new cluster C
15 **return** T

It is worth mentioning that UPGMA is simply another hierarchical clustering algorithm that 'dates' vertices of the constructed tree. A special type of rooted tree, which is known as ultrametric, is produced by UPGMA. The distance from the root to any leaf in ultrametric trees is the same, which is then connected with the 'neighboring leaves' idea. In 1987, an ingenious neighbor joining algorithm was designed by Naruya Saitou and Masatoshi Nei for phylogenetic tree reconstruction. In the case of additive trees, the neighbor joining algorithm somehow magically obtains pairs of neighboring leaves and proceeds by substituting such pairs with the leaves' parent [1]. However, neighbor joining works well for additive distance matrices. At the same time, it also works well with many others. Neighbor joining does not assume the existence of a molecular clock. It is gurranteed that the clusters that are merged in the course of tree reconstruction are not only close to each other but also are far apart from the rest.

11.9 Other Algorithms

We cover some typical algorithms that are used in bioinformatics in the above sections. We briefly describe other algorithms and provide pointers to the vast literature on this topic.

11.9.1 Branch-and-Bound Algorithm

As the various alternatives in a brute force algorithm are esplored, it is found that we can omit a large number of alternatives, a technique that is often called branch-and-bound, or pruning [1]. Suppose you were exhaustively searching the first floor and heard the phone ringing above your head. You could immediately rule out the need to find the basement or the first floor. What may have taken three hours may now only take one, depending on the number of space that can be ruled out.

11.9.2 Greedy Algorithm

Many algorithms are iterative procedures that choose among a number of alternatives at each literation. A good example is that a cashier can view the change problem as a series of decisions he or she has to make: which coin (among d denominations) to return first, which to return second, and so on. Some of these alternatives may produce correct solutions while others may not. In the case of greedy algorithm, the 'most attractive' alternative was chosen at each iteration, for example, the largest denomination possible [1]. USCHANGE used quarters, then dimes, then nickels, and finally pennies (in that order) to make change for M. By greedily choosing the largest denomination first, the algorithm avoided any combination of coins that included fewer than three quarters to make change for an amount larger than or equal to 75 cents. It should also be mentioned that the generalization of this greedy strategy, BETTERCHANGE, obtained incorrect results when certain new denominations were considered. In the case of telephone, the greedy algorithm would simply be to walk in the direction of the telephone's ringing until it was found. The problem here is that there may be a wall between you and the phone, preventing you from finding it. Unfortunately, in most cases of realistic problems, these sorts of difficulties were frequently observed. A greedy method will seem 'obvious' and natural in a number of studies, but will be subtly wrong.

11.9.3 Divide-and-Conquer Algorithm

As we know, one problem may be too big to be solved. However, two problems that are half the size may be significantly easier. In these cases, divide-and-conquer algorithms indicate their potential by doing just that: splitting the problem into smaller subproblems, solving the subproblems independently, and combining the solutions of subproblems into a solution of the original problem. Actually, the condition is often more complicated than this. Therefore, after splitting one problem into subproblems, a divide-and-conquer algorithm usually splits these subproblems into even smaller sub-subproblems until it reaches a point at which it no longer needs to recurse. A key step in many divide-and-conquer algorithms is the recombining of solutions to subproblems into a solution for a larger problem. This merging process usually needs a considerable amount of time. Examples of this technique can be seen from Jones and Pevzner [1].

11.9.4 Machine Learning

For the phone search problem, another way is to collect statistics over the course of a year about where you leave the phone, observing where the phone tends to end up most of the time. If the phone was left in the bathroom 80% of the time, in the bedroom 15% of the time, and in the kitchen 5% of the time, then a sensible time-saving strategy would be to begin the search in the bathroom, continue to the bedroom, and carry out in the kitchen. Machine learning algorithms often base their strategies on the computational analysis of previously collected data.

11.9.5 Randomized Algorithm

In many conditions, we may toss a coin to make a decision. For example, we may decide to start our search on the first floor if the coin comes up heads or on the second floor if the coin comes up tails. If we also happen to have a die, then after deciding on the second floor, we could roll it to decide in which of the six rooms on the second floor to start our search. Although tossing coins and rolling dice may be a fun approach to find the phone, it is certainly not the intuitive thing to do, nor is it at all clear whether it gives you any algorithmic advantage over a deterministic algorithm. Randomized algorithms really help solve practical problems, and some of them indicate a competitive advantage over deterministic algorithms. Randomized algorithms make random decisions throughout their operation [1]. At first glance, making random decisions does not seem particularly helpful. However, Comte de Buffon, an eighteenth-century French naturalist, proved the opposite by developing an algorithm to accurately compute by randomly dropping needles on a sheet of paper with parallel lines. The fact that a randomized algorithm undertakes a nondeterministic sequence of operations often means that, unlike deterministic algorithms, no input can reliably produce worst-case results. We usually use randomized algorithms in hard problems where an exact, polynomial-time algorithm is unknown.

11.9.6 Genetic Algorithm

Given their source of inspiration, i.e. evolution, which is at the heart of our domain, evolutionary algorithms indicate a special flavor. Evolutionary algorithms are a broad class of optimization algorithms that try to simulate in some way the inner workings of evolution, as we (think we) understand it [48]. One component common to all these algorithms is the generation of random perturbations, or mutations, and the presence of a fitness function that is used to eveluate the quality of a given point and filter out mutations that are not useful. In this sense, some other methods such as random descent methods and even simulated annealing can be regarded as special cases of evolutionary algorithms. One of the broadest subclasses of evolutionary algorithms is the genetic algorithms.

Genetic algorithms and the related field of artificial life push the evolutionary analogy one step further by simulating the evolution of populations of points in fitness space. Furthermore, in addition to mutations, new points are generated by a number of other operations mimicking genetic operators and sexual reproduction, such as crossover. It is mentioned that genetic algorithms are particularly flexible and make possible the evolution of complex objects, such as computer programs. However, they are still very slow even on current computers, although it is surely improved year by year.

Applications of genetic algorithms to problems in biology can be found in many studies [48, 75-76]. For example, Pugacheva et al. [76] aimed to show that amino acid sequences have a latent periodicity with insertions and deletions of amino acids in unknown positions of the analyzed sequence. Genetic algorithm, dynamic programming and random weight matrices were used to develop a new mathematical algorithm for latent periodicity search. A multiple alignment of periods was calculated with help of the direct optimization of the position-weight matrix without using pairwise alignments. The developed algorithm was applied to analyze amino acid sequences of a small number of proteins. The results showed the presence of latent periodicity with insertions and deletions in the amino acid sequences of such proteins, for which the presence of latent periodicity was not previously known. The origin of latent periodicity with insertions and deletions is discussed.

11.10 Concluding Remarks

This chapter introduces typical algorithmic techniques in bioinformatics such as Graph Theory, Dynamic Programming, Hidden Markov Model, Neural Network and Clustering analysis. It covers most of the current topics in bioinformatics. Different methods and the corresponding algorithms are provided. Furthermore, algorithms were illustrated by using some detailed examples. In addition, there are still some other algorithms that can be used individually and/or together with other algorithms for solving bioinformatics problems. For example, Simulated Annealing algorithm is a probabilistic technique for approximating the global optimum of a given function. Gradient descent algorithm, which is a first-order iterative optimization algorithm for finding the minimum of a function, is often combined with Hidden Markov Model or Neural Network algorithm. Expectation Maximization algorithm can be used for detecting & analyzing of gene chip image signal. Reconstructing full-length transcript isoforms from sequence fragments (such as ESTs) is a major interest and challenge for bioinformatic analysis of pre-mRNA alternative splicing. Xing et al. [77] introduced a probabilistic formulation of the isoform reconstruction problem, and provided an expectation-maximization (EM) algorithm for its maximum likelihood solution.

References

[1] Jones N C, Pevzner P A. *An Introduction to Bioinformatics Algorithms.* Cambridge: MIT Press, 2004.

[2] Gopi S, Singh A, Suresh S, Paul S, Ranu S, Naganathan A N. Toward a quantitative description of microscopic pathway heterogeneity in protein folding. *Physical Chemistry Chemical Physics*, 2017, 19(31): 20891-20903.

[3] Li Z, Liu Z, Zhong W, Huang M, Wu N, Xie Y, Dai Z, Zou X. Large-scale identification of human protein function using topological features of interaction network. *Sci. Rep.*, 2016(6): 37179.

[4] Mandoiu II, Zelikovsky A. *Bioinformatics algorithms: techniques and applications.* Hoboken: John Wiley & Sons, Inc., 2008.

[5] Niknam N, Khakzad H, Arab S S, Naderi-Manesh H. PDB2Graph: A toolbox for identifying critical amino acids map in proteins based on graph theory. *Comput. Biol. Med.*, 2016(72): 151-159.

[6] Milo R, Shen-Orr S, Itzkovitz S, Kashtan N, Chklovskii D, Alon U. Network motifs: Simple building blocks of complex networks. *Science*, 2002(298): 824-827.

[7] Shen-Orr S, Milo R, Mangan S, Alon U. Network motifs in the transcriptional regulation network of escherichia coli. *Nat. Genet.*, 2002(31).

[8] Middendorf M, Ziv E, Wiggins C H. Inferring network mechanisms: The Drosophila melanogaster protein interaction network. *Proc. Nat. Acad. Sci.* USA, 2005(102): 3192-3197.

[9] Huang C H, Chen T H, Ng K L. Graph theory and stability analysis of protein complex interaction networks. *IET. Syst. Biol.*, 2016, 10(2): 64-75.

[10] Stavrakas V, Melas I N, Sakellaropoulos T, Alexopoulos L G. Network reconstruction based on proteomic data and prior knowledge of protein connectivity using graph theory. *PLoS. One.*, 2015, 10(5): e012841.1.

[11] Ryslik G A, Cheng Y, Cheung K H, Modis Y, Zhao H. A graph theoretic approach to utilizing protein structure to identify non-random somatic mutations. *BMC Bioinformatics.* 2014(15): 86.

[12] Biswas A, Ranjan D, Zubair M, He J. A Dynamic Programming Algorithm for Finding the Optimal Placement of a Secondary Structure Topology in Cryo-EM Data. *J. Comput. Biol.*, 2015, 22(9): 837-843.

[13] Sankoff D. The early introduction of dynamic programming into computational biology. *Bioinformatics.* 2000(16): 41.

[14] Sabzekar M, Naghibzadeh M, Sadri J. Efficient dynamic programming algorithm with prior knowledge for protein β-strand alignment. *J. Theor. Biol.*, 2017(417): 43-50.

[15] Sabzekar M, Naghibzadeh M, Eghdami M, Aydin Z. Protein β-sheet prediction using an efficient dynamic programming algorithm. *Comput. Biol. Chem.*, 2017(70): 142-155.

[16] Cormen T H, Leiserson C E, Rivest R L, Stein C. *Introduction to Algorithms.* 2nd ed. Cambridge: MIT Press, 2001.

[17] Needleman S B, Wunsch C D. A general method applicable to the search for similarity in the amino acid sequence of two proteins. *J. Mol. Biol.*, 1970(48): 443.

[18] Smith T F, Waterman M S. Identification of common molecular subsequences. *J. Mol. Biol.*, 1980(147): 195.

[19] Jiang T, Lin G, Ma B, Zhang K. Ageneral edit distance between RNA structures. *J. Comput. Biol.*, 2002(9): 371.

[20] Holm L, Sander C. Searching protein structure databases has come of age. *Proteins*, 1994(19): 165.

[21] Wei Z, Zhu D, Wang L. A Dynamic Programming Algorithm For (1,2)-Exemplar Breakpoint Distance. *J. Comput. Biol.*, 2015, 22(7): 666-676.

[22] Siederdissen C H Z, Prohaska S J, Stadler P F. Algebraic Dynamic Programming over general data structures. *BMC Bioinformatics.* 2015(19): S2.

[23] Alterovitz G, Ramoni M. *Knowledge-Based Bioinformatics From Analysis to Interpretation.* Hoboken: John Wiley & Sons Inc., 2010.

[24] Dey D K, Ghosh S, Mallick B K. *Bayesian Modeling in Bioinformatics.* Boca Raton: Chapman & Hall/CRC, 2010.

[25] Mallick B K, Gold D, Baladandayuthapani V. *Bayesian Analysis of Gene Expression Data.* Hoboken: John Wiley & Sons Inc., 2009.

[26] Wilkinson D J. Bayesian methods in bioinformatics and computational systems biology. *Brief. Bioinformatics*, 2007, 8(2): 109–116.

[27] Gilks W R, Richardson S, Spiegelhalter D. *Markov Chain Monte Carlo in Practice: Interdisciplinary Statistics.* Boca Raton: Chapman & Hall/CRC, 1995.

[28] Thorne T. NetDiff-Bayesian model selection for differential gene regulatory network inference. *Sci. Rep.*, 2016(6): 39224.

[29] Spyrou C, Stark R, Lynch A G, Tavare S. BayesPeak: Bayesian analysis of ChIP-seq data. *BMC Bioinformatics*, 2009(10): 299.

[30] Zhang W, Zhu J, Schadt E E, Liu J S. A Bayesian partition method for detecting pleiotropic and epistatic eQTL modules. *PLoS. Comput. Biol.*, 2010, 6(1): e1000642.

[31] Zhang Y, Liu J S. Bayesian inference of epistatic interactions in case-control studies. *Nat. Genet.*, 2007, 39(9): 1167–1173.

[32] Carlin B P, Louis T A. *Bayesian Methods for Data Analysis.* Boca Raton: CRC Press, 2009.

[33] Albert J. *Bayesian Computation with R.* New York: Springer-Verlag, 2009.

[34] Siebert M, Söding J. Bayesian Markov models consistently outperform PWMs at predicting motifs in nucleotide sequences. *Nucleic. Acids. Res.*, 2016, 44(13): 6055-6069.

[35] Wang X, Gu J, Hilakivi-Clarke L, Clarke R, Xuan J. DM-BLD: differential methylation detection using a hierarchical Bayesian model exploiting local dependency. *Bioinformatics*, 2017a, 33(2): 161-168.

[36] Wang T, Chen Y P, MacLeod I M, Pryce J E, Goddard M E, Hayes B J. Application of a Bayesian non-linear model hybrid scheme to sequence data for genomic prediction and QTL mapping. *BMC Genomics.*, 2017b, 18(1): 618.

[37] Ejlali N, Faghihi M R, Sadeghi M. Bayesian comparison of protein structures using partial Procrustes distance. *Stat. Appl. Genet. Mol. Biol.*, 2017, 16(4): 243-257.

[38] Jeong H, Qian X, Yoon B J. Effective comparative analysis of protein-protein interaction networks by measuring the steady-state network flow using a Markov model. *BMC Bioinformatics*, 2016, 17(Suppl 13): 395.

[39] Fu C, Deng S, Jin G, Wang X, Yu Z G. Bayesian network model for identification of pathways by integrating protein interaction with genetic interaction data. *BMC Syst. Biol.*, 2017, 11(Suppl 4): 81.

[40] Xiong J. *Essential Bioinformatics.* New York: Cambridge University Press, 2006.

[41] Polanski A, Kimmel M. *Bioinformatics.* Berlin Heidelberg: Springer-Verlag, 2007.

[42] Nguyen T, Habeck M. A probabilistic model for detecting rigid domains in protein structures. *Bioinformatics*, 2016, 32(17): i710-i717.

[43] Zhou S, Wang Q, Wang Y, Yao X, Han W, Liu H. The folding mechanism and key metastable state identification of the PrP127-147 monomer studied by molecular dynamics simulations and Markov state model analysis. *Phys. Chem. Chem. Phys.*, 2017, 19(18): 11249-11259.

[44] Plattner N, Doerr S, De Fabritiis G, Noé F. Complete protein-protein association kinetics in atomic detail revealed by molecular dynamics simulations and Markov modelling. *Nat. Chem.*, 2017, 9(10): 1005-1011.

[45] Mackay L, Zemkova H, Stojilkovic S S, Sherman A, Khadra A. Deciphering

the regulation of P2X4 receptor channel gating by ivermectin using Markov models. *PLoS. Comput. Biol.*, 2017, 13(7): e1005643.

[46] Zhao L, Lascoux M, Waxman D. An informational transition in conditioned Markov chains: Applied to genetics and evolution. *J. Theor. Biol.*, 2016(402): 158-170.

[47] Baldi P, Chauvin Y, Hunkapillar T, McClure M. Hidden Markov models of biological primary sequence information. *Proc. Natl. Acad. Sci.* USA, 1994(91): 1059-1063.

[48] Baldi P, Brunak S. *Bioinformatics: The Machine Learning Approach.* Cambridge: MIT Press, 2001.

[49] Eddy S R. Hidden Markov models. *Curr. Opin. Struct. Biol.*, 1996(6): 361-365.

[50] Gerhold D, Caskey C T. It's the genes! EST access to human genome content. *Bioessays*, 1996(18): 973-981.

[51] Henikoff S, Henikoff J G. Protein family classification based on searching a database of blocks. *Genomics*, 1994(19): 97-107.

[52] Vijayabaskar M S. Introduction to Hidden Markov Models and Its Applications in Biology. *Methods. Mol. Biol.*, 2017(1552): 1-12.

[53] Francesco V D, Garnier J, Munson P J. Protein topology recognition from secondary structure sequences-Applications of the hidden Markov models to the alpha class proteins. *J. Mol. Biol.*, 1997(267): 446-463.

[54] Sonnhammer E L L, Eddy S R, Durbin R. Pfam: a comprehensive database of protein domain families based on seed alignments. *Proteins*, 1997(28): 405-420.

[55] Kundu S. Mathematical basis of improved protein subfamily classification by a HMM-based sequence filter. *Math. Biosci.*, 2017(293): 75-80.

[56] Lampros C, Papaloukas C, Exarchos T, Fotiadis D I. HMMs in Protein Fold Classification. In: Westhead D., Vijayabaskar M. (eds) Hidden Markov Models. *Methods in Molecular Biology*, vol. 1552. New York: Humana Press, 2017.

[57] Kamal M S, Chowdhury L, Khan M I, Ashour A S, Tavares JMRS, Dey N. Hidden Markov model and Chapman Kolmogrov for protein structures prediction from images. *Comput. Biol. Chem.*, 2017(68): 231-244.

[58] Jablonowski K. Hidden Markov Models for Protein Domain Homology Identification and Analysis. *Methods. Mol. Biol.*, 2017(1555): 47-58.

[59] Tsaousis G N, Theodoropoulou M C, Hamodrakas S J, Bagos P G. Predicting alpha helical transmembrane proteins using HMMs. *Methods. Mol. Biol.*, 2017(1552): 63-82.

[60] Voshol G P, Vijgenboom E, Punt P J. The discovery of novel LPMO families with a new Hidden Markov model. *BMC. Res. Notes.*, 2017, 10(1): 105.

[61] Wang T, Yun J, Xie Y, Xiao G. Finding RNA-Protein Interaction Sites Using HMMs. *Methods. Mol. Biol.*, 2017(1552): 177-184.

[62] Pereira M B, Wallroth M, Kristiansson E, Axelson-Fisk M. HattCI: fast and accurate attc site identification using Hidden Markov Models. *J. Comput. Biol.*, 2016, 23(11): 891-902.

[63] Malekpour S A, Pezeshk H, Sadeghi M. MGP-HMM: Detecting genome-wide CNVs using an HMM for modeling mate pair insertion sizes and read counts. *Math. Biosci.*, 2016(279): 53-62.

[64] Ferles C, Beaufort W S, Ferle V. Self-Organizing Hidden Markov Model Map (SOHMMM): biological sequence clustering and cluster visualization. *Methods. Mol. Biol.*, 2017(1552): 83-101.

[65] Stormo G D, Schneider T D, Gold L, Ehrenfeucht A. Use of the 'perceptron' algorithm to distinguish translational initiation sites in e. coli. *Nucl. Acids Res.*, 1982(10): 2997-3011.

[66] Rumelhart D E, Hinton G E, Williams R J. Learning internal representations by error propagation. In: Rumelhart D E, McClelland J L, and the PDP Research Group, editors, *Parallel distributed processing: Explorations in the microstructure of cognition*, volume 1: Foundations, 318-362. Cambridge: MIT Press, 1986.

[67] Wang Y B, You Z H, Li X, Jiang T H, Chen X, Zhou X, Wang L. Predicting protein-protein interactions from protein sequences by a stacked sparse autoencoder deep neural network. *Mol. Biosyst.*, 2017, 27; 13(7): 1336-1344.

[68] Du X, Sun S, Hu C, Yao Y, Yan Y, Zhang Y. DeepPPI: boosting prediction of protein-protein interactions with Deep Neural Networks. *J. Chem. Inf. Model.*, 2017, 57(6): 1499-1510.

[69] Dutta S, Madan S, Parikh H, Sundar D. An ensemble micro neural network approach for elucidating interactions between zinc finger proteins and their target DNA. *BMC Genomics.*, 2016, 22; 17(Suppl 13): 1033.

[70] Le N Q, Ho Q T, Ou Y Y. Incorporating deep learning with convolutional neural networks and position specific scoring matrices for identifying electron transport proteins. *J. Comput. Chem.*, 2017, 38(23): 2000-2006.

[71] Cang Z, Wei G W. TopologyNet: Topology based deep convolutional and multi-task neural networks for biomolecular property predictions. *PLoS. Comput. Biol.*, 2017, 13(7): e1005690.

[72] Brunak S, Engelbrecht J, Knudsen S. Prediction of human mRNA donor and acceptor sites from the DNA sequence. *J. Mol. Biol.*, 1991(220): 49-65.

[73] Snyder E E, Stormo G D. Identification of protein coding regions in genomic DNA. *J. Mol. Biol.*, 1995(248): 1-18.

[74] Umarov R K, Solovyev V V. Recognition of prokaryotic and eukaryotic promoters using convolutional deep learning neuralnetworks. *PLoS. One.*, 2017, 12(2): e0171410.

[75] Parsons R, Johnson M E. DNA sequence assembly and genetic programming-new results and puzzling insights. In: Rawlings C, Clark D, Altman R,

Hunter L, Lengauer T, and Wodak S, editors. *Proceedings of the Third International Conference on Intelligent Systems for Molecular Biology*, 277-284. Menlo Park: AAAI Press, 1995.

[76] Pugacheva V, Korotkov A, Korotkov E. Search of latent periodicity in amino acid sequences by means of genetic algorithm and dynamicprogramming. *Stat. Appl. Genet. Mol. Biol.*, 2016, 15(5): 381-400.

[77] Xing Y, Yu T, Wu YN, et al. An expectation-maximization algorithm for probabilistic reconstructions of full-length isoforms from splice graphs. *Nucleic Acids Res.*, 2006, 34(10): 3150-3160.

Chapter 12 An Introduction to R

Lishu Zhang[1]

R has becoming the most widely used software in bioinformatics. The purpose of this chapter is to provide basic guidance for biologists on how to use R. For readers, no prior knowledge of programming is required or assumed. We hope this chapter can motivate and help more biologists to use R in their scientific career.

12.1 What's R

R is a language and environment for statistical computing and graphics. It is a GNU project which is similar to the S language and environment which was developed at Bell Laboratories (formerly AT&T, now Lucent Technologies) by John Chambers and colleagues. R can be considered as a different implementation of S. There are some important differences, but much code written for S runs unaltered under R.

R was first written as a research project by Ross Ihaka and Robert Gentleman, and is now under active development by a group of statisticians called 'the R core team', with a home page at www.r-project.org. R is available free of charge and is distributed under the terms of the Free Software Foundation's GNU General Public License. You can download the program from the Comprehensive R Archive Network (CRAN). Ready-to-run 'binaries' are available for Windows, Mac OS X, and Linux. The source code is also available for download and can be compiled for other platforms.

R provides a wide variety of statistical (linear and nonlinear modelling, classical statistical tests, time-series analysis, classification, clustering ...) and graphical techniques, and is highly extensible. Currently, the Comprehensive R Archive Network (CRAN) package repository features more than ten thousand available R extension packages.

---

1. Zhang Lishu, College of Life Sciences and Bioengineering, School of Science, Beijing Jiao Tong University, Beijing, China, 100044.

12.2 How to Install R

R is free software. It is hosted on many different servers around the world ('mirrors') and can be downloaded from any of them. A list of all the download mirrors should be available through the homepage of R (https://www.r-project.org) (Fig. 12.1). Next, we will go through the R installation procedure.

① Look at the 'Getting Started' section on the home page of R and click on 'download R' and choose a mirror in the new page (such as 'https://cloud.r-project.org/'). For faster downloads, a server closer to your physical location should be chosen.

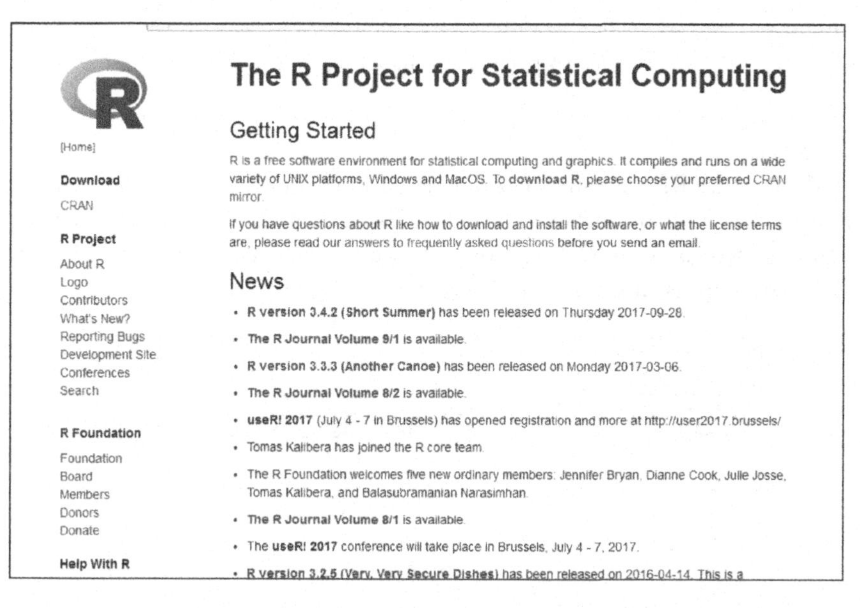

Fig. 12.1 The homepage of R

② Once you have chosen a mirror, at the top of your screen should be a list of links of R for each operating system (Fig. 12.2). Choose the operating system (assuming you work on a windows computer, choose 'windows') and in the new page choose 'base', on the next page, you should see a link saying something like 'Download R X.X.X for Windows' (where X.X.X gives the version of R, eg. R 3.4.2). Click on this link.

③ You may be asked if you want to save or run a file 'R-X.X.X-win.exe', you can choose 'Save' and save the executable file on your computer. Then double-click on the downloaded executable file to run it. The R Setup Wizard will appear in a window.

④ Follow the Setup Wizard until the installation has finished. The default

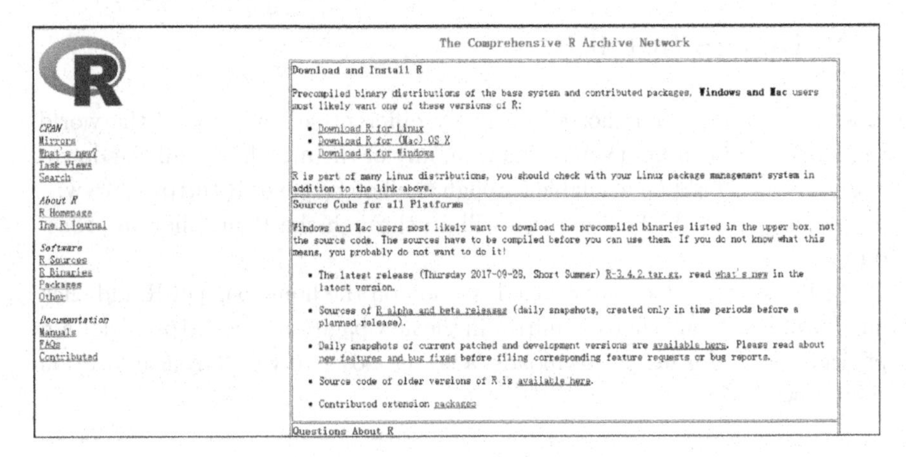

Fig. 12.2 The download page of R

settings are fine.

⑤ To start R, you can either double-click on the 'R' icon on the desktop of your computer or through the 'Start' button at the bottom left of your computer screen, followed by choosing 'All programs', and selecting 'R', then the R console (as shown in Fig. 12.3) should pop up.

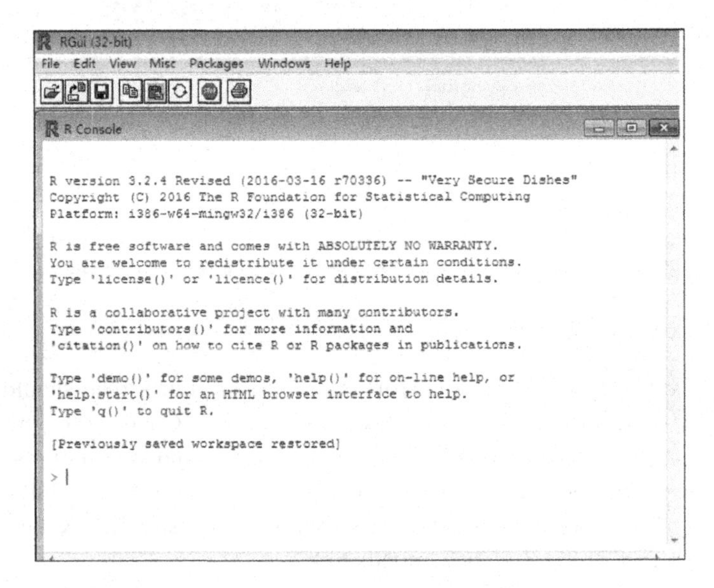

Fig. 12.3 R Console

12.3 RGui

As part of the process of downloading and installing R, you get the standard graphical user interface (GUI), called RGui. RGui gives you some tools to manage your R environment — most important, a console window. The console is where you type instructions, or scripts, and generally get R to do useful things for you.

When you open RGui for the first time, you see the R Console screen, which lists some basic information such as your version of R and the licensing conditions.

Below all this information is the R prompt, denoted by a > symbol. The prompt indicates where you type your commands to R; you see a blinking cursor to the right of the prompt (Figure 12.3).

12.3.1 Issue a Simple Command in the R Console

Now, let us use the Console to issue a very simple command to R. Type the following to calculate the sum of some numbers:

```
> 2+2+1
```

R responds immediately to your command, calculates the total, and displays it in the console:

```
>2+2+1
[1] 5
```

The answer is 5. R gives you one other piece of information: The [1] preceding 5 indicates that the value 5 is the first element in your answer. It is, in fact, the only element in your answer.

12.3.2 Close the R Console

To quit your R session, type the following code in the console:

```
> q()
```

At this point you will be asked whether you want to save the data from your R session. On some systems this will bring up a dialog box, and on others you will receive a text prompt to which you can respond Yes, No or Cancel (a single letter abbreviation will do) to save the data before quitting, quit without saving, or return to the R session. Data which is saved will be available in future R sessions. Click No, because you have nothing to save. This action closes your R session (as well as RGui, if you've been using RGui as your code editor).

Note the parentheses after the q. In R, to call a function you type the name followed by the arguments in parentheses. If the function takes no arguments you just type the name followed by left and right parenthesis. If you forget the parentheses and type just the name of the function, R will list it.

The R Console allows command editing. You will find that the left and right arrow keys, home, end, backspace, insert, and delete work exactly as you would expect. You also get a command history: the up and down arrow keys can be used to scroll through recent commands. Thus, if you make a mistake all you need to do is press the up key to recall your last command and edit it.

12.3.3 Getting Help

Even with good introductory books on R, you'll need to use the R Help files. The R Help files provide detailed information about the use of different functions and their peculiarities. R has excellent built-in help for every function that explains how to use that function.

To search through the Help files, you'll use one of the following functions:

?: Displays the Help file for a specific function. For example,

```
> ?read.table()
```

and then displays the Help file for the read.table function.

??: Searches for a word (or pattern) in the Help files. For example,

```
> ??read
```

returns the names of functions that contain the word read in either the function names or their descriptions.

RSiteSearch(): Performs an online search with an internet search engine. This search engine allows you to perform a search of the R functions, package vignettes and the R-help mail archives. For example,

```
>RSiteSearch("linear models")
```

does a search at website for the search term "linear models".

12.3.4 Working Directory

Your working directory is the folder on your computer in which you are currently working. When you ask R to open a certain file, it will look in the working directory for this file, and when you tell R to save a data file or figure, it will save it in the working directory. Before you start working, please set your working directory to where all your data and script files are or should be stored. Type in the command window: setwd("directoryname").

For example:

```
>setwd("D:/Users/R/")
```

R is case sensitive, so make sure you write capitals where necessary. Let us use "getwd()" to see the working directory and you can see R has set the working directory according your command.

```
> getwd()
[1] "D:/Users/R"
```

12.3.5 Scripts

R is an interpreter that uses a command line based environment. This means that you have to type commands. Sometimes you will want to type the commands in files, the so called scripts, and then run these commands all at once. These scripts have typically file names with the extension .R (e.g., user.R). You can open an script window to edit these files by clicking "File" and "New" or "Open file...". You can run the commands in your script either copy-paste the commands into the command window or highlight the commands you want to execute and then click the button indicated in Figure 12.4. You can always run the whole script with the console command "source", so e.g. for the script in the file user. R you type:

```
> source("user.R")
```

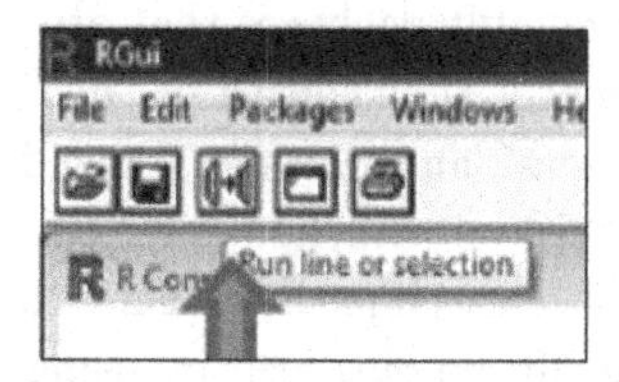

Fig. 12.4 Button for running command

For those who do more R-programing, some other integrated development environment (IDE) for R (such as R Studio interface) is recommended. R Studio includes a console, syntax-highlighting editor that supports direct code execution, as well as tools for plotting, history, debugging and workspace management. To install R Studio, go to http://www.rstudio.org/.

12.4 How to Install R Extention Packages

Although a team of statisticians and programmers maintain R, many independent groups submit contributed packages. This is a very extensive resource of a large amount of programs and will not be install by default. To find out about contributed package please click on the link "packages" at the left of the http://cran.r-project.org. This will link to a page with lots of information on contributed packages (Figure 12.5).

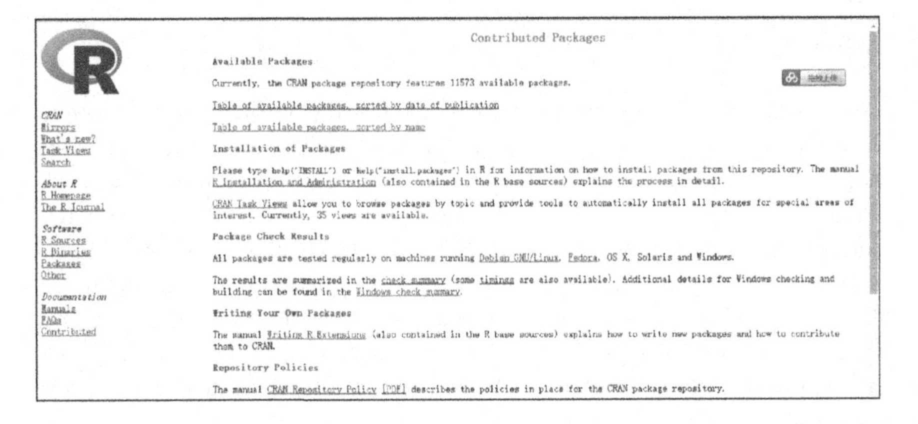

Fig. 12.5 Page for R Packages

There is a small set of default packages that are attached every time R is started interactively. Currently, they consist of "methods" "stats" "graphics" "grDevices" "utils" "datasets" and "base". When R is run in batch mode, not all of these packages will be attached. You can find out what packages are currently attached to the search path by using the function "search", and more detailed information can be obtained by using "sessionInfo".

```
> search()
[1] ".GlobalEnv"        "package:stats"     "package:graphics"
[4] "package:grDevices" "package:utils"     "package:datasets"
[7] "package:methods"   "Autoloads"         "package:base"
```

Many functions can generate information about packages that have already been installed on the user's system. A vector listing the base names of packages that are currently attached can be obtained using ".packages". The return value of ".packages" is invisible, so we first assign it to a temporary variable, as in the example below.

```
> z=.packages()
> z
```

```
[1] "stats"     "graphics"  "grDevices" "utils"
    "datasets"  "methods"
[7] "base"
```

There are several ways to install extension packages: ① from the GUI, select "Packages → Install Packages", note only the packages from the selected repositories will be shown; ② from command line type "install.packages ("package name")"; ③ download a compressed copy of the package from the R website and install from local zip or .tar.gz file.

Let us install an R package called "rmeta" package by the first way. You should choose "Install package(s)···" from the "Packages" menu at the top of the R console (Figure 12.6). This will ask you what website you want to download the package from. It will also bring up a list of available packages that you can install, and you should choose the package that you want (here is "rmeta") to install from that list. This will install the "rmeta" package.

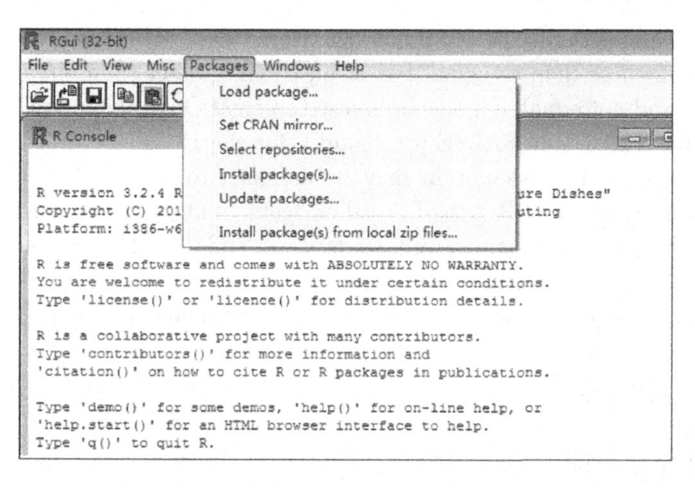

Fig. 12.6 "Packages" menu at R console

The "rmeta" package is now installed. Whenever you want to use the "rmeta" package after this, after starting R, you first have to load the package by typing into the R console:

```
> library("rmeta")
```

The procedure above can be used to install the majority of R packages. However, the Bioconductor set of bioinformatics R packages need to be installed by a special procedure (please refer to section 12.10).

12.5 Expressions and Assignments

As we have exercised in section 12.3, R works like a calculator, you type an expression and get the answer:

```
> 1+2
[1] 3
```

The standard arithmetic operators are “+”, “-”, “*”, and “/” for add, subtract, multiply and divide, and “^” for exponentiation, so 2^3=8. These operators have the standard precedence, with exponentiation highest and addition/subtraction lowest, but you can always control the order of evaluation with parentheses. You can use mathematical functions, such as “sqrt”, “exp”, and “log”. For example:

```
> log(0.3/(1-0.3))
[1] -0.8472979
```

R also understands the relational operators “<=” “<” “==” “>” “>=” and “!=” for less than or equal, less than, equal, greater than, greater than or equal, and not equal. These can be used to create logical expressions that take values TRUE and FALSE (or T and F for short).

The results of a calculation may be assigned to a named object. The assignment operator in R is “<-”, read as “gets”, but by popular demand R now accepts the equal sign as well, so “x <- 2” and “x = 2” both assign the value 2 to a variable (technically an object) named x.

Typing a object name prints its contents. The name “pi” is used for the constant π. Thus,

```
> s <- pi/3
> s
[1] 1.047198
```

assigns $\pi/3$ to the variable “s” and then prints the result.

12.6 Data Structure

Like in many other programs, R organizes numbers in scalars (a single number – 0-dimensional), vectors (a row of numbers, also called arrays – 1-dimensional) and matrices (like a table – 2-dimensional). All variables (scalars, vectors, matrices, etc.) created by R are called objects. The “s” you defined above was a scalar. R is designed to work with vectors and matrices as well.

12.6.1 Vector

To define a vector with the numbers 3, 4 and 5, you need the function "c", which is short for concatenate (paste together).

```
> b=c(3,4,5)
> b
[1] 3 4 5
```

The simplest way to create a vector is the following command:

```
> vector1 = 1:9
> vector1
[1] 1 2 3 4 5 6 7 8 9
```

The colon ":" operator creates a sequence of numbers from the left hand value to the right hand value, iterating in increments of 1. Another example:

```
> 1.2:9
[1] 1.2 2.2 3.2 4.2 5.2 6.2 7.2 8.2
```

Some functions are useful in creating vectors. You can also use the "seq" function to create a sequence given the starting and stopping points and an increment. For example, here are eleven values between 0 and 1 in steps of 0.1:

```
> seq(0, 1, 0.1)
[1] 0.0 0.1 0.2 0.3 0.4 0.5 0.6 0.7 0.8 0.9 1.0
```

The "rep" function can be used for repeat or replicate. For example, "rep(3,4)" replicates the number three four times. The first argument can be a vector, so "rep(x,3)" replicates the entire vector x three times. If both arguments are vectors of the same size, then each element of the first vector is replicated the number or times indicated by the corresponding element in the second vector. Consider this example:

```
> rep(1:3, 2)
[1] 1 2 3 1 2 3
> rep(1:3, c(2,2,2))
[1] 1 1 2 2 3 3
```

The first call repeats the vector 1:3 twice. The second call repeats each element of 1:3 twice, and "rep(1:3, rep(2,3))" will return the same results.

When adding vectors, R adds the first element of one vector to the first element of the other, the second element of one vector to the second element of the other, etc. For example:

```
> vector3 = 1:9
> vector3
```

```
[1] 1 2 3 4 5 6 7 8 9
> vector4 = 9:1
> vector4
[1] 9 8 7 6 5 4 3 2 1
> vector3 + vector4
[1] 10 10 10 10 10 10 10 10 10
```

The first element of this new vector was 9 + 1, the second 8 + 2, etc.

If x and y are vectors of different lengths, the shorter one is recycled as many times as necessary to add something to each element of the larger vector.

```
> vector5 = 1:3
> vector5
[1] 1 2 3
> vector3=1:9
> vector3
[1] 1 2 3 4 5 6 7 8 9
> vector3 + vector5
[1] 2 4 6 5 7 9 8 10 12
```

In this case the first element is 1 + 1, the second element is 2 + 2, the third element is 3 + 3, but then the smaller vector loops back to its first element to add to the fourth element of the larger vector, giving the calculation 1 + 4 for the fourth element.

When the larger vector is not a multiple of the smaller vector the process is the same, except R will give a warning message informing you that the larger vector is not an integer multiple, so not all elements of the smaller vector were added to the larger vector an equal number of times.

Multiplying, dividing, exponentiating, and subtracting vectors works in the same, element-by-element fashion.

```
> vector3 * vector5
[1] 1 4 9 4 10 18 7 16 27
```

12.6.2 Matrices

R also understands matrices and higher dimensional arrays. To define a matrix, use the function "matrix":

```
>matrix(c(1, 2, 3, 4, 5, 6, 7, 8, 9), nrow = 3)
          [ ,1 ]  [ ,2 ]  [ ,3 ]
[ 1, ]       1       4       7
```

```
[ 2, ]     2       5       8
[ 3, ]     3       6       9
```

As you can see, the matrix() function takes the data you input and the number of rows you input (nrow) and makes a matrix by filling down each column from the left to the right. You can also specify the number of columns (ncol):

```
>matrix(1:8, ncol = 2)
[ ,1 ]  [ ,2 ]
[ 1, ]  1       5
[ 2, ]  2       6
[ 3, ]  3       7
[ 4, ]  4       8
```

Elements of a matrix can be addressed in the usual way "[row,column]". When you want to select a whole row, you leave the spot for the column number empty (the other way around for columns of course).

```
> mat1<-matrix(c(1, 2, 3, 4, 5, 6, 7, 8, 9), nrow = 3)
> mat1
          [,1]  [,2]  [,3]
[1,]         1     4     7
[2,]         2     5     8
[3,]         3     6     9
> mat1[1,2]
[1] 4
> mat1[1,]
[1] 1 4 7
> mat1[,2]
[1] 4 5 6
```

Alternatively, data can be changed. For example, to remove the second column from the matrix one would use the following command:

```
> mat1[,-2]
      [,1]  [,2]
[1,]   1     7
[2,]   2     8
[3,]   3     9
```

R takes the matrix, removes the second column, and shifts everything over.

If you want to change the actual values of the data just access the part of the matrix you want to change and use the "=" operator as follows:

```
> mat1[1, 1] = 15
> mat1
      [,1] [,2] [,3]
[1,]   15    4    7
[2,]    2    5    8
[3,]    3    6    9
```

12.6.3 Data Frames

Time series are often ordered in data frames. A data frame is a matrix with names above the columns. This is nice, because you can call and use one of the columns without knowing in which position it is.

Let us construct a data frame. The columns have the names x, y and z.

```
t = data.frame(x = c(11,12,14), y = c(19,20,21),
                              z = c(10,9,7))
> t
   x  y  z
1 11 19 10
2 12 20  9
3 14 21  7
```

To retrieve data in a cell, we would enter its row and column coordinates in the single square bracket "[ ]" operator. The two coordinates are separated by a comma. In other words, the coordinates begins with row position, then followed by a comma, and ends with the column position. The order is important.

Here is the cell value from the first row, third column of t.

```
> t [1,3]
[1] 10
```

Moreover, we can use the row and column names instead of the numeric coordinates.

```
> t [1,"x"]
[1] 11
```

And we can select the whole row or whole column of t,

```
> t [,"z"]
[1] 10  9  7
> t [2,]
x y z
```

```
2 12 20 9
```

The number of data rows in the data frame is given by the "nrow" function. And the number of columns of a data frame is given by the "ncol" function.

```
> nrow(t)
[1] 3
> ncol(t)
[1] 3
```

12.6.4 Lists

Another basic structure in R is a list. The main advantage of lists is that the "columns" (they're not really ordered in columns any more, but are a collection of vectors) don't have to be of the same length, unlike matrices and data frames. Lists are great when you want to store multi-dimensional objects into one object.

Let's construct a list named "list1" by giving names and values.

```
> list1 <- list (name="Tom", age=30, scores=c(85, 76, 90))
> list1
$name
[1] "Tom"
$age
[1] 30
$scores
[1] 85 76 90
```

We can retrieve a list slice with the single square bracket "[ ]" operator.

```
> list1["name"]
$name
[1] "Tom"
```

In order to reference a list member directly, we have to use the double square bracket "[[ ]]" operator. The following object list1[["scores"]] is a copy of scores, but is not a slice containing scores. We have modified its content directly, but the list slice "scores" is unaffected.

```
> list1[["scores"]]+10
[1] 95 86 100
> list1["scores"] # to see if scores is unaffected
$scores
[1] 85 76 90
```

12.7 Importing Data Into R

R has many functions that allow you to import data from other applications. The following table (Table 12.1) lists some of the useful text import functions, what they do, and examples of how to use them.

Table 12.1 Text import functions

Function	What It Does	Example
read.table()	Reads any tabular data where the columns are separated (for example by commas or tabs). You can specify the separator (for example, commas or tabs), as well as other arguments to precisely describe your data	read.table(file="myfile", sep="t", header=TRUE)
read.csv()	A simplified version of read.table() with all the arguments preset to read CSV files, like Microsoft Excel spreadsheets	read.csv(file="myfile")
read.csv2()	A version of read.csv() configured for data with a comma as the decimal point and a semicolon as the field separator	read.csv2(file="myfile", header=TRUE)
read.delim()	Useful for reading delimited files, with tabs as the default separator	read.delim(file="myfile", header=TRUE)
scan()	Allows you finer control over the read process when your data isn't tabular	scan("myfile", skip = 1, nmax=100)
readLines()	Reads text from a text file one line at a time	readLines("myfile")
read.fwf	Read a file with dates in fixed-width format. In other words, each column in the data has a fixed number of characters	read.fwf("myfile", widths=c(1,2,3))

Let's study the usage of read.table as an example.

The basic syntax behind R "read.table" function to read the data from a text file is as shown below.

```
read.table(file, header = FALSE, sep = "", quote = "\"'",
           dec = ".", numerals = c("allow.loss",
           "warn.loss", "no.loss"),
           row.names, col.names, as.is = !stringsAsFactors,
           na.strings = "NA", colClasses = NA, nrows = -1,
           skip = 0, check.names = TRUE,
           fill = !blank.lines.skip,
           strip.white = FALSE, blank.lines.skip = TRUE,
           comment.char = "#",
           allowEscapes = FALSE, flush = FALSE,
           stringsAsFactors = default.stringsAsFactors(),
           fileEncoding = "", encoding = "unknown", text,
```

```
skipNul = FALSE)
```

There are some explanation for the syntax of function "read.table" in the Table 12.2, and more help can be accessed through R help files by typing "?read.table" in R console.

Table 12.2 Syntax of function read.table

file	The name of the file(within" " or a variable of mode character), possibly with its path (using symbol "/",not "\"), or a remote access to a file of type URL(HTTP://...)
header	FALSE or TRUE to indicate if a file contains the names of the variables on its first line
sep	The field separator used in the file, for instance sep="\t" means that it is tab seperated
quote	The characters used to cite the variables of mode character
dec	The character used for the decimal point
row.names	A vector of row names
col.names	A vector of optional names for the variables. The default is to use "V" followed by the column number
as.is	Controls the conversion of character variables as factors(if FALSE) or keeps them as character(if TRUE)
na.strings	A value given to missing data(converted as NA)
colClasses	A vector of classes to be assumed for the columns
nrows	the maximum number of rows to read in
skip	the number of lines of the data file to skip before beginning to read data
check.names	Logical. If TRUE then the names of the variables in the data frame are checked to ensure that they are syntactically valid variable names
fill	If TRUE and all lines do not have the same number of variables, "blanks" are added

In addition to these options to read text data, the package "foreign" allows you to read data from other popular statistical formats, such as SPSS. To use these functions, you first have to load the built-in foreign package, with the following command:

```
> library("foreign")
```

The Table 12.3 lists the functions to import data from SPSS, Stata, and SAS.

Table 12.3 Functions of import data

Function	What It Does	Example
read.spss	Reads SPSS data file	read.spss("myfile")
read.dta	Reads Stata binary file	read.dta("myfile")
read.xport	Reads SAS export file	read.xport("myfile")

12.8 Exporting Data

There are three main types of output from R. Two of the main types of output are the same as the two main types of input — delineated files such as ".csv" and R object files such as ".rdata". Any R data type can be written to a ".csv" or ".rdata" file and these data files are stored in your current R working directory. The third type of output is graphical output, and the most common form of graphical output is the ".pdf" file. Other file types such as ".jpg", ".png", ".bmp", and ".tiff" are available as well.

Below, let's see an example for R data exporting. If we export data to a tab delimited text file, we can use "write.table" function. "write.table" prints its required argument x (after converting it to a data frame if it is not one nor a matrix) to a file or connection. The basic syntax behind this function is as follows:

```
write.table(x, file = "", append = FALSE, quote = TRUE, sep = "",
            eol = "\n", na = "NA", dec = ".", row.names = TRUE,
            col.names = TRUE, qmethod = c("escape", "double"),
            fileEncoding = "")
```

12.9 Loops/Statements

Although R is built around operations on matrices, it still contains a powerful set of tools called loops and statements that you will find in many other programming languages.

The first and perhaps most important loop is called the "for-loop". Below is an example illustrating how a for-loop works (please input the following commands into the script window and then run them):

```
for( i in 1:3 ) {
print(i)
print(i + 5)
}
```

The output should be as follows:

```
[1] 1
[1] 6
[1] 2
[1] 7
[1] 3
[1] 8
```

As you can see, the for-loop works by selecting a value for i (in this case the values of i it iterates through are contained in the vector 1:3), then goes through the code from top to bottom. First, $i = 1$, and the loop calls "print(i)". The function print( ), true to its name, outputs whatever is within its parentheses. Thus, the number 1 is output first. Then, the function print( ) is called again, this time as "print(i + 5)", outputting the number 6, or 1 + 5. Then, the process starts over for $i = 2$ and finally for $i = 3$, after which the loop finishes. Only commands within the { } are run.

Next, let's look at a commonly used statement — the "if-statement". The following commands illustrate a usage:

```
matrix1<-matrix(1:9,nrow=3)
for(i in 1:3) {
  if(matrix1[i,1] >= 2) {
  matrix1[i,1] = 0
  }
}
print(matrix1)
```

The output should look like the following:

```
        [ ,1 ] [ ,2 ] [ ,3 ]
[ 1, ]     1      4      7
[ 2, ]     0      5      8
[ 3, ]     0      6      9
```

The commands above combine the for-loop and the if-statements. These R commands tell the computer to look at the first element of each row in matrix1 and check if it is greater than or equal to 2, then, if that element is greater than or equal to 2, that value is set to 0.

The last main statement that we will cover in this section is the "ifelse-statement". This statement is similar to the if-statement but allows you to control what happens for both the true and false Boolean values created. To illustrate:

```
> ifelse(matrix1 > 5, 1, 0)
        [ ,1 ] [ ,2 ] [ ,3 ]
[ 1, ] 0       0      1
[ 2, ] 0       0      1
[ 3, ] 0       1      1
```

The "ifelse( )" function is checking which values of matrix1 are larger than 5, and then it is assigning those values that are TRUE to 1, and the values that are FALSE to 0. The important part about this statement is to remember that the parameter order is what you want to set true values to, followed by what you want to set false values to.

12.10 Bioconductor

12.10.1 Bioconductor Packages Overview

The Bioconductor (www.bioconductor.org) project provides many additional R packages for statistical data analysis in different life science areas, such as tools for microarray, next generation sequence and genome analysis. It is widely used by bioinformatics. There are more than 1300 R packages available in Bioconductor so far. This includes R packages such as "affy", "affyPLM", etc. To obtain a broad overview of available Bioconductor packages, you can refer to website http://www.bioconductor.org/packages/release/BiocViews.html#__Software. It is a very useful resource for a beginner to explore all available Bioconductor Packages (Figure 12.7).

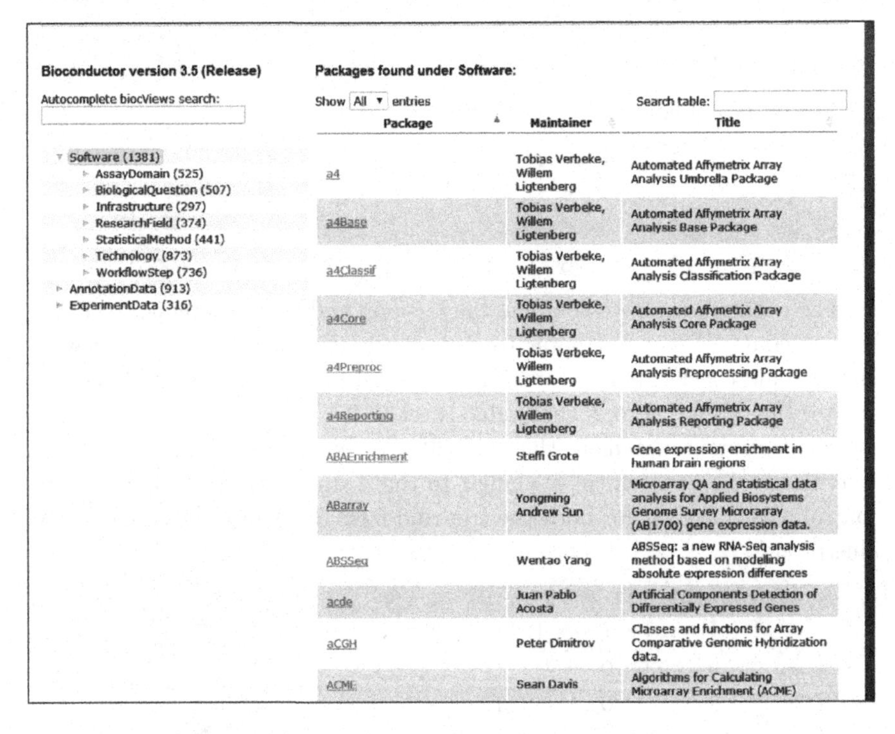

Fig. 12.7 Screenshot of Bioconductor Packages

12.10.2 Install Bioconductor Packages

It is recommended to install Bioconductor packages with the biocLite. R installation script. To install core packages, type the following in an R command window:

```
## try http:// if https:// URLs are not supported
>source("https://bioconductor.org/biocLite.R")
>biocLite()
```

Note that "## try http:// if https:// URLs are not supported" is just a comment, not a command, so it is not necessary and can be deleted. Comments starting with a hashmark ('#'), can be put almost anywhere in R.

This will install a core set of Bioconductor packages ("affy" "affydata" "affyPLM" "annaffy" "annotate" "Biobase" "Biostrings" "DynDoc" "gcrma" "genefilter" "geneplotter" "hgu95av2.db" "limma" "marray" "matchprobes" "multtest" "ROC" "vsn" "xtable" "affyQCReport"). This takes a few minutes.

At a later date, you may wish to install some extra Bioconductor packages that do not belong to the core set of Bioconductor packages. For example, to install the Bioconductor package called "yeastExpData", start R and type in the R console:

```
>source("http://bioconductor.org/biocLite.R")
> biocLite("yeastExpData")
```

Whenever you want to use a package after installing it, you need to load it into R by typing:

```
>library("yeastExpData")
```

We can use the basic R help functions to get additional information about R packages and their functions, such as we can use library(help=affy) to List all functions/objects of "affy" package (Figure 12.8).

12.11 Further Resources

From this chapter you have got a bit insight into R. We really encourage you to start the usage of R step by step. There is now an extensive and rapidly growing literature on R which will give you enough help on your work. Here are some useful websites for your reference.

• CRAN Home Page: http://cran.r-project.org/, useful site for updating R and finding more information on packages and current R events.

• R Journal: http://journal.r-project.org/, an academic journal that has articles directly relating to R and new R packages.

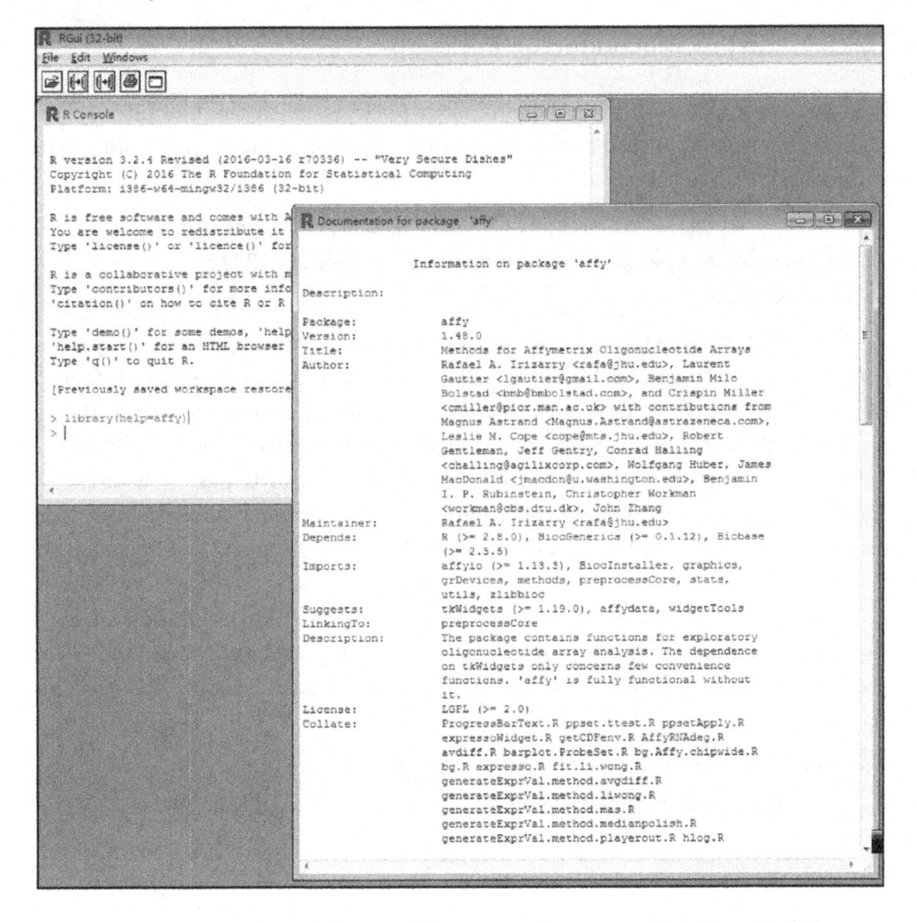

Fig. 12.8 Use library(help=affy) to get information of package "affy"

- R Manuals: http://cran.r-project.org/ > Documentation > Manuals.
- Bioconductor Home Page: http://www.bioconductor.org/.
- R & Bioconductor Manual: http://manuals.bioinformatics.ucr.edu/home/R_BioCondManual#TOC-Basics-on-Functions-and-Packages.

References

[1] Venables W N, Smith D M and the R Core Team. An Introduction to R. [2017-08-30]. https://cran.r-project.org/doc/manuals/r-release/R-intro.html.

[2] Robert Gentleman. *R Programming for Bioinformatics*. New York: CRC press, 2009

[3] Emmanuel Paradis. R for Beginners. (2005-09-12) [2017-08-30]. https://cran.r-project.org/doc/contrib/Paradis-rdebuts_en.pdf.

[4] http://manuals.bioinformatics.ucr.edu/home/R_BioCondManual#TOC-Basics-on-Functions-and-Packages.

[5] Bioconductor. Open Source Software of Bioinformatics. [2017-08-30]. http://www.bioconductor.org.

[6] The Comprehensive R Archive Network. [2017-08-30]. http://cran.r-project.org.

[7] Aedin Culhane. Basic Introduction to R and Bioconductor. (2011-05-23) [2017-08-30]. http://bcb.dfci.harvard.edu/~aedin/courses/R/CDC/basic-introduction-to-r-and-bioconductor.pdf.

[8] Paul Torfs, Claudia Brauer. A (very) short introduction to R. (2014-03-03) [2017-08-30]. https://cran.r-project.org/doc/contrib/Torfs+Brauer-Short-R-Intro.pdf.

[9] Germán Rodríguez. Introducing R. 2012. [2017-08-30]. http://data.princeton.edu/R.

Index

M

N

O

P

Q

R

S

T

U

V

W

Printed in the United States
By Bookmasters